FLORA ZAMBESIACA

Flora terrarum Zambesii aquis conjunctarum

VOLUME TWELVE: PART TWO

FLORA ZAMBESIACA

MOZAMBIQUE

MALAWI, ZAMBIA, ZIMBABWE

BOTSWANA

VOLUME TWELVE: PART TWO

Edited by
J.R. TIMBERLAKE & E.S. MARTINS

on behalf of the Editorial Board:

D.J. MABBERLEY
Royal Botanic Gardens, Kew

M.A. DINIZ
*Centro de Botânica, Instituto de Investigação
Científica Tropical, Lisboa*

J.R. TIMBERLAKE
Royal Botanic Gardens, Kew

Published by the Royal Botanic Gardens, Kew,
for the Flora Zambesiaca Managing Committee
2009

First published in 2009 by
Royal Botanic Gardens, Kew,
Richmond, Surrey, TW9 3AB, UK
www.kew.org

Distributed on behalf of the Royal Botanic Gardens, Kew in North America by the University of Chicago Press, 1427 East 60th Street, Chicago, IL 60637, USA

ISBN 978-1-84246-193-8
eISBN 978-1-84246-624-7

British Library Cataloguing in Publication Data
A catalogue record for this book is available from the British Library

Typesetting by Christine Beard
Publishing, Design and Photography
Royal Botanic Gardens, Kew

Printed in the UK by Marston Book Services Ltd
Printed in the USA by The University of Chicago Press

For information or to purchase all Kew titles please visit
www.kewbooks.com or email publishing@kew.org

Kew's mission is to be the global resource in plant and fungal knowledge and the world's leading botanic garden.

Kew receives half of its running costs from Government through the Department for Environment, Food and Rural Affairs (Defra). All other funding needed to support Kew's vital work comes from members, foundations, donors and commercial activities including book sales.

CONTENTS

FAMILIES INCLUDED IN VOLUME 12, PART 2   *page* vi

LIST OF NEW NAMES PUBLISHED IN THIS PART   vi

GENERAL SYSTEMATIC TEXT   12,2: **1**

INDEX TO VOLUME 12, PART 2   12,2: **203**

FAMILIES INCLUDED IN VOLUME 12, PART 2

ALISMATACEAE

LIMNOCHARITACEAE

HYDROCHARITACEAE

NAJADACEAE

APONOGETONACEAE

JUNCAGINACEAE

POTOMOGETONACEAE

ZOSTERACEAE

ZANNICHELLIACEAE

CYMODOCEACEAE

DIOSCOREACEAE

BURMANNIACEAE

PANDANACEAE

VELLOZIACEAE

COLCHICACEAE

LILIACEAE sensu stricto

SMILACACEAE

LIST OF NEW NAMES PUBLISHED IN THIS PART

	Page
Ottelia luapulana Symoens sp. nov.	12,2: **36**
Ottelia lisowskii Symoens sp. nov.	12,2: **38**
Afrothismia zambesiaca Cheek sp. nov.	12,2: **141**
Gymnosiphon afro-orientalis Cheek sp. nov.	12,2: **146**

ALISMATACEAE

by E.S. Martins & L. Catarino

Perennial aquatic or marsh herbs, usually deciduous and sometimes stoloniferous, rarely annual, with latex. Roots thin, numerous, fibrous. Rhizome short, woody and irregular or tuberous. Leaves basal, exstipulate; petioles usually long, expanded below and forming an open sheath; leaf-blades erect, rarely floating or submerged, entire, linear, linear-lanceolate, ovate, hastate, sagittate or subreniform with parallel veins and an apical 'water gland' (hydathode). Inflorescence usually long pedunculate, compound, with whorls of flowers or branches, racemose, paniculate or spike-like, rarely pseudo-umbellate or reduced to one flower. Bracts at the base of each whorl 2–3, bracteoles sometimes present. Flowers regular, hypogynous, bisexual or unisexual. Sepals 3, herbaceous, persistent. Petals 3, soon deciduous, rarely absent. Stamens 3, 6, 9 or more; filaments filiform or flattened below; anthers 2-celled, dehiscing by longitudinal and lateral slits; pollen grains globular. Staminodes present in some species. Carpels 3–many, free or joined at the base, arranged in a whorl or spirally on a flat or convex receptacle, unilocular; style terminal or ventral with apical stigma; ovules 1, rarely 2 or many, basal and erect or on the suture. Sterile carpels sometimes present in unisexual male flowers. Fruit compound of achenes, sometimes with floating chambers, or rarely of basally dehiscing follicles. Seeds oblong, indented laterally, smooth, wrinkled or ridged, without endosperm; embryo horseshoe-shaped.

A family predominantly found in temperate and warm regions of the northern hemisphere, with 13 genera and about 90 species; 6 genera and 7 species in the Flora Zambesiaca area. Species show great morphological variability, particularly in the size and shape of leaves and in overall plant size. This had led some authors to consider the extreme forms frequently found in the tropics as species or varieties. Vegetative reproduction from vegetative buds in the inflorescence is a feature of some species.

1. Flowering plants up to 5(15) cm tall, usually with stolons; inflorescence of 1–3 flowers in 1 whorl · **5. Ranalisma**
 – Flowering plants more than 20 cm tall, without stolons; inflorescence of more than 2 whorls · 2
2. Inflorescence unbranched, spike-like, flowers sessile or subsessile; leaves with a small linear or linear-lanceolate blade, sometimes reduced to a long linear petiole · **6. Wiesneria**
 – Inflorescence branched, or if unbranched then the flowers pedicellate; leaves with a well developed blade, never reduced to the petiole · · · · · · · · · · · · · · 3
3. Leaf-blades cuneate or rounded at base; achenes laterally compressed · · · · · 4
 – Leaf-blades cordate or sagittate at base; achenes not laterally compressed · · · 5
4. Leaves 3–5-nerved from the base; flowers unisexual; achenes bunched on receptacle · **2. Burnatia**
 – Leaves (5)7–9-nerved from the base; flowers bisexual; achenes in a whorl on receptacle · **1. Alisma**
5. Flowers bisexual; leaf-blades deeply cordate · · · · · · · · · · · · · · · · · **3. Caldesia**
 – Flowers polygamous; leaf-blades sagittate · · · · · · · · · · · · · · · **4. Limnophyton**

1

1. ALISMA L.

Alisma L., Sp. Pl.: 342 (1753); Gen. Pl., ed.5: 160 (1754).

Perennial aquatic or marsh herbs with a woody irregular rhizome. Leaves erect; leaf-blade linear-lanceolate to ovate, acute at the apex, rounded to cordate but sometimes decurrent at the base. Inflorescence a compound pyramidal panicle; branches and flowers in bracteate whorls of 4 or more. Flowers pedicellate, bisexual. Sepals 3. Petals 3, larger than sepals. Stamens 6, in pairs opposite the petals, with filiform filaments. Carpels numerous, free, in a single whorl, laterally compressed; style ventral, short, erect or curved; ovules solitary in each carpel, basal. Achenes numerous in a whorl on the receptacle, ridged dorsally.

A genus of about 9 species mainly found in the northern hemisphere. Now cosmopolitan through recent introductions. One species in the Flora area.

Alisma plantago-aquatica L., Sp. Pl.: 342 (1753). —Wright in F.T.A. **8**: 207 (1901). — Carter in F.T.E.A., Alismataceae: 5, fig.3 (1960). —Wild in Kirkia **2**: 36, t.13A (1961). —Obermeyer in F.S.A. **1**: 96, fig.29.1 (1966). —Symoens & Billiet in F.A.C., Alismataceae: 8, pl.3 (1975). —Gibbs Russell in Kirkia **10**: 435 (1977). — Cook, Aq. Wetl. Pl. Sthn. Africa: 49, fig.28 (2004). Type: Europe (LINN 473.1 holotype). FIGURE 12.2.1.

Perennial aquatic herb. Roots thin, densely covering the rhizome. Leaves with petiole 10–30(42) cm long; limb 6–15(20) × 2–8(10) cm, ovate to oblong-lanceolate, rounded to subcordate at base, acute at apex, with 5–9 nerves, not all radiating from apex of petiole, the cross-veins dense, parallel, arising from the midnerve. Inflorescence a much branched pyramidal panicle with about 6 whorls of branches, the lower compound, each whorl with 4–6 branches and flowers, the ultimate bearing 1–6(9) umbellate flowers; peduncle 35–50(100) cm high, terete; bracts membranous, lanceolate, acuminate, those of the lowest whorl 10–20 mm long; bracteoles several, smaller. Pedicels 10–30 mm long, filiform. Sepals c.2 mm long, ovate, green. Petals c.4 mm long, obovate, white or pink. Stamens c.2 mm long; anthers c.0.75 mm long, narrowly elliptic. Carpels c.20, ovate, laterally compressed; style c.1 mm long, erect or slightly curved; stigma papillose. Achenes c.2 mm long, broadly ovate, 2–3-furrowed dorsally, pale brown or yellowish.

Zimbabwe. C: Harare Dist., 6.4 km from Harare (Salisbury) on Beatrice road, first bridge on dirt road to Highfields Township, fl.& fr. 27.iii.1968, *Kelly* 37 (K, LISC, PRE, SRGH).

Cosmopolitan; known from the temperate regions of Europe, Africa, Asia, Australia, North America and South America (Chile). In Africa from Morocco to Egypt and southwards via the East African highlands to Tanzania. Introduced in Angola (Huíla), Zimbabwe and South Africa. In wet soils or shallow water of marshes and river banks; 900–2280 m in tropical East Africa.

Conservation notes: Introduced species; not threatened.

The seeds are buoyant because of subepidermal air-tissue and may float for many months. They are eaten by birds and pass unharmed through the alimentary canal.

2. BURNATIA Micheli

Burnatia Micheli in A. & C. De Candolle, Monogr. Phan. **3**: 81 (1881).

Perennial swamp or aquatic herbs, dioecious by abortion, with tuberous rhizome densely covered with ± fleshy roots. Leaves radical, erect, with long petiole; leaf-blade linear-lanceolate to ovate, the base decurrent to rounded. Inflorescence overtopping the leaves, a compound panicle with whorls of 3 branches or 3 flowers, sometimes reduced and few-flowered; bracts 3

Fig. 12.2.1. ALISMA PLANTAGO-AQUATICA. 1, habit (× ⅓); 2, part of inflorescence (× 6); 3, fruit (× 9); 4, achene, lateral view (× 12); 5, achene, dorsal view (× 12); 6, embryo (× 12). 1–2 from *Milne-Redhead & Taylor* 7036, 3–6 from *D.F. Smith*. Drawn by Margaret Stone. From F.T.E.A.

below the branches, lanceolate, membranous, much smaller below the flowers. Male flowers: sepals 3; petals 3, smaller than sepals; stamens 9, the outer 6 opposed to the sepals, the inner 3 opposed to the petals; abortive carpels about 12. Female flowers: sepals 3; petals 0 or minute; staminodes 0–2; carpels 8–20, free, closely arranged on a small receptacle; style ventral, very short, with stigma discoid, papillose; ovules basal, solitary. Achenes bunched on the stipitate receptacle, laterally compressed, with a narrow ± circular wing on each side following outline of the seed cavity.

A monospecific genus known only from tropical and subtropical Africa.

Burnatia enneandra Micheli in A. & C. De Candolle, Monogr. Phan. **3**: 81 (1881). — Wright in F.T.A. **8**: 213 (1901). —Hutchinson & Dalziel, F.W.T.A. **2**(2): 304 (1936). —Carter in F.T.E.A., Alismataceae: 13 (1960). —Podlech in Merxmüller, Prodr. Fl. SW Afrika, Alismataceae: 1 (1966). —Obermeyer in F.S.A. **1**: 99, fig.29.3 (1966). —Hepper in F.W.T.A., ed.2, **3**: 14, fig.319 (1968). —Symoens & Billiet in F.A.C., Alismataceae: 16, pl.5 (1975). —Symoens in Fl. Cameroun **26**: 22, pl.6 (1984). —Clarke & Klaassen, Water Pl. Namibia: 67 (2001). —Cook, Aq. Wetl. Pl. Sthn. Africa: 49, fig.29 (2004). Type: Sudan, Kordofan, near Arashkol Mtn., 17.x.1839, *Kotschy* 192 (BM holotype, K, G, M, P, W). FIGURE 12.2.**2**.

    Echinodorus schinzii Buchenau in Bull. Herb. Boiss. **4**: 413 (1896). Type: Namibia, Ovamboland, Omulonga, 20.iii.1890, *Rautanen* 51 (Z holotype)

    Rautanenia schinzii (Buchenau) Buchenau in Bull. Herb. Boiss. **5**: 855 (1897). —Wright in F.T.A. **8**: 212 (1901).

Perennial herb up to 1 m high, glabrous. Rhizome ovoid to globose 0.8–2 cm in diameter, sometimes poorly developed, covered with abundant somewhat fleshy roots. Leaves wholly or partially submerged; petiole (8)10–45(60) cm long, flattened; leaf-blade very variable, 10–25 × 0.5–4(7) cm, linear to narrowly lanceolate, or rarely ovate-lanceolate, decurrent to cuneate, or rarely rounded at base, subacute at apex, with 3, 5 or 7 nerves from base, the median up to 1.5 mm wide on the lower surface. Inflorescences 1–3 per plant; peduncle up to 85 cm long. Male inflorescence a panicle 15–30(40) cm long, with 1–5 whorls of branches; lowest bracts 9–35 mm long. Female inflorescence 5–15 cm long; lowest bracts 3–30(39) mm long. Flowers small, the bracts and sepals whitish or tinged violet. Male flowers: pedicels 6–10(20) mm long, filiform; sepals 2–3 mm long, ovate, concave; petals 1–2.5 mm long, narrow; stamens with filaments 1.5 mm long, flattened, whitish; anthers 1 mm long, whitish to yellowish; sterile carpels 1–1.5 mm long, compressed, greenish. Female flowers: sepals 1.5 mm long, ovate, concave; petals, if present, minute and scale-like; carpels 8–20, but often c.12 on a short torus, obovoid, compressed. Achenes 6–20, but usually c.12, 1.5 mm long, black.

Caprivi Strip. Recorded from E Caprivi by Clarke & Klaassen (2001). **Botswana.** N: Swamp at Mampi, S of Sepupa (Sepopa), 18°46'30"S, 22°14'15"E, fl. 6.v.1975, *P.A. Smith* 1383 (K, LISC, SRGH). **Zambia.** B: Machili, fl. 24.xii.1960, *Fanshawe* 6016 (K). N: Mbala Dist., Lumi Marsh, Kawimbe, 1740 m, fl. 29.xii.1959, *Richards* 12024 (K). W: 132 km S of Mwinilunga on road to Kabompo, fl. 25.i.1975, *Brummitt, Chisumpa & Polhill* 14128 (K, SRGH). C: Ndola Dist., Sacred Lake near St. Anthony's Mission, c.48 km SW of Luanshya, 1200 m, fl.& fr. 14.ii.1975, *Hooper & Townsend* 36 (K, SRGH). S: Kalomo, fl. 11.ii.1965, *Fanshawe* 9089 (K). **Zimbabwe.** N: Hurungwe Dist., Zambezi Valley, Menswa Pan, 520 m, fl.& fr. 26.ii.1953, *Wild* 4076 (K, LISC, PRE). W: Hwange Dist., Hwange (Wankie) Nat. Park, c.38.5 km WSW of Shapi Camp, 1030 m, fl. 22.ii.1967, *Rushworth* 101 (K, LISC). C: Chegutu Dist., Hippo Pools Dam, fl. 24.ii.1969, *Mavi* 978 (K, LISC, PRE). **Malawi.** C: Lisasadzi R., near bridge on main Kasungu road, 1075 m, fl. 15.i.1959, *Robson & Jackson* 1198 (K). S: Zomba Dist., S of Kachulu Harbour, fl.& fr. 6.iii.1985, *Patel* 2088 (MAL). **Mozambique.** N: Montepuez Dist., 35 km from Montepuez to Namuno, 480 m, fl.& fr. 31.xii.1963, *Torre & Paiva* 9794 (COI, EA, K, LISC, LMU, PRE). GI: Guijá Dist.,

Fig. 12.2.**2**. BURNATIA ENNEANDRA. 1, habit (male plant) (× ¹/₃); 2, part of male inflorescence (× 2); 3, male flower (× 9); 4, part of female inflorescence (× 2); 5, female flower (× 9); 6, achene, lateral view (× 9); 7, achene, dorsal view (× 9); 8, seed (× 9). 1 from *Dalziel* 260, 2–3 from *Rayner* 476, 4–5 from *Thomas* 3558, 6–7 from *Bally* 5237. Drawn by Margaret Stone. From F.T.E.A.

30 km from Guijá (Caniçado), on road to Saúte, fl.& fr. 12.ii.1948, *Torre* 7317 (COI, J, K, LISC, LMA, SRGH, WAG).

Widespread in tropical Africa. On the edges of slow-flowing rivers and streams, shallow lakes, water holes, dambos, and swamps; 100–1750 m.

Conservation notes: Widespread species; not threatened.

Podlech (1966) refers to collections from Ovamboland and from the Okavango-Gebiet in N Namibia, but none from Caprivi Strip. However, Obermeyer (1966) refers to *Killick & Leistner* 3113 from Caprivi, Katima Mulilo area "where it is locally common". These specimens were not seen.

Tubers edible; said to be sweet-smelling.

3. CALDESIA Parl.

Caldesia Parl., Fl. Ital. **3**: 598 (1860).
Alisma sect. *Caldesia* (Parl.) Hook. f. in Bentham & Hooker, Gen. Pl. **3**: 1005 (1883).

Annual or perennial aquatic and marsh herbs. Leaves all basal, floating or submerged, rarely erect and aerial; floating leaf-blades broadly elliptic to broadly ovate, or subreniform, rounded to obtuse at the apex, truncate to deeply cordate at base. Inflorescences with peduncle longer than the leaves, pyramidal, compound with whorls of 3 branches or 3 flowers, subtended by 3 lanceolate bracts and up to 3 inconspicuous bracteoles, sometimes reduced and few-flowered. Flowers pedicellate, bisexual. Sepals 3, spreading or reflexed. Petals 3, delicate. Stamens 6–9(11) with filiform or flattened filaments. Carpels 2–9(20), free, on a small flat receptacle; style ventral, persistent. Achenes swollen, smooth, ridged, tuberculate or warty.

A genus of 4 species from warm and temperate regions of Africa, Asia, Australia and Europe. Only 1 species in the Flora area.

Caldesia reniformis (D. Don) Makino in Bot. Mag. (Tokyo) **20**: 34 (1906). — Hutchinson & Dalziel, F.W.T.A. **2**(2): 304 (1936). —Troupin in Bull. Jard. Bot. État **25**: 221 (1955). —Carter in F.T.E.A., Alismataceae: 7, fig.4 (1960). —Hepper in F.W.T.A., ed.2, **3**: 11 (1968). —Symoens & Billiet in F.A.C., Alismataceae: 5, pl.2 (1975). —Gibbs Russell in Kirkia **10**: 435 (1977). —Cook, Aq. Wetl. Pl. Sthn. Africa: 50, fig.29 (2004). Type: Nepal, *Wallich* s.n. (K isotype). FIGURE 12.2.3.
Alisma reniforme D. Don, Prodr. Fl. Nepal: 22 (1825).

Perennial herb up to 1.5 m tall. Rhizome poorly developed, with latex; roots up to 35 cm long and 2 mm wide. Leaves all basal with floating blades; petiole from 10–30 cm to more than 150 cm long depending on water depth, transversely septate; leaf-blade 3–8(12) × 2–6(10) cm, broadly ovate to reniform, rounded at apex, deeply cordate at base, with 13–17 nerves radiant from apex of petiole, glossy, tinged with purple. Inflorescence emergent, up to 1.5 m high; whorls 3–8, the lowest always branched; bracts lanceolate or triangular, those of the lowest whorl 10–15 × 6–10 mm, thickened; bracteoles much shorter, membranous. Flowers with pedicels 1–3.5 cm long, sometimes replaced by vegetative buds. Sepals 3–5 × 2–3 mm, broadly ovate, green, spreading. Petals slightly larger than sepals, delicate, white to bluish. Stamens 6; filaments c.2 mm long, flattened and broadened at base; anthers basifixed, 1–1.2 mm long, yellow. Carpels 10–15, c.1 mm long, ovoid; style c.1 mm long, slightly arched. Achenes about 8, c.3 mm long, obovoid, smooth, with 3–5 longitudinal ribs and an apical beak formed by the persistent style; exocarp spongy and endocarp woody.

Botswana. N: Boteti (Botletle) R. at Samadupi Drift, 1000 m., fl.& fr. 23.i.1972, *Gibbs Russell & Biegel* 1380 (K, LISC). **Zambia**. B: Kalabo Dist., near resthouse Kalabo, fl. 13.xi.1959, *Drummond & Cookson* 6423 (K, LISC, PRE). N: Mbala Dist., Saisi Valley, 1520 m, fr. 10.x.1970, *Sanane* 1369 (K). C: Luangwa Valley, lagoon near

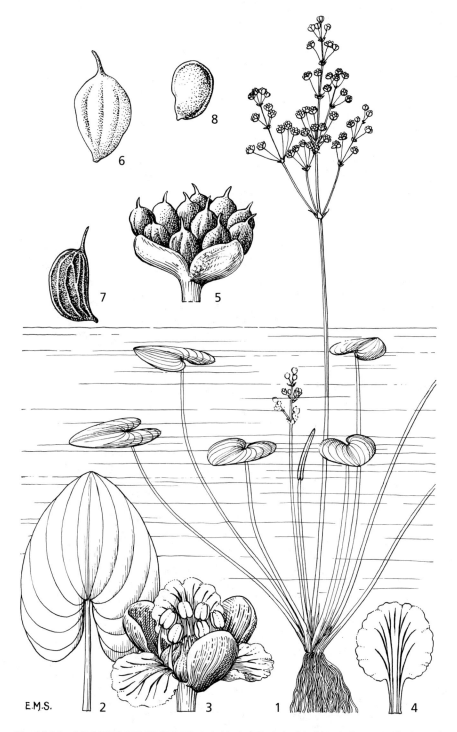

Fig. 12.2.**3**. CALDESIA RENIFORMIS. 1, habit ($\times \frac{1}{3}$); 2, leaf ($\times \frac{2}{3}$); 3, petal ($\times 6$); 4, petal ($\times 6$); 5, fruit ($\times 3$); 6, achene, dorsal view ($\times 6$); 7, achene, lateral view ($\times 6$); 8, seed ($\times 6$). 1–2, 7 from *Hancock & Chandler* 136, 3–4 from *Knowles* 44, 5–6, 8 from *Schweinfurth* ser. III, 222. Drawn by Margaret Stone. From F.T.E.A.

Kapamba R., fr. 13.xi.1965, *Astle* 4096 (K). **Zimbabwe**. C: Chirumanzu (Chilimanzi), fl. 12.viii.1964, *Wild & Drummond* 6575 (K, LISC). **Malawi**. C: Mchinji Dist., Ludzi Mission, Ludzi Dambo, fl. 23.xi.1983, *Salubeni & Patel* 3552 (LISC).

Widespread but not common in tropical Africa. Also known from Madagascar, India, Sri Lanka, Malaysia, China, Indochina, Japan, Indonesia, New Guinea and N Australia. In shallow lakes, lagoons, marshes and low current river margins; 900–1500 m.

Conservation notes: Widespread species; not threatened.

The species is closely related to the S European *C. parnassifolia* (Bassi) Parl., from which differs in having all its parts larger, leaf-blades rounded at the apex, and a more complex inflorescence.

4. LIMNOPHYTON Miq.

Limnophyton Miq., Fl. Ind. Bat. **3**: 242 (1856).

Annual or perennial emergent aquatic or marsh herbs. Leaves erect, glabrous, scabrid or pubescent, with a long petiole; leaf-blade sagittate to hastate, subacute to rounded at apex and with more or less acute basal lobes, or leaf-blade linear-lanceolate. Inflorescences with peduncle longer than leaves, racemes of 4–7 whorls of flowers, sometimes the lowest with 2–3 branches and the inflorescence paniculate, each whorl subtended by 3 bracts. Flowers regular, male in upper whorls, bisexual and male in the lower; bracts membranous, glabrous or pubescent; bracteoles 3 or more, smaller than bracts, glabrous or pubescent. Sepals 3, free, reflexed after anthesis, persistent. Petals 3, free, larger than sepals, orbicular or obovate, clawed, white. Stamens 6, with filaments broadened at base. Carpels 10–30, free, ovoid, on a small globose receptacle; style ventral; ovules solitary, basal. Achenes asymmetrical, obovoid, shortly stipitate, with 2 lateral air-chambers between the exocarp and endocarp.

A genus of 3 species of which *Limnophyton angolense* has a tropical African distribution, *L. fluitans* Graebn. is restricted to Cameroon and S Nigeria, and *L. obtusifolium* is widespread in tropical Africa, Madagascar and SE Asia.

Leaves glabrous or glabrescent; angle between the basal lobes more than 90°; bracts triangular; bracteoles 3–6; flowers 5–10 in each whorl · · · · · · · · **1**. *obtusifolium*
Leaves pubescent; angle between the basal lobes less than 90°; bracts lanceolate; bracteoles 10 or more; flowers 15 or more in each whorl · · · · · · · · **2**. *angolense*

1. **Limnophyton obtusifolium** (L.) Miq., Fl. Ind. Bat. **3**: 242 (1856). —Micheli in A. & C. De Candolle, Monogr. Phan. **3**: 39 (1881). —Durand & Schinz, Consp. Fl. Afr. **5**: 487 (1894) in part. —Wright in F.T.A. **8**: 209 (1901) in part. —Hutchinson & Dalziel, F.W.T.A. **2**(2): 303 (1936) in part. —Carter in F.T.E.A., Alismataceae: 9, fig.5 (1960). —Obermeyer in F.S.A. **1**: 97, fig.29.2 (1966). —Hepper in F.W.T.A., ed.2, **3**: 11, fig.318 (1968). —Ross, Fl. Natal: 56 (1972). —Symoens &

Fig. 12.2.4. LIMNOPHYTON OBTUSIFOLIUM. 1, leaf (× ¹/₃); 2, lower whorl of inflorescence in fruiting (× 1); 3, achene, dorsal view (× 0 4); 4, achene, lateral view (× 4); 5, transverse section of achene showing air-chambers (× 4). LIMNOPHYTON ANGOLENSE. 6, leaf (× ¹/₃); 7, undersurface of leaf (× 2); 8, lower whorl of inflorescence (× 1); 9, male flower (× 3); 10, bisexual flower (× 3); 11, achene in fresh state, dorsal view (× 4); 12, achene when dried, lateral view (× 4); 13, transverse section of achene, showing air chambers (× 4). 1, 3, 5 from *Wailly* 5387, 2 from *Welch* 375, 6–8 from *Brown* 136, 9, 11 from *Milne-Redhead* 4082, 10 from *Eggeling* 865, 12–13 from *Michelmore* 315. Drawn by Margaret Stone. From F.T.E.A.

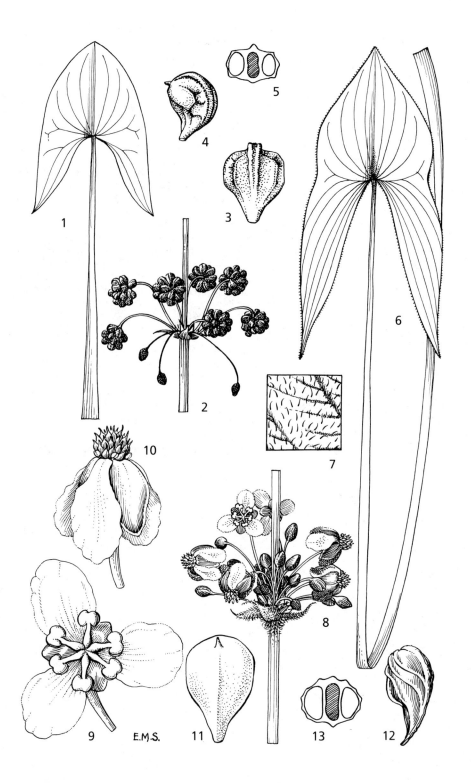

1

4

5

3

2

6

10

7

9

E.M.S.

11

8

13

12

Billiet in F.A.C., Alismataceae: 11 (1975). —Gibbs Russell in Kirkia **10**: 435 (1977). —Clarke & Klaassen, Water Pl. Namibia: 67 (2001). —Cooke, Aq. Wetl. Pl. Sthn. Africa: 51, fig.30 (2004). Type: Asia, *Plukenet* t.220, fig.7 in Herb. Sloane, vol. 97, fol.181 (BM lectotype). FIGURE 12.2.**4**.

   Sagittaria obtusifolia L., Sp. Pl.: 993 (1753).

Annual or perennial, aquatic or marsh herb up to 1(1.5) m high, with a small rhizome. Roots fibrous and spongy, with transverse septa. Leaves glabrous or glabrescent, pellucid-dotted; petioles up to 50 cm long with a triangular cross-section, spongy, the transverse septa distinct in juveniles; blade of mature leaves sagittate, wider than long, 5–10 cm along the midrib and 6–20 cm at maximum width, obtuse at apex, with the angle between basal lobes more than 90° and lobes often as long or longer than upper part of blade, acuminate; nerves 13–17(19), 3–4 pairs arising from apex of petiole and arching to the blade apex, the other curving downwards to tip of the lobes; blade of seedling leaves ovate, obovate or hastate with small lobes. Inflorescence a pyramidal panicle 20–50 cm long when fully developed (often much reduced in small plants), on a stout peduncle 30–80(100) cm long. Flowers in 4–7 dense whorls, the lower branched; bracts 10–20 × 10 mm, ovate or triangular, rounded at apex, scarious, reflexed with age; bracteoles 3–6, c.3 mm long; basal whorl with 6–10 bisexual flowers and 0–2 male flowers; pedicels 10–40 mm long, slender in male flowers, thickened to 2 mm in diameter and recurved in fruit. Sepals 5–6 × 3–4 mm in bisexual flowers, c.4 × 2 mm in male flowers, green. Petals 4–8 mm long, white. Achenes up to 30, forming a globose head 10–12 mm in diameter, each c.4 × 3 mm, obovoid, shortly stipitate, reticulately ridged, hard, pale-brown.

Caprivi Strip. Recorded from E Caprivi by Clarke & Klaassen (2001). **Botswana**. N: near Zibadianja and Savuti headwaters, fr. 22.vi.1975, *P.A. Smith* 1394 (K, SRGH). **Zambia**. B: Western Prov. (Barotseland), Sesheke, fl. i.1925, *Borle* s.n. (PRE). C: Mpika Dist., Mfuwe, fl. 3.v.1965, *B.L. Mitchell* 2777 (K, LISC). S: Namwala Dist., fr. 1934–35, *Gordon Read* 56 (K, PRE). **Zimbabwe**. W: Nkayi Dist., Dagamela, fl. 3.iv.1972, *Chiparawasha* 7 (SRGH). C: Sebungwe, Chicomba vlei, fl. 12.iii.1952, *Lovemore* 254 (K, LISC). E: Chipinge Dist., Chibuwe Pan, 700 m, fr. 1.vi.1972, *Gibbs Russell* 2064 (K, PRE, SRGH). S: Mwenezi Dist., Mwenezi (Nuanetsi) R., pan 4.8 km downstream from Malipati, W of river, fl. 25.iv.1961, *Drummond & Rutherford-Smith* 7522 (K, LISC, PRE). **Malawi**. S: Chikwawa Dist., Lengwe Nat. Park (Game Reserve), 105 m, fl. 7.iv.1970, *Hall-Martin* 782 (K, PRE). **Mozambique**. Z: between Quelimane and Nicuadala, fl.& fr. 25.vii.1942, *Torre* 4415 (COI, EA, K, LISC, LMA). T: near Mecumbura (Chissico), 8 km N of Zimbabwe border, fl.& fr. 30.iv.1964, *Wild* 6546 (BM, COI, PRE). MS: Maringa, Save R., 180 m, fl.& fr. 28.vi.1950, *Chase* 2478 (BM, COI, LISC). GI: Guijá Dist., near Guijá (Caniçado), fl. 5.v.1944, *Torre* 6578 (COI, K, LISC, LMU, MO, PRE, SRGH). M: Magude Dist., near Chobela, fl. 11.ii.1948, *Torre* 7298 (COI, K, LISC, PRE).

Widespread in tropical and subtropical Africa, from Mauritania to Ethiopia and Somalia and southwards through West Africa to Angola and Botswana, and through East Africa to Swaziland and South Africa. Also in Madagascar, India, Sri Lanka, Indochina and Malaysia. In swamps, small or large pools, and seasonal pans and waterholes; 0–1000 m.

Conservation notes: Widespread species; not threatened.

This species exhibits a large phenotypic diversity in size as well as in the shape of its leaves. Even very small plants with only juvenile-type leaves produce flowers and fruits.

2. **Limnophyton angolense** Buchenau in Engler, Pflanzenr. **16**: 23 (1903). —Carter in F.T.E.A., Alismataceae: 11, fig.5/6–13 (1960). —Hepper in F.W.T.A., ed.2, **3**: 11, fig. 318 (1968). —Symoens & Billiet in F.A.C., Alismataceae: 12, pl.4 (1975).

—Cook, Aq. Wetl. Pl. Sthn. Africa: 51, fig.30 (2004). Type: Angola, Kubango, *Baum* 364 (BM lectotype, B†, BR, COI, G, K, M, Z). FIGURE 12.2.**4**.

Limnophyton obtusifolium sensu auct. mult., non (L.) Miq. —Gilg in Warburg, Kunene-Samb.-Exped. Baum: 171 (1903). —Durand & Schinz, Consp. Fl. Afr. **5**: 487 (1894) in part as regards *Barter* 1532 & *Welwitsch* 3010. —Rendle, Cat. Afr. Pl. Welw. **2**(1): 93 (1899). — Wright in F.T.A. **8**: 209 (1901) in part as regards *Scott Elliot* 4972, *Barter* 1532, *Welwitsch* 3010 & *Carson* 40. —R.E. Fries, Wiss. Ergebn. Schwed. Rhod.-Kongo Exped. **1**: 188 (1916). — Hutchinson & Dalziel, F.W.T.A. **2**(2): 303 (1936) in part as regards *Scott Elliot* 4972, *Deighton* 357, *Thomas* 2520, *Linder* 285 & *Barter* 1532.

Perennial aquatic or marsh herb up to 1.5(2) m tall. Leaves with spongy angular petiole up to 120 cm long but usually shorter, pubescent; blade of mature leaves sagittate, usually constricted laterally at level of petiole apex, 15–33 cm in total length, 8–17 cm along the midrib and 5–20 cm wide at the constriction, pubescent especially on lower surface, rarely the older leaves glabrous; basal lobes well developed, usually longer than midrib, acuminate, with the angle between them less than 90°; nerves 17–25, 4–5(6) pairs arising from apex of petiole and arching to blade apex, the other descending to tip of the lobes. Inflorescence with 4–6 whorls, the lower sometimes branched; bracts lanceolate, acute to obtuse, the largest 15–25 × 5–10 mm, pubescent on back and margins; bracteoles 10–20, 1 mm long, pubescent. Flowers many in each whorl, the basal one with 7–13 bisexual flowers and 6–12 male flowers; pedicels 20–45 mm long, thin, weakly recurved in fruit. Sepals 5–6 × 3–4 mm in bisexual flowers, 4 × 2 mm in male flowers, green. Petals 4–8 mm long, delicate. Achenes numerous, shortly stipitate, up to 5 × 4 mm when dried, obovoid, with longitudinal ridges scarcely evident, smooth, red-brown.

Botswana. N: Isaaga R., tributary of the Moanachira R., 950 m, fl.& fr. 1.v.1972, *P.A. Smith* 200 (BM, K, LISC, PRE). **Zambia**. B: Mongu Dist., edge of Bulozi Plain below Mongu, fl. 9.xi.1959, *Drummond & Cookson* 6271 (LISC, PRE). N: Chinsali Dist., Ishiba Ngandu (L. Young, Shiwa Ngandu), 1350 m, fl.& fr. 17.i.1959, *Richards* 10710 (K). C: Lake Lusiwasi, 1560 m, fl. 7.i.1959, *Zinderen Bakker* 919 (K, PRE).

Scattered throughout tropical Africa from Guinea and Sierra Leone to Sudan and Uganda, and southwards to Angola and Zambia. In swamps and shallow lakes and lagoons; 900–1600 m.

Conservation notes: Widespread species; not threatened.

5. RANALISMA Stapf

Ranalisma Stapf in Hooker's Icon. Pl. **27**: t.2652 (1900).
Echinodorus sensu Wright in F.T.A. **8**: 211 (1901), non L.C. Rich. (1848).

Perennial stoloniferous marsh herbs, glabrous. Leaves erect or spreading; petioles long, sheathing at base; leaf-blades lanceolate to narrowly ovate or linear, cordate to cuneate at base. Inflorescence pseudo-umbellate with 1–3 flowers and 2 spathaceous and membranous bracts, the flowers sometimes replaced by vegetative buds which develop into new plants. Flowers bisexual, with convex receptacle. Sepals 3, herbaceous, persistent and reflexed at fruiting stage. Petals 3, white, larger than sepals, deciduous. Stamens (6)9(12) with filiform filaments; anthers ovoid, basifixed. Carpels numerous, free, congested on receptacle; styles terminal, hooked; ovules solitary, basal. Achenes many, spirally arranged on elongated receptacle, strongly compressed laterally, winged, with a terminal beak formed by the persistent style, glandular. Seed oblong, compressed.

A genus of 2 species, one in tropical SE Asia and the Malay Peninsula (*R. rostratum* Stapf), the other in tropical Africa. The flowers are often replaced by vegetative buds and the inflorescence bends over to the substrate, giving rise to vegetative reproduction.

Ranalisma humile (Kunth) Hutch. in F.W.T.A. **2**: 303, fig.280 (1936). —Carter in
　　F.T.E.A., Alismataceae: 2, fig.1 (1960). —Hepper in F.W.T.A. ed.2, **3**: 9, fig.317
　　(1968). —Symoens & Billiet in F.A.C., Alismataceae: 4, pl.1 (1975). Type:
　　Senegal, near Richard-Tol, *Lelièvre* (B† holotype). FIGURE 12.2.5.
　　　Alisma humile Kunth, Enum. Pl. **3**: 154 (1841).
　　　Echinodorus humilis (Kunth) Buchenau in Pringsh. Jahrb. **7**: 28 (1868). —Wright in F.T.A.
　　　8: 211 (1901).
　　　Sagittaria humilis (Kunth) Kuntze, Rev. Gen. Pl. **3**: 326 (1891).

　　Small water-loving herb, 3–5 cm tall if growing on exposed mud, up to 15 cm tall if growing
in water, with short stolons and abundant thin roots. Leaves erect or spreading; petiole 1–10 cm
long, flattened; leaf-blade 1–3 × 0.3–1.1(1.5) cm and lanceolate to narrowly obovate when
emerged, up to 10 × 0.5 cm and linear-lanceolate when submerged, acute to obtuse at apex,
obtuse to cuneate or slightly decurrent at base, with 3(5) radiant nerves arising from apex of
petiole. Inflorescence of 1 or rarely 2 flowers on a peduncle about as long as the petiole; bracts
3–5 mm long; vegetative buds sometimes present, replacing the flowers. Sepals 3 × 4 mm, ovate,
green. Petals 6 × 4 mm, white to pinkish, delicate. Stamens (6)9(12), with filaments 2.5 mm
long and anthers up to 1mm, basifixed. Carpels many (up to 70), obovoid, compressed; style 1
mm long. Achenes densely packed in a globose head 5–8(10) mm in diameter, 2–2.5 × 1.5–2
mm, obliquely obovoid, with the persistent style forming a beak 1 mm long.

　　Zambia. B: Mongu Dist., Zambezi R., Sandaula pontoon, fl.& fr. 12.xi.1959,
Drummond & Cookson 6372 (K, LISC). N: Mbala (Abercorn), lagoon at confluence of
Lufubu and Mulugu R., 6.x.1956, *Richards* 6380 (K).

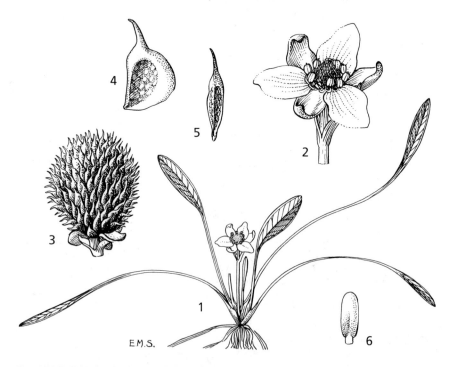

Fig. 12.2.5. RANALISMA HUMILE. 1, habit (× 1); 2, flower (× 3); 3, fruit (× 3); 4, achene,
lateral view (× 6); 5, achene, dorsal view (× 6); 6, seed (× 6). 1–2 from *Richards* 6380, 3–6 from
Milne-Redhead 5027. Drawn by Margaret Stone. From F.T.E.A.

Few specimens recorded from the Flora area, but probably undercollected. Widespread from Senegal to the Sudan, and from Gabon, Congo, Tanzania and Angola. In moist sand by rivers, wetlands, marshes and river margins; 1000–1500 m. Conservation notes: Widespread species, but apparently localised in the Flora area; probably Lower Risk near threatened.

6. WIESNERIA Micheli

Wiesneria Micheli in A. & C. De Candolle, Monogr. Phan. **3**: 82 (1881).

Perennial aquatic herbs, weakly rooting or floating, with rhizome scarcely developed. Leaves wholly submerged or partly emerged, up to 1 m long; petioles up to 60 cm, depending on water depth and flow, terete; blades submerged or floating, linear to narrowly lanceolate, usually shorter than petiole. Inflorescences emergent, unbranched and spike-like, with whorls of sessile to shortly pedicelled flowers, lower whorls of female flowers, the upper of male flowers; peduncle shorter than the leaves; bracts 3, united. Vegetative buds sometimes present in lower and middle whorls. Male flowers up to 8 in each whorl, shortly pedicellate; bracteoles ovate; sepals 3, ovate, persistent; petals 0–3, smaller than sepals; stamens 3 with flattened filaments; abortive carpels 2–3, gland-shaped. Female flowers 3 in each whorl, subsessile to shortly pedicellate, ebracteolate; sepals 3, ovate; petals 0–3, very small; staminodes 3, subulate; carpels 3–6, free; styles terminal, persistent, with papillose stigma; ovules solitary, basal. Achenes ovoid, swollen, ridged or smooth, with small air-chambers.

A genus of 3 species, 2 from tropical Africa and Madagascar and a third, *W. triandra* (Dalzell) Micheli, from E India.

Some authors give the generic name as *Wisneria*, but since it was named after Wiesner, a biology teacher in Vienna, the correct name is *Wiesneria*.

Leaves with a definite blade distinct from the petiole; female flowers subsessile or
  with a pedicel up to 2 mm long; achenes ridged · · · · · · · · · · **1.** *schweinfurthii*
Leaves without a definite blade, or only little wider than the petiole; female flowers
  with a pedicel 2.5–4 mm long; achenes smooth · · · · · · · · · · · · · · · **2.** *filifolia*

1. **Wiesneria schweinfurthii** Hook. f. in Bentham & Hooker, Gen. Pl. **3**: 1007 (1883). —Wright in F.T.A. **8**: 214 (1901). —Hutchinson & Dalziel, F.W.T.A. **2**(2): 303 (1936). —Hepper in F.W.T.A., ed.2, **3**: 10 (1968). —Symoens & Billiet in F.A.C., Alismataceae: 18, pl.6 (1975). —Cook, Aq. Wetl. Pl. Sthn. Africa: 51, fig.31 (2004). Types: Sudan, Gir, *Schweinfurth* 2157 (K, P syntypes) and Jur, Jur Ghattas, *Schweinfurth* 2304 (B, K syntypes).
    Wiesneria sparganiifolia Graebn. in Bot. Jahrb. Syst. **48**: 402 (1912).

Aquatic herb with abundant, septate roots 1–2 mm in diameter, sometimes anchored in other floating plants. Leaves sheathing at base; petiole up to 70(100) cm long and 0.4-0.6 cm in diameter, swollen, terete, septate, submerged; blade 18–25 × 0.7–1(1.5) cm, linear to narrowly lanceolate, submerged or floating, usually distinguishable from petiole, subobtuse at apex, 1–3-nerved, sometimes lacking. Inflorescences 1 to several on each plant; peduncle up to 60 cm long, slender; spikes 5–20 cm long, with 6–15 whorls, internodes first short but expanding from below with age. Flowers white, pale mauve, or pinkish, sessile or shortly pedicellate, male flowers in the upper whorls and female in the lower ones; vegetative buds sometimes present in lower and middle whorls; outer bracts without a stiff midrib. Bracts united into a membranous sheet 2–3 mm long, entire or sometimes 3-lobed. Male flowers up to 8 in each whorl; pedicels 1.5–1.8 mm long; sepals 3, ovate, unequal; anthers 0.4–0.5 mm long, basifixed. Female flowers 3 in each whorl; pedicels up to 2 mm long, longer in fruiting stage, angular; sepals 3, ovate, unequal, the largest up to 4 mm long; petals 0–3, shorter than sepals; carpels 3–6, ovoid, styles 1.5–2 mm long. Achenes 2–6, ovoid, 3–3.5 × 2.2–2.5 mm, ± ridged and often warty, with a beak 1.5–2 mm long.

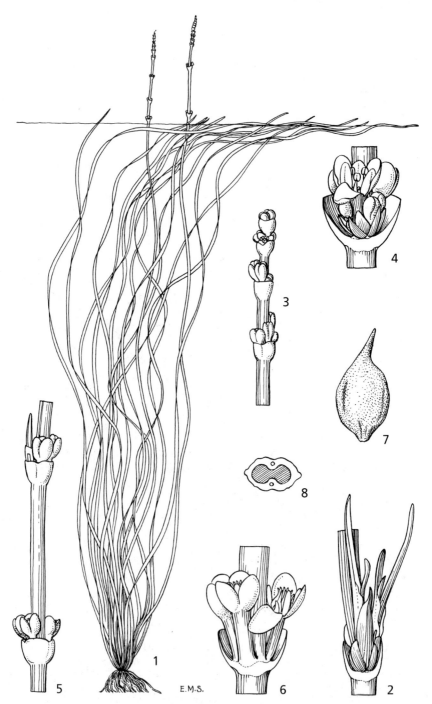

Fig. 12.2.**6**. WIESNERIA FILIFOLIA. 1, habit (× ¹/₃); 2, vegetative buds from a lower whorl of inflorescence (× 3); 3, upper part of inflorescence with male flowers (× 1¹/₂); 4, whorl of male flowers with bract cut away to show bracteoles and flower buds (× 6); 5, lower whorls of inflorescence with female flowers (× 1¹/₂); 6, whorl of female flowers with bract cut away (× 3); 7, achene, lateral view (× 6); 8, transverse section of achene showing air-chambers (× 6). 1 from *Parker*, 2–4 from *Lowe* 267, 5–8 from *Hancock* 7a. Drawn by Margaret Stone. From F.T.E.A.

Botswana. N: Okavango Delta, Boro R., fr. 28.i.1975, *Hiemstra* 40 (K). **Zambia**. W: Kasempa Dist., 7 km E of Chizera, fl.& fr. 27.iii.1961, *Drummond & Rutherford-Smith* 7460 (K, LISC). N: Mporokoso Dist., Mweru Wantipa road to Bulaya, 1050 m, fl. 12.iv.1957, *Richards* 9174 (K).

Widespread but apparently not common, in tropical West and Central Africa. In rivers, lagoons, water channels and shallow pans; to 1050 m.

Conservation notes: Widespread species; not threatened.

2. **Wiesneria filifolia** Hook. f. in Bentham & Hooker, Gen. Pl. **3**: 1007 (1883). — Carter in F.T.E.A., Alismataceae: 11, fig. 6 (1960) as '*Wisneria filifolia*'. —Symoens & Billiet in F.A.C., Alismataceae: 20 (1975). —Cook, Aq. Wetl. Pl. Sthn. Africa: 52, fig.32 (2004). Type: C Madagascar, *Baron* 591 (K holotype). FIGURE 12.2.**6**.

Aquatic herb, sometimes anchored in other floating plants. Leaves sheathing at base; petiole (10)30–80 cm long, swollen, ± 3-angled, septate, submerged; leaf-blade scarcely distinguishable from petiole, up to 24 × 0.15–0.3 cm, linear, usually partly submerged. Inflorescences 1 to several on each plant; peduncle 30–125 cm long, less strong than petioles; spikes 5–20 cm long, with 5–11(15) whorls, the internodes first short but expanding up to 3–5 cm from below with age. Flowers white or pinkish, shortly pedicellate, male flowers in the upper whorls and female in the lower; vegetative buds sometimes present in lower and middle whorls; outer bracts sometimes with a thickened and stiff midrib. Bracts united into a membranous sheet 2–4.5 mm long, entire or sometimes 3-lobed. Male flowers 4–8 in each whorl; pedicels 2 mm long; sepals 3, ovate, spreading, unequal; anthers 1 mm long. Female flowers 3 in each whorl; pedicels 3–4.5(5) mm long, longer at fruiting stage, angular; sepals 3, ovate, 3–3.5 mm long, unequal; petals 1–3, minute; carpels 3–6, ovoid, styles 1–1.5 mm long. Achenes 2–6, ovoid, 3–4 × 2.2–2.5 mm, smooth, slightly indented each side, with a beak 1–1.5 mm long.

Botswana. N: Thamalakane R., above confluence with Boro R., 930 m, fl. 22.i.1972, *Biegel & Russell* 3715 (K, LISC). **Zambia**. N: Mbala Dist., Lake Chila, 1590 m, fl. 25.ix.1963, *Richards* 18193 (K). **Zimbabwe**. E: Nyanga Dist., Van Niekerk Ruins, pool of Ngarawe R., st. 2.viii.1950, *Wild* 3519 (K). W: Nyamandhlovu Dist., Nyamandhlovu Pasture Res. Station, st. 13.ii.1955, *Plowes* 1776 (K).

Also known from Congo, Burundi, Uganda, Tanzania, Angola and Madagascar. In rivers, lagoons, pans and in ± stagnant deep (0.5–1.5 m) waters; 930–1700 m.

Conservation notes: Widespread species; not threatened.

LIMNOCHARITACEAE

by E.S. Martins

Annual or perennial aquatic, marsh or swamp herbs, glabrous, lactiferous, with well developed aerenchyma. Rhizome short, erect. Roots short, fibrous. Leaves radical, petiolate and sheathing at base or cauline and in groups along the stem, lower ones linear, upper ones petiolate; leaf blade entire, linear, lanceolate, oblanceolate or ovate to orbicular, cuneate to truncate at base, acute to rounded at apex tipped by a hydathode, with ± parallel curved nerves. Inflorescence erect, umbellate or in axillary fascicles, rarely solitary flowers; bracts 2 or 3; bracteoles several. Flowers actinomorphic, bisexual. Sepals 3, imbricate, persistent. Petals 3, delicate. Stamens 3–9 or more, free; filaments flattened; anthers attached by the base, 2-celled, dehiscing longitudinally by lateral slits. Staminodes present or absent. Carpels 6–∞, superior, sessile, free or cohering only at the base, unilocular; stigmas sessile; ovules ∞, anatropous or campylotropous, scattered over the ovary wall on reticulately branched placentas. Fruiting carpels many-seeded, dehiscing by the ventral suture. Seeds without endosperm; embryo horseshoe-shaped.

Family with 3 genera and c.11 species, confined to the tropics and subtropics.

BUTOMOPSIS Kunth

Butomopsis Kunth, Enum. Pl. **3**: 164 (May 1841). —Micheli in A. & C. de Candolle, Monogr. Phan. **3**: 87 (1881).
Tenagocharis Hochst. in Flora **24**: 369 (June 1841).

Annual or perennial aquatic or swamp herbs. Leaves arising from base, petiolate, erect; leaf-blade narrowly obovate to oblanceolate. Peduncle erect, longer than leaves, angular. Inflorescence in 1 or (in larger plants) 2 umbellate whorls; bracts 2, and several bracteoles at base of each whorl. Flowers long-pedicelled, regular. Pedicels angular, 3-edged to the apex. Sepals 3, erect. Petals 3, white, withering and disintegrating into a mucilaginous mass after anthesis. Stamens c.9; filaments flattened to base; anthers narrowly oblong; staminodes lacking. Carpels 5–8, cohering at base. Fruiting carpels with their tips exserted from calyx. Seeds small, numerous, broadly oblong, dark brown to blackish, smooth, shining; embryo horseshoe-shaped.

Genus with a single species, widespread in tropical Africa, tropical and subtropical Asia and northern Australia.

Butomopsis latifolia (D. Don) Kunth, Enum. Pl. **3**: 165 (May 1841). —Symoens in Symoens, Hooper & Compère, Stud. Aquat. Vasc. Pl.: 304 (1982). —Cook, Aq. Wetl. Pl. Sthn. Africa: 172–173 (2004). Type: Nepal, *Hamilton* s.n. (location unknown). FIGURE 12.2.7.
Butomus latifolius D. Don, Prodr. Fl. Nep.: 22 (1825).
Butomopsis lanceolata (Roxb.) Kunth, Enum. Pl. **3**: 165 (1841). —Wright in F.T.A. **8**: 214 (1901). Type: Nepal, *Wallich* s.n. (K holotype?).
Tenagocharis latifolia (D. Don) Buchenau in Abh. Naturwiss. Vereine Bremen **2**: 2 (1869). —Engler, Pflanzenw. Afrikas **2**: 105, fig.95 (1908). —Hutchinson in F.W.T.A. **2**: 298 (1936). —Carter in F.T.E.A, Butomaceae: 1, fig.1 (1960). —Hepper in F.W.T.A., ed.2, **3**: 5, fig.315 (1968).

Annual or perennial herb, 15–130 cm high, with milk-like latex. Petiole 6–13(44) cm long with a well-developed sheathing base; leaf-blade 3–8(24) × 0.5–1.4(5.5) cm, narrowly oblanceolate to narrowly obovate, cuneate at base, rounded to bluntly acuminate at apex with a large hydathode underside, entire, glabrous; nerves 3–7. Peduncles 2, rarely 3, on each plant, 7–15(80) cm long. Flowers 3 on weak plants, up to 20 in each whorl in strong specimens with 2 whorls; pedicels 1.5–7(11) cm long, erect, angular but 3-edged above; bracts 8–22 × 3–6 mm, lanceolate or triangular-lanceolate, membranous; bracteoles several, membranous, smaller than bracts. Sepals 5–7 × 4–5 mm (up to 11 × 6 mm in fruit), broadly ovate, obtuse to rounded at apex, green-greyish, scarious at margins, reticulately veined. Petals white, very delicate, slightly smaller than sepals. Stamens (8)9(12); filaments c.2 mm long, elongating to 4 mm at maturity, hyaline; anthers 1.5–2.5 mm long. Carpels 6–9, c.5 × 2 mm; stigmas sessile. Fruiting carpels 3–9, up to 13 mm long. Seeds 0.4–0.5 × 0.25 mm, minutely warty.

Botswana. N: Chosoroga (Tsotsoroga) pan , fl.& fr. 18.v.1977, *P.A.Smith* 2025 (K, SRGH). **Zambia**. B: Machili, fl.& fr. 1.vii.1963, *Fanshawe* 7892 (K). N: Kasama Dist., 10.5 km E of Kasama, fl.& fr. 6.v.1962, *Robinson* 5161 (K, PRE). S: Namwala Dist., Umbuzu, on Ngabo Kwichila road, fr. 18.iv.1963, *van Rensburg* 2040 (K). **Zimbabwe**. W: Hwange Dist., Hwange Nat. Park, pan 6.4 km W of Shumba Camp, 1030 m, fl.& fr. 29.iv.1970, *Rushworth* 2500 (LISC, PRE, SRGH). **Mozambique**. T: Near Mucumbura (Chissico), 8 km N of Zimbabwe border, fl. 30.iv.1964, *Wild* 6547 (K, LISC, PRE).

Found across tropical Africa from Mauritania, Senegal, Guinea-Bissau, Guinea, Ivory Coast, Ghana, Mali, Burkino-Faso, Niger, Nigeria, Cameroon, Chad, Sudan,

Fig. 12.2.7. BUTOMOPSIS LATIFOLIA. 1, plant in fruit (× 1); 2, flower with a sepal and petal pulled down (× 3); 3, young carpel opened out (× 9); 4, young fruit (× 3); 5, ripe carpel after dehiscence (× 3); 6, seed (× 60). 1 from *Lea* 220, 2–4 from *H.B. Johnston* IV.82, 5–6 from *Lind* 260. Drawn by Margaret Stone. From F.T.E.A.

Uganda and Tanzania; also in Asia from Nepal, India and Bangladesh and in W Australia. In moderately deep water of pans and swamps and on the dried-up periphery, in moist sandy or loamy soils and mud; 600–1100 m.

Conservation notes: Widespread species; not threatened.

Juveniles are generally submerged, while adult plants are emergent. The species survives unfavourable periods as seed.

HYDROCHARITACEAE

by J.J. Symoens[1]

Freshwater or marine aquatic herbs, monoecious, dioecious, gynodioecious or trioecious (male, female and hermaphrodite forms on different plants), perennial, rarely annual, completely or partly submerged, some species floating and forming mats. Roots mostly simple, rarely branched (*Hydrocharis, Limnobium*). Stems elongate or contracted, often rhizomatous or stoloniferous, or erect, simple or branched. Leaves in a basal rosette or on the stem, alternate, opposite or whorled, submerged or floating, sometimes partly emergent, petiolate or sessile, leaf-base usually sheathing the stem; stipules absent or membranous if present; intravaginal scales mostly present, situated in leaf axils. Inflorescence usually a few-flowered cyme or single flower, subtended by a spathe consisting primarily of 2 opposite, free or united bracts, rarely reduced to a single bract; spathes sessile to long-pedunculate, often ribbed or winged. Flowers 1–many per spathe, pedicellate or sessile, bisexual or unisexual (then often with rudiments of the other gender), regular, sometimes slightly irregular (*Maidenia, Nechamandra, Vallisneria*), sometimes cleistogamous. Hypanthium (perianth tube) often present in hermaphrodite and female flowers, ± elongate above ovary apex and exserted above the spathe at anthesis. Sepals 3, free, green or whitish, often reflexed, falling early or persisting in fruit, rarely absent; petals mostly 3, free, white or coloured, often showy, sometimes rudimentary or absent. Stamens 1–many, often of unequal height, in 1–5 (usually 3-merous) whorls, often staminodial; filaments filiform or flattened, rarely absent; anthers basifixed or dorsifixed, mostly bithecate and 4-sporangiate, sometimes 1-thecate and 2-sporangiate, opening dorsally or along sides by longitudinal slits; staminodes sometimes present in pistillate flowers. Carpels (2)3–6(20 or more), ± united, surface glabrous; ovary inferior, ovoid to linear, 1-locular, placentation parietal, placentas sometimes protruding nearly to centre of ovary, rarely basal; ovules 1–many, anatropous or orthotropous, mostly with 2 integuments, rarely one (*Thalassia*); styles and stigmas as many as carpels, stigmas entire or more often 2-lobed or bifid, papillate; rudimentary pistillode often present in staminate flowers. Fruit a capsule or berry, dry or ± fleshy, globose, ovoid or cylindric, ripening under water, opening regularly or breaking up irregularly by decay of pericarp. Seeds 1–200 (or more), usually minute, ellipsoidal, fusiform or oblong, without or with scant endosperm; testa smooth or striate, ribbed, warty or tuberculate, sometimes densely covered with unicellular hairs; embryo mostly straight, sometimes curved (*Halophila*).

A family of 17 genera and c.80 species widespread in temperate and tropical regions; only 3 genera are marine, the remainder occurring in fresh or slightly brackish water. In the Flora Zambesiaca area there are 7 genera. *Egeria densa* Planch. has been introduced into East and South Africa.

Miki (Bot. Mag. (Tokyo) **51**: 472–480, 1937) emphasized remarkable similarities in the vegetative and reproductive structures of *Najas* and the submerged Hydrocharitaceae and suggested that *Najas* was derived from the Hydrocharitaceae by simplification. From studies of seed-coat structure, Shaffer-Fehre (Bot. J. Linn.

---

[1] The author wishes to thank Prof. J. Léonard (Brussels) for comments on the manuscript and Dr D. Simpson (Kew) and Dr S. Bandeira (Maputo) for useful information.

Soc. **107**: 169–188 & 189–209, 1991) confirmed this relationship and formally included the genus *Najas* within the Hydrocharitaceae, a view also confirmed from *rbcL* and *matK* gene sequence data by Les *et al.* (Mol. Phylogen. Evol. **2**: 304–314, 1993; Syst. Bot. **22**: 443–463, 1997) and by Tanaka *et al.* (J. Plant Res. **110**: 329–337, 1997). The inclusion of *Najas* brings the number of species in Hydrocharitaceae to around 115. Although Najadaceae is treated separately in this Flora, as it was in other regional African floras, the arguments supporting its inclusion in Hydrocharitaceae should not be ignored.

As is frequent in aquatic plants, many species show a wide range of phenotypic variability. As far as possible, fresh material should be used for identification since the flowers are mostly very delicate and damaged during desiccation. Specimens freshly collected or preserved in alcohol are sometimes necessary for a safe decision about the unisexual or bisexual character of flowers and the distinction of stamens and staminodes.

1. Freshwater plants; perianth mostly with 3 sepals and 1-3 petals, the latter sometimes rudimentary · 2
   – Marine plants; perianth segments 3 · 6
2. Leaves regularly arranged on elongate stems; spathes sessile or nearly so · · · 3
   – Leaves usually in a basal rosette or spaced along a stolon; spathes pedunculate, rarely sessile ·4
3. Leaf apex with 2 spine cells; intravaginal scales entire to papillate; male flowers with 3 fertile stamens and 3 erect feathery staminodes; styles bifid almost to base · **1. Lagarosiphon**
   – Leaf apex acute and terminating in a single spine cell; intravaginal scales fringed with finger-like orange-brown hairs; male flowers with 3 stamens without staminodes; styles simple · **2. Hydrilla**
4. Petals conspicuous and broad · **3. Ottelia**
   – Petals linear or very reduced · 5
5. Leaf bases not clasping stems; petals band-shaped to filiform, scarcely wider than sepals; styles 3, simple · **4. Blyxa**
   – Leaf bases clasping stem; male flowers numerous, minute, free from spathe when mature, single petal rudimentary or absent; female flowers on long flexible peduncles, petals 3, translucent; styles 2–3, each with 2 papillose lobes · **5. Vallisneria**
6. Plant delicate, horizontal stems (if present) not more than 2 mm thick, creeping, not wholly buried in substrate; leaves in pairs, sessile or petiolate, without a distinct sheath; blade elliptic, oblong or linear, less than 8 cm long; flowers with 3 tepals, the female extremely reduced · · · · · · · · · · **7. Halophila**
   – Plant robust, rhizomes 2–10 mm thick, buried in substrate; leaves in 2 rows, sessile, distinctly sheathing; blade linear, ± falcate, often more than 10 cm long; flowers with 3 tepals or 3 sepals + 3 petals · 7
7. Rhizome 10–15 mm wide, densely covered with persistent fibrous strands of decayed leaves; roots numerous, crowded, without root-hairs; male spathe with numerous flower buds, becoming detached just before flowering; female peduncle up to 50 cm long, contracting spirally after pollination; flowers with 3 sepals and 3 tepals · **6. Enhalus**
   – Rhizome 2–3 mm wide, without fibres; roots 1 per node, with root-hairs; male spathe with only one flower, shed after anthesis; female peduncle 1–1.5 cm long, elongating to 2–4 cm after pollination; flowers with 3 tepals · · · · · **8. Thalassia**

1. LAGAROSIPHON Harv.

Lagarosiphon Harv. in J. Bot. (Hooker) **4**: 230 (1841). —Symoens & Triest in Bull. Jard. Bot. Belg. **53**: 441–488 (1983).

Freshwater herbs, perennial, submerged except for the flowers at anthesis, dioecious. Roots unbranched. Stems elongate, cylindrical, the basal part persistent, rhizomatous with distant axillary branches, densely leaved at nodes, internodes lengthening afterwards. Leaves alternate, subopposite or subverticillate, sessile, linear to linear-lanceolate, often recurved; margins green or with several longitudinal rows of translucent fibres, serrulate or denticulate; apex with 2 spine cells. Intravaginal scales 2, axillary, minute. Inflorescences axillary, solitary, sessile and unisexual. Male spathes consisting of 2 united bracts, ovate or obovate, compressed or cup-shaped, toothed; receptacle with numerous stalked flower buds that break off and rise to the water surface where they open. Male flowers with a perianth of 2 whorls, each with 3 segments, reflexed at anthesis, the outer slightly narrower; stamens 3, filaments expanding and stretching out parallel to water surface with anthers at right angles; staminodes 2–3, erect, longer than stamens, papillate and usually coloured above, joined at top, acting as a sail. Female spathes 1-flowered, comprising 2 united bracts, narrowly ovate to oblong or cylindrical, entire or toothed. Female flowers sessile, but hypanthium exserted near apex of spathe, lengthening so that the flower bud reaches the water surface; perianth segments 6, all alike, or the outer slightly smaller, spreading and floating; staminodes 3, minute, filiform; ovary unilocular, with 3 parietal placentas; styles 3, each divided above into 2 long papillate, often brightly coloured stigmas; ovules 5–30, orthotropous. Capsule ovate, extended into a beak which protrudes and tears the spathe-valves; pericarp honeycombed, becoming mucilaginous and bursting irregularly and so dispersing the seeds. Seeds 5–30, ellipsoidal, with short stipe at base, tapering at apex, densely ribbed or honeycombed, buoyant at first, sinking later.

A genus of 9 species with 8 in Sub-Saharan Africa and 1 endemic to Madagascar. *L. major* is adventive, often invasive, in Europe and New Zealand.

All *Lagarosiphon* species can be identified without flowers. A lens (× 12 magnification) is required to see details of the leaf margins, but a microscope is necessary to see the epidermal cells and details of the central area. Polarized light makes the observation of fibre rows easier.

1. Stems slender, 0.3–2 mm in diameter; leaves thin, transparent, (0.25)0.4–1.5(2.2) mm wide; epidermal cells rectangular; central area mostly bordered by 2 parallel fibre lines (suggesting tramway rails) · · · · · · · · · · · · 2
- Stems thick, up to 3 mm in diameter; leaves thick, opaque, mostly 2–4.4 mm broad; epidermal cells equal-sided; central area not bordered by fibre lines · · 4
2. Leaf margins green, without rows of sclerenchyma fibres; central area narrow or broad, midrib mostly distinct; leaf teeth on protuberances at least at leaf base · 3
- Leaf margins almost translucent, with 2–6 rows of sclerenchyma fibres; central area narrow, midrib faint; leaf teeth not on protuberances · · · · · · **3.** *muscoides*
3. Leaves alternate, opposite or occasionally whorled; central area broad, usually with 2 longitudinal air-spaces divided by transverse septa · · · · · · · **1.** *cordofanus*
- Leaves regularly whorled; central area narrow, without transverse septa · **2.** *verticillifolius*
4. Leaf margins with 2 rows of sclerenchyma fibres, each side with more than 50 teeth; leaf teeth relatively short, blunt, pointing upwards and not on protuberances · **4.** *major*
- Leaf margins green, without sclerenchyma fibres, bearing less than 50 teeth on each side; leaf teeth long, sharp, spreading or recurved, on broad triangular protuberances · **5.** *ilicifolius*

1. **Lagarosiphon cordofanus** Casp. in Pringsheim. Jahrb. Wiss. Bot. **1**: 504, t.29, figs.75–78 (1858). —Ridley in J. Linn. Soc., Bot. **22**: 234 (1886). —Wright in F.T.A. **7**: 4 (1897). —Symoens & Triest in Bull. Jard. Bot. Belg. **53**: 456, fig.5 (1983). —Simpson in F.T.E.A., Hydrocharitaceae: 2, fig.1 (1989). —Clarke & Klaassen, Water Pl. Namibia: 70 (2001). —Cook, Aq. Wetl. Pl. Sthn. Afr.: 143 (2004). Type: Sudan, 'Arasch-Cool', *Kotschy* 170 (B† holotype, G lectotype selected by Symoens & Triest (1983), BM, K, L, MPU, P, Z).

Udora cordofana Hochst. in Sched. Pl. Exs. Kotschyi, It. Nubic., no.170 (1841), nomen nudum.

Lagarosiphon nyassae Ridl. in J. Linn. Soc., Bot. **22**: 234 (1886). —Durand & Schinz, Consp. Fl. Afr. **5**: 2 (1894). —Gürke in Engler, Pflanzenw. Ost-Afr. **C**: 95 (1895). —Burkill in Johnston, Brit. Cent. Afr.: 269 (1897). —Wright in F.T.A. **7**: 3 (1897). —Peter in Repert. Spec. Nov. Regni Veg. Beih. **40**(1): 122 (1929). —Binns, First Check List Herb. Fl. Malawi: 53 (1968). Type: Malawi, Lake Malawi (Nyassa), 1877, *Laws* s.n. (BM holotype, BM, K).

Lagarosiphon tenuis Rendle in J. Linn. Soc., Bot. **30**: 380, t.31 (1895). —Wright in F.T.A. **7**: 3 (1897). Type: Kenya, Machakos Dist., E Ngulia, Kenani, 1893, *Gregory* s.n. (BM holotype).

Lagarosiphon crispus Rendle in J. Linn. Soc., Bot. **30**: 381, t.31 (1895). —Wright in F.T.A. **7**: 4 (1897). —T.C.E. Fries in R.E. Fries, Wiss. Ergebn. Schwed. Rhod.-Kongo-Exped.:189 (1916). —Obermeyer in Bothalia **8**: 143, figs.1 & 2 (1964); in F.S.A. **1**: 106, figs. 32 & 33 (1968). —Podlech in Merxmüller, Prodr. Fl. SW Afr., fam.142: 2 (1966). —Gibbs Russell in Kirkia **10**: 437 (1977). —Hall-Martin & Drummond in Kirkia **12**: 155, 157 (1980). Type: Tanzania, 'Zanzibar coast', Ugui, 1886, *W.E. Taylor* s.n. (BM holotype).

Lagarosiphon fischeri Gürke in Engler, Pflanzenw. Ost-Afr. **C**: 95 (1895). —Wright in F.T.A. **7**: 4 (1897). Type: Tanzania, Masai plains, *Fischer* 116 (B† holotype).

Lagarosiphon tsotsorogensis Bremek. & Oberm. in Ann. Transv. Mus. **16**: 401 (1935). Type: Botswana, Tsotsoroga Pan, 17.vi.1930, *van Son* in TRV 28853 (PRE holotype, BM).

Lagarosiphon sp. sensu Hall-Martin & Drummond in Kirkia **12**: 157 (1950).

Perennial herb. Stems filiform, up to 2.5 m long and 0.5–2 mm wide. Leaves alternate, subopposite or locally whorled, spreading, linear, (5)7–20(29) × 0.3–1(1.5) mm, soft, thin, transparent, apex tapering to very acute; epidermal cells predominantly elongate; margins green, without sclerenchyma fibres, each with (15)25–50(66) teeth situated on triangular protuberances, somewhat curved, pointing upwards, sharp; central area broad, rarely narrow, with larger cells, usually with 2 longitudinal air-spaces divided by transverse septa and 2 parallel fibre lines. Intravaginal scales colourless, ovate or narrowly ovate, apex with 1–4 papillae. Male inflorescences with spathe containing 7–14 flower buds, valves ovate to broadly ovate, 1.3–2.7 × 1–2.1 mm, bearing 3–16 teeth; male flowers whitish, perianth segments c.1 mm long; stamens c.2 mm long, anthers with pointed tip. Female inflorescences with spathe valves ovate to narrowly ovate, 1.7–2.8 × 0.7–1.2 mm, entire or bearing up to 10 teeth; female flowers whitish, perianth segments obovate, 0.7–1 × c.0.5 mm; ovary with 10–40 ovules. Capsule ovoid, tapered above, 3–4 × 1.5–2 mm, pericarp honeycombed when dried. Seeds narrowly ellipsoid, 1–1.5 × 0.3–0.5 mm.

Botswana. N: Shishikola Pan, ♀ fl. 24.vi.1975, *P.A. Smith* 1396 (B, BR, LISC, PRE, SRGH). **Zambia**. N: Kaputa Dist., Mweru Wantipa, ♀ fl. 9.iv.1957, *Richards* 9110 (BR, K). C: near Kabwe (Broken Hill), Chirukutu, ♀ fl. viii.1911, *R.E. Fries* 289 (K, UPS, Z). E: c.7 km from Mfuwe on road to Nyamaluma, ♀ fl. 28.ii.1988, *Phiri* 2284 (K, UZL). S: Namwala, Mulela floodplain, ♀ fl. 28.iv.1964, *van Rensburg* 2905 (K). **Zimbabwe**. N: Hurungwe Dist., Zambezi Valley, Mhenzu (Menswa) Pan, ♀ fl. 26.ii.1953, *Wild* 4069 (BM, K). W: mentioned by Gibbs Russell (1977). E: Chipinge Dist., between Manzwire and Maduku, ♀ fl. 31.i.1975, *Gibbs Russell* 2712 (BR, K, PRE). S: 27 km ESE of Chirundu Bridge, ♀ fl. 31.i.1958, *Drummond* 5404 (BM, BR, EA, K, LISC, P, PRE). **Malawi**. N: Livingstonia, Lake Malawi (Nyassa), ♀ fl. 1877, *Laws* 7 (BM, K). C: Lilongwe Game Reserve, iii.1969, *Hall-Martin* in PRE 57646 (PRE). S: Lengwe Nat. Park (Game Reserve), st. 7.iv.1970, *Hall-Martin* 781 (K, PRE). **Mozambique**. N: 20 km from Nampula, road to Meconta, ♀ fl. 6.x.1936, *Torre* 889 (COI, LISC).

Widespread in the Sudano-Zambezian Region; also in Cameroon, Sudan, Ethiopia, Somalia, Congo (Katanga), Rwanda, Uganda, Kenya, Tanzania, Angola, Namibia and the northern provinces of South Africa. In lakes, dams, ponds, swamps and on floodplains in permanent or temporary pools, also in slow-flowing freshwater with water depth 0.15–2 m; 100–1050 m.

Conservation notes: Widespread species; not threatened.

Lagarosiphon cordofanus is a very variable species, well characterized by the leaves having a broad central area, usually with 2 longitudinal air-spaces and 2 parallel fibre lines, and marginal teeth situated on triangular protuberances. Some specimens from the Bangweulu–Luapula area with linear, narrow leaves (1–1.2 mm wide) and a distinct midrib, but without the two fibre lines, could represent a hybrid between *L. cordofanus* and *L. ilicifolius*.

2. **Lagarosiphon verticillifolius** Oberm. in Bothalia **8**: 142, fig.1 (1964). —Ross, Fl. Natal: 57 (1973). —Gibbs Russell in Kirkia **10**: 437 (1977). —Symoens & Triest in Bull. Jard. Bot. Belg. **53**: 461, fig.6 (1983). —Clarke & Klaassen, Water Pl. Namibia: 70 (2001). —Cook, Aq. Wetl. Pl. Sthn. Afr.: 143 (2004). Type: South Africa, KwaZulu-Natal, Hlabisa, Hluhluwe Game Reserve, 21.iv.1955, *Ward* 2551 (PRE holotype, B). FIGURE 12.2.8.

Lagarosiphon muscoides Harv. var. *major* sensu Schinz & Junod in Bull. Herb. Boiss. **7**: 889 (1899). —Pedro in Bol. Soc. Estudos Moçamb. **24**: 19 (1954), non Harvey.

Stems filiform, 0.5–1 mm wide. Leaves mostly regularly whorled (5–7), occasionally subverticillate, spreading, linear, 6.5–15.5(20) × 0.4–1.2 mm, soft, thin, transparent, apex very acute; epidermal cells predominantly elongate, rectangular; marginal cells almost without chlorophyll, each side with 26–58 teeth on triangular protuberances, somewhat curved at least at leaf base, pointing upwards, sharp, central area with slightly larger cells between two distinct lines of sclerenchyma fibres, no transverse septa. Male inflorescences with spathe containing 10–20 flower buds, valves ovate or narrowly ovate, 2.3–3.4 × 1–1.6 mm, entire or with up to 9 teeth; male flowers c.1.6 mm wide, perianth whitish (not yet fully described). Female inflorescences with spathe valves narrowly ovate, tubular above, 2.2–4.5 × 0.7–1 mm, entire or with up to 14 teeth; female flower with whitish perianth, petals slightly broader and more obtuse than sepals, 0.75–1.3 mm long; ovary with 6–12 ovules. Capsule narrowly ovoid, tapered above, 5–6.5 × 1.5–2.1 mm, containing 4–6 seeds. Seeds narrowly elliptic, 1.5–2 × 0.5–0.7 mm.

Zimbabwe. S: Mwenezi Dist., SW of Mateke Hills, Malangwe R., ♀ fl. 6.v.1958, *Drummond* 5627 (K, LISC) & ♂ fl. 6.v.1958, *Drummond* 5628 (K, LISC, PRE). **Mozambique**. Z: Maganja da Costa, road to Namacurra, Muijaiana stream, ♀ fl. 27.i.1966, *Torre & Correia* 14220 (LISC). MS: Gorongosa Nat. Park, Xivulo, ♂ fl. 6.vii.1972, *Ward* 7763 (PRE). GI: near Massangena, ♀ fl. 27.vii.1973, *Correia & Marques* 3089 (LMU, WAG). M: Maputo (Delagoa) Bay, Lake Rikatla, 25°46'S; 32°37'E, ♀ fl. x.1893, *Junod* 430 (BR, Z); Marracuene (Vila Luísa), R. Incomati, 5.v.1968, *Balsinhas* 125-7 (COI).

Also in South Africa (northern provinces and KwaZulu-Natal) and Swaziland; possibly in Namibia (Bushmanland). Pans, dams and streams; 30–620 m.

Conservation notes: Widespread species; not threatened.

3. **Lagarosiphon muscoides** Harv. in J. Bot. **4**: 230, t.22 (1841). —Wright in F.T.A. **7**: 3 (1897); in F.C. **5**: 1 (1912). —Dinter in Repert Spec. Nov. Regni Veg. **18**: 435 (1922). —Wager in Trans. Roy. Soc. S. Afr. **16**: 192, 194 (1928). —Obermeyer in Bothalia **8**: 140, fig.1 (1964); in F.S.A. **1**: 103, fig.32 (1966). —Podlech in Merxmüller, Prodr. Fl. SW Afr., fam.142: 2 (1966). —Ross, Fl. Natal: 57 (1973). —Gibbs Russell in Kirkia **10**: 437 (1977). —Symoens & Triest in Bull. Jard. Bot.

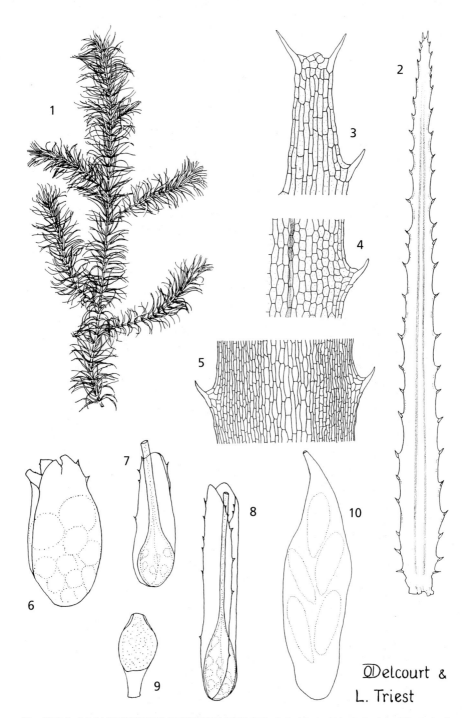

Fig. 12.2.**8**. LAGAROSIPHON VERTICILLIFOLIUS. 1, habit (× ¹/₂); 2, leaf (× 22); 3, leaf apex (× 125); 4, leaf margin (× 90); 5, leaf, central part (× 90); 6, male spathe (× 14); 7, 8, female spathe (× 14); 9, ovule and funicle (× 35); 10, fruit (× 10). 1–5 from *Ward* 2518; 6 from *Ward* 2549; 7 from *Correia & Marques* 1319; 8–9 from *Correia & Marques* 3089; 10 from *Codd* 6069. Drawn by O. Delcourt and L. Triest. Reproduced with permission from Bull. Jard. Bot. Nat. Belg. (1983).

Belg. **53**: 463, fig.8 (1983). —Simpson in F.T.E.A., Hydrocharitaceae: 4 (1989). —Clarke & Klaassen, Water Pl. Namibia: 70 (2001). Type: South Africa, Eastern Cape, Albany, *Zeyher* s.n. (TCD lectotype selected by Symoens & Triest 1983).

Hydrilla dregeana C. Presl in Abh. Böhm. Ges. Wiss., ser.5, **3**: 542 (1845). Type: South Africa, Eastern Cape, Port Elizabeth, Swartkops R., 1840, *Drège* 2276c (K).

Hydrilla muscoides (Harv.) Planch. in Ann. Sci. Nat., Bot. sér.3, **11**: 79 (1849).

Lagarosiphon schweinfurthii Casp. in Bot. Zeitung (Berlin) **28**: 88 (1870). —Ridley in J. Linn. Soc., Bot. **22**: 234 (1886). —Wright in F.T.A. **7**: 3 (1897). —Hepper in F.W.T.A., ed.2, **3**: 9 (B1968). Type: Sudan, Bongoland, Gir, 27.vii.1879, *Schweinfurth* 2158 (G lectotype selected by Symoens & Triest (1983), K, P, W).

Stems filiform, 0.5–1.5 mm wide. Leaves alternate, subopposite or locally whorled, spreading, linear, (4.8)6–15(20) × 0.5–1.5 mm, soft, thin, transparent, apex attenuate to very acute; epidermal cells predominantly elongate; margins with 2–6 rows of sclerenchyma fibres, each side with 28–86 teeth, not on protuberances, pointing upwards, straight and sharp; central area with slightly larger cells between two distinct lines of sclerenchyma fibres, midrib faint, no transverse septa. Intravaginal scales 0.25 × 0.1 mm, colourless, ovate to lanceolate, entire to papillate. Male inflorescences with spathe containing up to 40 flower buds, valves ovate to broadly ovate, 3–4 × 1.7–2.8 mm, bearing 3–30 teeth; male flowers white to pink; sepals 1 × 0.5 mm, petals 0.7–1 × 0.33 mm; stamens 1–1.7 mm long, staminodes 2–4.5 mm long. Female inflorescences with spathe valves ovate to narrowly ovate, 3.8–4.5 × 0.9–2 mm, with 7–41 teeth; female flowers with sepals 1.5–2 × 1.1 mm, petals 1.2–1.5 × 0.9 mm; ovary with 4–8 ovules, stigmas red. Capsule narrowly ovoid, tapered above, 4.3–8(10) × 1.3–2.2(3) mm broad. Seeds narrowly ellipsoid, 2–2.5 × 0.5 mm.

Caprivi Strip. Mentioned from E Caprivi by Clarke & Klaassen (2001). **Botswana**. N: Okavango R., between Mohembo and Shakawe, ♂ fl. 28.iv.1975, *Gibbs Russell* 2851 (BR, K). SE: Content Farm, ♀ fl. 10.iii.1978, *Hansen* 3372 (C, GAB, K, PRE, SRGH). **Zimbabwe**. W: Hwange Dist., Hwange Nat. Park, Caterpillar Pan, ♀ fl. 21.xi.1968, *Rushworth* 1280 (BR, K, P). C: Marondera Dist., Chikokorana Pan, ♂ & ♀ fl. 29.iv.1972, *Gibbs Russell* 1982 (BR, K, PRE). S: Beitbridge Dist., Tshiturapadsi (Chiturupadzi), Dip Camp area, 90 km E of Beitbridge, 19.iii.1967, *Mavi* 249 (PRE). **Malawi**. S: Zomba Mt., v.1970, *Vorster* 2145 (PRE).

Widely distributed in the Sudano-Zambezian region, Burkina Faso, Mali, Chad, Central African Republic, Sudan, Uganda, Tanzania, Namibia, Lesotho and South Africa (from Limpopo Province to Eastern Cape). Dams, pans, also in slow flowing water, associated with *Najas horrida* and *Potamogeton octandrus*; 400–1500 m.

Conservation notes: Widespread species; not threatened.

The forms *typica*, *brevifolia* and *longifolia* described by Wager (1928) have no taxonomic value, transplantation experiments having shown that differences in leaf length and number of teeth is influenced by light and temperature conditions. In optimum conditions, newly grown leaves are larger and bear more teeth.

4. **Lagarosiphon major** (Ridl.) Moss in Wager in Trans. Roy. Soc. S. Afr. **16**: 193, figs.4,6 (1928). —Wild in Kirkia **2**: 39, t.13d (1961). —Obermeyer in Bothalia **8**: 145, fig.1 (1964); in F.S.A. **1**: 106, fig.32 (1966). —Ross, Fl. Natal: 57 (1973). — Gibbs Russell in Kirkia **10**: 437 (1977). —Symoens & Triest in Bull. Jard. Bot. Belg. **53**: 476, fig.13 (1983). —Cook, Aq. Wetl. Pl. Sthn. Afr.: 144, fig.144 (2004). Type: South Africa, Umdizine R., 1880, *Cooper* 17 (BM lectotype selected by Symoens & Triest (1983), BOL, G, K, PRE, W, Z).

Lagarosiphon muscoides Harv. var. *major* Ridl. in J. Linn. Soc., Bot. **22**: 233 (1886). — Wright in F.C. **5**(3): 2 (1912).

Stems firm, c.3 mm wide. Leaves alternate, occasionally whorled, spreading or recurved, linear, 6.5–25 × 2–4.4 mm, firm, mostly opaque, apex mostly obtuse, rarely acute or acuminate;

epidermal cells predominantly equal-sided; margins with 2 (or 3 at leaf base) rows of sclerenchyma fibres, almost translucent, each side bearing 50–100 teeth, not on protuberances, relatively short, pointing upwards, blunt; central area narrow, consisting of some mesophyll, mostly with air lacunae and aerenchyma, no sclerenchyma fibres. Male inflorescences with spathe containing c.50 flower buds, valves ovate, 3–5 × 1.9–2.7 mm, bearing 3–18 small teeth; male flowers with pinkish perianth, sepals 1.25 × 0.66 mm, petals 1 × 0.35 mm; stamens 1.25 mm long, staminodes c.2 mm long. Female inflorescence with spathe valves ovate to narrowly ovate, 3.5(5.5) × 1.4(2.5) mm, bearing 2–3 small teeth; female flowers with pinkish perianth segments c.1.25 mm long, sepals somewhat broader than petals; ovary containing 10–12 ovules, stigmas purplish. Capsule ovoid to narrowly ovoid, tapered above, 4–5 × c.1 mm. Seeds narrowly ellipsoid, 2 × 0.7 mm.

Botswana. N: cited by Barnes & Turton, List Fl. Pl. Botswana: 37 (1986), no material seen. **Zimbabwe**. W: Matobo Dist., Matopos dam, st. 14.i.1958, *Rattray* in *GHS* 82664 (BM, K, SRGH). C: Harare Dist., Hunyani Poort dam, st. 29.x.1953, *Wild* 4146 (BM, K, PRE). E: Nyanga Dist., 8 km ENE of Nyanga, Troutbeck Lake, 31.xii.1973, *Bamps, Symoens & Vanden Berghen* 499 (BR).

Also in South Africa from the northern provinces to the Eastern Cape. Widely introduced, often invasive and troublesome in Europe and New Zealand. Dams and lakes; 900–2000 m.

Conservation notes: An invasive introduced species.

5. **Lagarosiphon ilicifolius** Oberm. in Bothalia **8**: 145, fig.1 (1964); in F.S.A. **1**: 108, fig.32 (1966). —Podlech in Merxmüller, Prodr. Fl. SW Afr., fam.142: 2 (1966). —Gibbs Russell in Kirkia **10**: 437 (1977). —Symoens & Triest in Bull. Jard. Bot. Belg. **53**: 483, fig.16 (1983). —Simpson in F.T.E.A., Hydrocharitaceae: 5 (1989). —Clarke & Klaassen, Water Pl. Namibia: 70 (2001). —Cook, Aq. Wetl. Pl. Sthn. Afr.: 143, fig.143 (2004). Type: Botswana, Lake Ngami, Toteng, 30.ix.1954, *Story* 4727 (PRE holotype). FIGURE 12.2.**9**.

Lagarosiphon muscoides var. major sensu Schinz in Bull. Herb. Boiss. **4** (app.3): 10 (1896), non Ridley.

Hydrilla verticillata sensu Gibbs in J. Linn. Soc., Bot. **37**: 471 (1906). —Eyles in Trans. Roy. Soc. S. Afr. **5**: 293 (1916). —T.C.E. Fries in R.E. Fries, Wiss. Ergebn. Schwed. Rhod.-Kongo-Exped.: 188 (1916). —Wild in Clark, Victoria Falls Handb.: 134, 138 (1952), non Royle.

Stems robust, 1.2–3 mm wide. Leaves alternate, occasionally subopposite or even whorled, mostly recurved, ovate to narrowly ovate, 4.5–13.5 × 2–3.7 mm, fairly firm, opaque, apex acute; epidermal cells predominantly equal-sided; margins with outer row of abaxial epidermal layer almost translucent, without sclerenchyma fibres, each side with 12–48 teeth on broad triangular protuberances, strong and sharp, many of them spreading or recurved; central area narrow, midrib faint, somewhat prominent below, cells somewhat larger and narrower, with neither septa or sclerenchyma fibres. Intravaginal scales colourless, acuminate, c.0.35 × 0.17 mm. Male inflorescence with up to 30 flower buds per spathe, valves ovate, 4–4.2 × 2–3 mm, pale violet or lilac, with 16–25 teeth; male flower not yet described. Female inflorescence with spathe valves ovate to narrowly ovate, 2–4.3 × 0.8–1.8 mm, pale violet or lilac, with 6–23 teeth; female flower with perianth pale lilac, c.1.6 mm wide; ovary with 6–9 ovules, stigma purple. Capsule ovoid to narrowly ovoid, tapered above, 3.5–7 × 1.2–1.5 mm, containing 6–9 seeds. Seeds narrowly ellipsoid, 1.5–2 × 0.5 mm.

Caprivi Strip: E Caprivi, Kamatanda, Lake Liambezi, ♀ fl. 26.ii.1975, *Edwards* 4325 (K, PRE). **Botswana**. N: Toteng, NE tip of Lake Ngami, fr. 20.ix.1954, *Story* 4727 (PRE); near Taun, Taoghe R., ♂ fl. 31.vii.1974, *P.A. Smith* 1069 (COI, SRGH). SE: Gaborone–Lobatse Road, Thlowane R., ♀ fl. 19.xii.1978, *Hansen* 3553 (BM, C, WAG). **Zambia**. B: Mongu Dist., Luanginga R., 7 km NW of Sandaula, ♀ fl. 17.xi.1959,

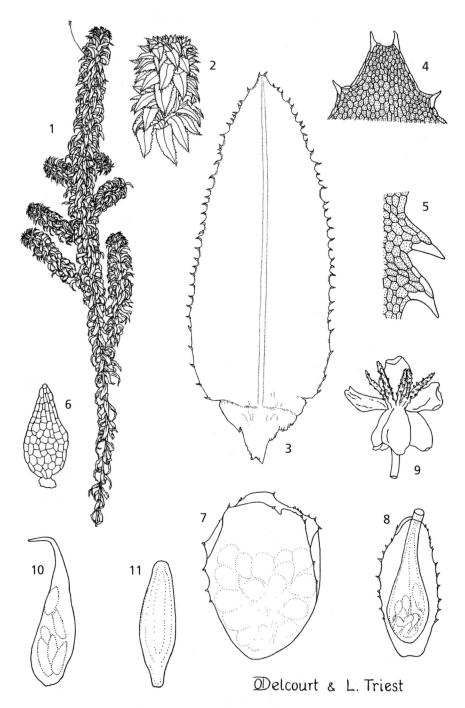

Fig. 12.2.**9**. LAGAROSIPHON ILICIFOLIUS. 1, female plant (× ¹/₂); 2, top of plant (× 1¹/₂); 3, leaf (× 16); 4, leaf apex (× 60); 5, leaf margin (× 107); 6, squamule (× 70); 7, male spathe (× 10); 8, female spathe (× 10); 9, female flower (× 11); 10, fruit (× 6); 11, seed (× 20). 1 from *Schmitz* 7049; 2–5 from *Symoens* 9257a; 6, 8 from *Menezes* 3461; 7, 9 from *H.& E.Hess* 52/1885; 10–11 from *Story* 4727. Drawn by O. Delcourt and L. Triest. Reproduced with permission from Bull. Jard. Bot. Belg. (1983).

Drummond & Cookson 6586 (BM). N: Luwingu Dist., Matongo Is., st. 19.ii.1959, *Watmough* 280 (K). S: Livingstone, Zambezi R., st. 21.vii.1941, *Greenway* 6252 (K, PRE). **Zimbabwe.** N: Binga, Lake Kariba, near mouth of Mosangwe R., ♀ fl. 4.ix.1973, *Gibbs Russell* 2591 (BR, K, PRE). W: Zambezi R., Victoria Falls, st. ix.1905, *Gibbs* 136 (BM).
Mostly distributed in the Zambezian area – Congo (Katanga), Angola, Namibia, and also in Lake Victoria. Lakes, dams, ponds and rivers, associated with *Ceratophyllum demersum, Nymphaea caerulea, Najas horrida, Hydrilla verticillata,* water depth up to 3 m; 500–1150 m.
Conservation notes: Widespread species; not threatened.

2. HYDRILLA Rich.

Hydrilla Rich. in Mém. Cl. Sci. Math. Inst. Nat. France **1811**(2): 9, 61, 73, 75 (1814). —Cook & Lüönd in Aquatic Bot. **13**: 485–504 (1982).

Perennial or annual freshwater herbs, gynodioecious, submerged except for flowers at anthesis. Roots unbranched; stems elongate, branched at distant intervals, stoloniferous below, erect and spreading above, forming turions (vegetative buds) on stolons, erect stems and branches. Leaves opposite and scale-like or whorled and foliate, sessile; margins serrate or toothed, apex with a single spine cell. Intravaginal scales mostly 2 in leaf axils, but 4 or more when axillary buds develop. Male inflorescences with solitary spathes in leaf axils, often 1 on each leaf, comprising 2 pear-shaped united bracts with subulate appendages at end; 1-flowered, opening to liberate the flower bud which rises to the surface where it opens. Male flowers with pedicel c.0.5 mm long, perianth of 2 whorls of 3 segments each; sepals supporting the floating flower, petals shorter and narrower than sepals; stamens 3, pollen projected explosively above water surface; staminodes absent, pistillodium absent. Female inflorescences with sessile spathes, membranous, bifid at top, 1(2)-flowered. Female flowers with inferior ovary, subsessile, the long hypanthium carrying bud to water surface where perianth segments open, forming an underwater funnel but open to air above; sepals oblong-ovate, petals shorter and much narrower than sepals; staminodes 3, minute; styles 3, simple, much shorter than perianth segments, ovules few, anatropous. Fruit cylindrical, smooth or armed, indehiscent, ripening under water. Seeds few, fusiform, testa smooth.

A monospecific genus comprising *H. verticillata,* widely distributed in Asia and Australia, perhaps also native around the African Great Lakes as well as in N Europe; recently introduced in several African, European and American countries.

Hydrilla verticillata (L. f.) Royle, Ill. Bot. Himal. Mts. **1**: 376 (1839). —Ridley in J. Linn. Soc., Bot. **22**: 236 (1886). —Cook & Lüönd in Aquatic Bot. **13**: 485, figs.1–5 (1982). —Simpson in F.T.E.A., Hydrocharitaceae: 7, fig.2 (1989). — Cook, Aq. Wetl. Pl. Sthn. Afr.: 142 (2004). Type: India, *Linnean Herb.* no.11066-1 (LINN lectotype). FIGURE 12.2.**10**.
   Serpicula verticillata L.f., Suppl. Pl.: 416 (1781).
   Hydrilla verticillata var. *brevifolia* Casp. in Pringsheim, Jahrb. Wiss. Bot. **1**: 495 (1858). — Gürke in Engler, Pflanzenw. Ost-Afr. **C**: 95 (1895). Types: India, E India, *Willdenow* 17636 (2 syntypes); *Royle* s.n. (P syntype); Coromandel, 1826, *Bélanger* 28 (syntype); *Wallich* 5048 (LINN syntype); *Stocks* in *Herb. Hooker* (K syntype).

Annual or perennial submerged aquatic herb. Roots long and simple, adventitious, arising at nodes. Stems either creeping and stoloniferous or erect, terete, slender, 0.3–0.8 mm wide, up to 2(3) m long, usually reddish, freely branched near base, usually sparingly branched above. Leaves opposite on stolons or at base of stem or branch, ovate or widely ovate, rarely more than 4 mm long, but whorled (3–12) on the erect stems and branches, mostly linear to narrowly

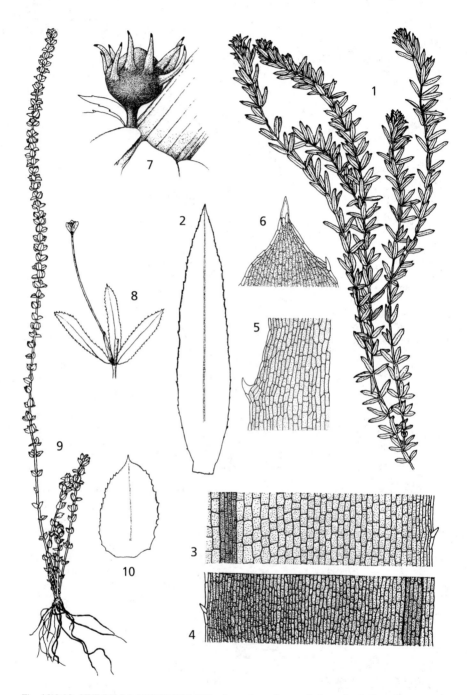

Fig. 12.2.**10**. HYDRILLA VERTICILLATA. Long-leaved type: 1, branched plant fragment (×
¹/₂); 2, leaf (× 8); 3, detail of upper leaf surface (× 66); 4, detail of lower leaf surface (× 66); 5,
leaf margin (× 40); 6, leaf apex (× 40); 7, male spathe (× 15); 8, flowering female spathe (× 2).
Short-leaved type: 9, habit (× ¹/₂); 10, leaf (× 8). 1 from *Bamps* 3171, 2–4 from *Ghesquière* 5009,
5–7 from *Gomes e Sousa* 3403, 8 from *Cyril* in EA 13883, 9–10 from *Symoens* 14760. Drawn by A.
Fernandes. Reproduced with permission from Flore d'Afrique Centrale (Meise).

lanceolate, sometimes widely ovate, (7)12–20 mm long (rarely more), 1.5–4 mm wide, soft, green, often with reddish-brown spots and stripes, wide and partly embracing stem at base, acute at apex; midrib distinct, sometimes with unicellular spines underneath; margins with 1–3 rows of translucent cells and unicellular spines, regularly spaced, pointing upwards, mostly visible to the naked eye. Intravaginal scales narrowly triangular to lanceolate, c.0.5 mm long, translucent, with finger-like orange-brown hairs. Male infloresences on peduncle up to 1 mm long, spathes comprising 2 pear-shaped or globose united bracts, c.1.5 mm wide, with 8–16 subulate appendages; male flowers with pedicel c.0.5 mm long, sepals strongly convex at anthesis, 1.5–3 mm × 2 mm, green at base, reddish higher, petals elongate, ± spathulate, whitish or reddish. Female spathes solitary or rarely 2 in leaf axils, sessile, cylindrical but tapering outwards, translucent or reddish-brown, disintegrating as fruit ripens; female flowers with filiform hypanthium, 15–100 mm long, sepals 2–4 mm long, translucent, streaked with red or white, petals transparent, occasionally with reddish streaks, staminodes colourless, ovary pink, 2–3 mm long, containing up to 10 ovules, stigmas 0.75–1 mm long, whitish to pinkish-purple. Fruit cylindrical, 5–10(15) × 1.5 mm, smooth or with lateral linear processes, developing under water. Seeds 2–5(7), borne in a linear sequence, c.2.5 mm long, brown.

Zambia. N: Mpulungu, Lake Tanganyika, st. 20.x.1947, *Greenway & Brenan* 8244 (BM). **Mozambique**. M: Marracuene Dist., Incomati valley, ♀ 10.iv.1946, *Gomes e Sousa* 3403 (BM, COI, K, LISU).

Widely distributed in Asia, New Guinea and Australia, probably also native in NE Europe. In Africa probably native in and around the central African lakes (Congo, Rwanda, Burundi, Uganda, Kenya, Tanzania, N Zambia), but a recent introduction elsewhere (Canary Is., Ivory Coast, Mozambique, Réunion and Mauritius); absent from Madagascar. In lakes, swamps, irrigation ditches and slowly flowing waters up to 3 m deep, sometimes invasive; 90–800 m (to 1950 m in East Africa).

Conservation notes: Introduced across most of its range in the Flora area; not threatened.

Hydrilla verticillata shows much variation in robustness, the number of leaves in each whorl, leaf length and shape, the number of rows of marginal cells, degree of ramification of stems, length of internodes, etc. Isoenzyme analysis and random amplified DNA analysis show the existence of several population groups of *Hydrilla* according to geographical origin. The plants from the East African Great Lakes have short, leathery, ovate and mostly recurved leaves, and are always sterile. They can be confused with *Lagarosiphon ilicifolius*, especially when both species grow together. They can be distinguished by examining the lower leaves on stems and branches – in *Hydrilla* leaves are in whorls while in *L. ilicifolius* they are spirally arranged. Moreover, the leaf apex bears only 1 spine in *Hydrilla* and 2 in *Lagarosiphon*.

3. OTTELIA Pers.

Ottelia Pers., Syn. Pl. **1**: 400 (1805). —Cook, Symoens & Urmi-König in Aquatic Bot. **18**: 263–274 (1984).

Damasonium Schreb. in Linnaeus, Gen. Pl., ed.8: 242 (1789), nomen illegit., non Miller.

Boottia Wall., Pl. Asiat. Rar. **1**: 51 (1830).

Freshwater herbs, annual or perennial, monoecious, dioecious or trioecious (♀, ♂ and hermaphrodite). Roots unbranched; stems usually erect, simple, contracted and corm-like, or sometimes creeping, rhizomatous and irregularly branched. Leaves usually arising from base, without stipules: juvenile leaves linear to elliptic or ovate, not or hardly differentiated into petiole and blade; adult leaves sessile or petiolate; petiole usually sheathing at base, submerged, sinuous, sometimes laterally winged, armed or unarmed; blade submerged or floating, ovate, elliptic to suborbicular, or lanceolate, strap-like or even graminiform, base attenuate, wedge-shaped, rounded or cordate, apex obtuse, acute or acuminate, margins entire, denticulate or prickly, sometimes undulate to curled; up to 15 prominent longitudinal veins connected by cross-veins, the latter often oblique giving the blade the appearance of a quilt with rhomboidal patches. Intravaginal scales sometimes present, 2–10 per leaf. Inflorescences with long peduncles, spathes solitary, ovoid to narrowly cylindrical, sometimes compressed, with wings, ribs, spines or warts, rarely smooth, 2–6-lobed at apex, 1–60-flowered. Flowers bisexual or unisexual. Male flowers up to 60 per spathe, stalked, remaining attached at anthesis. Bisexual and female flowers 1–2 per spathe (outside the Flora area sometimes up to 4 or even 12), sessile, sometimes cleistogamous. Sepals 3, free, linear, oblong or ovate, green with membranous margins, persistent above fruit in bisexual and female flowers. Petals 3, free, oblong, obovate or suborbicular, large, white or coloured, showy, but delicate and short-lived. Stamens (3)6–15 (occasionally more), in whorls of 3, some (all in female flowers) staminodial, filaments often flattened, anthers basifixed or slightly dorsifixed, dehiscing laterally. Carpels 3–15(20 or more), united into an ovoid, ellipsoid or narrowly cylindrical ovary, placentation laminar-diffuse, with placentas sometimes protruding nearly to ovary centre, ovules numerous (to 200 or more), anatropous; styles 3–15(20), each divided into 2 linear, papillate, stigmatic arms, pistillode usually 3-lobed in male flowers. Fruit a somewhat fleshy capsule, ovoid or cylindrical, attenuate at apex, regularly dehiscent or opening irregularly by decay of pericarp. Seeds numerous, small, oblong or fusiform, sometimes with a short apical projection, testa mostly covered with flaccid unicellular hairs, somewhat mucilaginous forming an abundant pulp shed shortly after seeds are liberated from the fruit.

A genus of about 25 species of the warm regions of the Old World with 1 species in Brazil and 1 species naturalized in Europe. Nine species recognised in the Flora Zambesiaca area.

Ottelia alismoides (L.) Pers. has been mentioned from Lake Malawi (Wright in F.T.A. **7**: 7, 1897) and from Inhaca Is., Mozambique (Mogg in Macnae & Kalk, Nat. Hist. Inhaca Is., ed.1: 140, 1958). However, as its presence in the Flora Zambesiaca area seems improbable and could not be confirmed by any herbarium specimen, it has not been included here. *O. alismoides* is easily recognisable with its large membranaceous leaves, approximately as long as wide, mostly cordate, and the spathes have 3–12 longitudinal wings (see Cook & Lüönd in Aquatic Bot. **20**: 133, 1984).

Key to Ottelia *species with mature leaves*

1. Flowers bisexual · 2
   – Flowers unisexual · 3
2. Plants smooth or bearing only minute teeth (0.1–0.25 mm long) on leaf margins and spathes · **1.** *ulvifolia*
   – Plants with leaf blades prickly on main veins and often on margins; spathes coarsely prickly or tuberculate · **9.** *verdickii*
3. Leaves sharply delineate into petiole and blade · 4
   – Leaves not sharply delineated into petiole and blade, linear or lanceolate · · ·5
4. Base of leaf blade rounded or slightly cordate, never cuneate or attenuate; blade 3–13.5 × 2– 7 cm, ovate to nearly circular, rarely elliptic · · · · · · · · · · · **2.** *exserta*
   – Base of leaf blade mostly cuneate, only non-cuneate in blades more than 14 cm long; blades (8)11–22(42) × 6–13.5 cm, elliptic, elliptic-oblong or elliptic-lanceolate, rarely ovate · **3.** *fischeri*
5. Leaves more than (1.7)2 cm wide · 6
   – Leaf blades 0.2–1.5(2) cm wide · 7
6. Male spathes very elongate, length/width ratio 7–11.6, not winged; flowers yellow · **4.** *luapulana*
   – Male spathes oblong-lanceolate, length/width ratio 3.5–5, slightly (c.1 mm) winged; flowers white · **5.** *lisowskii*
7. Leaves thick, triangular in cross-section; margins and midrib with coarse prickles · **7.** *muricata*
   – Leaves thin, flat, mostly translucent; margins and midrib smooth or with very fine teeth · 8

8. Leaves grass-like or very narrowly lanceolate · · · · · · · · · · · · · · · · **6.** *cylindrica*
– Leaves strap-shaped · **8.** *kunenensis*

1. **Ottelia ulvifolia** (Planch.) Walp. in Ann. Bot. Syst. **3**: 510 (1852). —Ridley in J. Linn. Soc., Bot. **22**: 238 (1886). —Dandy in J. Bot. **72**: 138 (1934). —Wild in Clark, Victoria Falls Handb.: 138 (1952). —Brenan in Mem. New York Bot. Gard. **9**: 80 (1954). —Pedro in Bol. Estudos Moçamb. **24**: 19 (1954). —Obermeyer in F.S.A. **1**: 109, fig.35 (1966). —Podlech in Merxmüller, Prodr. Fl. SW Afr., fam.142: 3 (1966). —Hepper in F.W.T.A., ed.2, **3**: 7, t.6 (1968). —Gibbs Russell in Kirkia **10**: 437 (1977). —Munday & Forbes in J. S. Afr. Bot. **45**: 3 (1979). — Simpson in F.T.E.A., Hydrocharitaceae: 15, fig.6 (1989). —Clarke & Klaassen, Water Pl. Namibia: 72 (2001). —Cook, Aq. Wetl. Pl. Sthn. Afr.: 147 (2004). Type: Madagascar, *Lyell* 149 (K lectotype).

   Damasonium ulvifolium ('*ulvaefolium*') Planch. in Ann. Sci. Nat., Bot. sér.3, **11**: 81 (1849).
   Ottelia lancifolia A. Rich., Tent. Fl. Abyss. **2**: 280, t.95 (1850). —Eyles in Trans. Roy. Soc. S. Afr. **5**: 293 as *O. paucifolia* A. Rich. (1916). —T.C.E. Fries in R.E. Fries, Wiss. Ergebn. Schwed. Rhod.-Kongo-Exped.: 190 (1916). —Binns, First Check List Herb. Fl. Malawi: 53 (1968). Type: Ethiopia, Shire, *Quartin-Dillon* (P holotype).
   Ottelia lancifolia var. *fluitans* Ridl. in J. Linn. Soc., Bot. **22**: 238 (1886). —Wright in F.T.A. **7**: 7 (1897). Type: Angola, Pungo Andongo, Pedra Songue, 1857, *Welwitsch* 6468 (BM holotype).
   Ottelia vesiculata Ridl. in J. Linn. Soc., Bot. **22**: 237 (1886). —Eyles in Trans. Roy. Soc. S. Afr. **5**: 293 (1916). Type: Angola, Huíla, Lopollo & Mumpulla, iv.1860, *Welwitsch* 6467 (BM holotype, M).
   Ottelia plantaginea Ridl. in J. Linn. Soc., Bot. **22**: 238 (1886). —Wright in F.T.A. **7**: 7 (1897). —Suessenguth & Merxmüller, Contrib. Fl. Marandellas Dist.: 73 (1951) as *O.* cf. *plantaginea*. Type: Angola, Huíla, Catumba, iv.1860, *Welwitsch* 6469 (BM holotype).
   Ottelia latifolia De Wild., Pl. Nov. Hort. Then. **5**: 153, t.33 (1905). Type: Mozambique, Morrumbala Mt., xii.1900, *Luja* 392 (BR sheet 1 holotype, BR sheet 2).
   Ottelia obtusifolia T.C.E. Fr. in R.E. Fries, Wiss. Ergebn. Schwed. Rhod.-Kongo-Exped.: 190, fig.18a (1916). Type: Zambia, Ndola, viii.1911, *R.E. Fries* 503 (UPS holotype).
   Ottelia australis Bremek. in Ann. Transv. Mus. **15**: 235 (1933). Type: South Africa, Pietersburg Dist., Vivo vlei, 20.i.1931, *Bremekamp & Schweickerdt* 203 (PRE holotype).
   Ottelia vernayi Bremek. & Oberm. in Ann. Transv. Mus. **16**: 401 (1935). Type: Botswana, Maun, Chobe R., 2.vi.1930, *van Son* in *TRV* 28852 (PRE holotype, BM).

Annual or perennial, submerged herb, monoecious, smooth or bearing only minute, unicellular teeth (0.1–0.25 mm long) on leaf margins, peduncles and spathes. Roots to 28 cm long, white; stem erect, contracted, corm-like. Leaves numerous, often in thick tufts, petiole distinct, 6–30(100) × 0.1–0.6 cm, or undifferentiated from a gradually tapering blade; blade submerged, floating only at very low water levels, ovate-lanceolate to linear-lanceolate, (3.5)8–40(45) × (1)2–4.5(11.5) cm, acute to broadly obtuse, thin and often translucent, pale green to brownish-green, often marked with irregular purple zebra-markings, (5)7–13 prominent longitudinal veins with many finer parallel veins connected by cross-veins; margins ± undulate, smooth or very finely denticulate. Inflorescences with ± 3-angled peduncles, 14–50(60) × 0.3–0.7 cm, smooth or with very minute teeth, loosely spirally twisted and retracted after anthesis; spathes elliptic-oblong, ovate or lanceolate, sometimes nearly circular, compressed, more rarely subcylindrical, (15)25–45(60) × 4–20(30) mm, green, turning brown in fruit, smooth or minutely prickly, (3)5–15(19) visible veins on each face, mostly with 2 wings (up to 4 mm broad), ± undulate, paler and sometimes nearly translucent; apex mostly opening in 2 lobes, sometimes deeply cleft at maturity. Flowers 1 (rarely 2) per spathe, bisexual, emerging just above the water, falling early; sepals 3, lanceolate to linear, 8–15(20) × (1.6)2–3(4) mm, ± translucent, pale green to pale pinkish-brown, darker at tip, with 1–3(7) darker green or brown veins; petals 3, obovate, (10)12–22(30) × 8–10 mm, yellow, white or white with yellow base, ± crinkled; stamens (3)6, 4–8 mm long, anthers oblong; ovary with (3)6 carpels, narrowly ellipsoidal, pale green, styles (3)6, each split into 2 long, orange-yellow stigmas. Fruit ovoid to oblong-cylindrical, 20–40 × 12 mm, opening by decay of pericarp, hypanthium exserted by 3–5(10) mm above spathe. Seeds numerous, small, oblong to fusiform, 2–2.5 × 0.5–0.7 mm, embedded in a pulpy mass; testa dark brown.

Caprivi Strip: E Caprivi, Sangwali, 6.x.1970, *Vahrmeijer* 2168 (K). **Botswana**. N: Mababe R., fl. 17.vi.1978, *P.A. Smith* 2459 (K, SRGH). **Zambia**. B: c.16 km NE of Mongu, fl. 18.xi.1959, *Drummond & Cookson* 6595 (K, SRGH). N: Lubunda, Mubende R., bridge on Luapula Valley road, fl.& imm.fr. 1.iv.1961, *Symoens* 8461 (BR, BRVU, K). W: Mufulira Dist., near Mufulira, fl. 28.ix.1947, *Greenway & Brenan* 8123 (BM). E: Lundazi Dist., 37 km from Lumimba School to Lundazi, after Mbuzi Wildlife Gate, fl.& fr. 14.viii.1979, *Chisumpa* 546 (K, NDO). S: Choma Dist., Mapanza to Pemba road, c.13 km N of Muhamwa R., fl. 23.iv.1963, *van Rensburg* 2082 (K). **Zimbabwe**. N: Hurungwe Dist., Mwami (Miami), fl. iv.1926, *Rand* 39 (BM). W: Matobo Dist., Matopos Hills, fl. iv.1904, *Eyles* 34 (BM). C: Marondera (Marandellas), fl. 28.iii.1943, *Dehn* 758 (M). E: Chipinge Dist, near Chibuya (Chibuwe), fl. 1.vi.1972, *Gibbs Russell* 2065 (BR, M, MO, SRGH). S: Mentioned by Gibbs Russell (1977). **Malawi**. N: Nkhata Bay Dist., Kabunduli road, fl. 10.vii.1969, *Pawek* 2606 (K). C: Nkhota Kota (Kota Kota), fl.& fr. viii.1946, *Vernay* 17398 (NY). S: Machinga Dist., W of T.A. Nkhokwe, fl. 28.iv.1982, *Patel* 858 (BR, K, MAL). **Mozambique**. N: Lichinga (Vila Cabral), fl. 23.v.1934, *Torre* 125 (BM, COI). Z: Morrumbala Mt., fl.& imm.fr. xii.1900, *Luja* 392 (BR). T: Fíngoè, Rio Mucanha, fl. 31.v.1962, *Gomes e Sousa* 4763 (COI, K, M, UPS). MS: Inhamadzi (Inyamadzi) Valley, fr. 8.ix.1906, *Swynnerton* 781 (B, BM, K). M: Inhaca Is. (see Munday & Forbes 1979, no material seen; probably extinct due to severe reduction of swamps up to and during the 20th century).

Widespread over most of tropical Africa from Senegal to Sudan and Ethiopia southward to Namibia, Botswana and the northern provinces of South Africa; also in Madagascar. Mainly in watercourses, also in lakes, dams, ponds, backwaters, swamps, irrigation channels, and ephemeral river pools, water depth 0.1–3 m, on clay or sandy bottoms; 90–1800 m, but to 2700 m outside the Flora area. Flowers during the rainy season.

Conservation status: Widespread taxon; not threatened.

Ottelia ulvifolia is very variable in total height, leaf shape and size and spathe-shape. Specimens from the Flora area have flowers with yellow petals, while those from the forest regions of the Congo and neighbouring countries have white petals.

Ottelia obtusifolia, only known from its holotype with indistinct wings and obtuse leaves, and *Ottelia vernayi* with minutely scabrid leaves, peduncles and spathes, are within the variation range of *O. ulvifolia*. Delicate specimens with ascending lanceolate leaves, 6.5–20 × 0.4–1.3 cm, are not uncommon in Tanzania, Zambia and south to central Zimbabwe.

2. **Ottelia exserta** (Ridl.) Dandy in J. Bot. **72**: 137 (1934). —Obermeyer in F.S.A. **1**: 109, fig.34 (1966). —Podlech in Merxmüller, Prodr. Fl. SW Afr., fam.142: 3 (1966). — Ross, Fl. Natal: 57 (1973). —Gibbs Russell in Kirkia **10**: 437 (1977). —Hall-Martin & Drummond in Kirkia **12**: 157 (1980). —Simpson in F.T.E.A., Hydrocharitaceae: 17 (1989). —Cook, Aq. Wetl. Pl. Sthn. Afr.: 146 (2004). Types: Mozambique, Shire R. above Shamo, viii.1859, *Kirk* (K syntype); Chupanga (Shupanga), 5.i.1860, *Kirk* (K syntype); Madagascar, *Forbes* 2 (BM syntype). FIGURE 12.2.11.

    Boottia exserta Ridl. in J. Linn. Soc., Bot. **22**: 240, t.13 (1886). —Durand & Schinz, Consp. Fl. Afr. **5**: 4 (1894). —Gürke in Engler, Pflanzenw. Ost-Afr. **C**: 95 (1895). —Wright in F.T.A. **7**: 8 (1897).

Boottia macrantha C.H. Wright in F.T.A. **7**: 9 (1897). Type: Mozambique, Quelimane, viii.1887, *Scott* s.n. (K holotype).

Boottia mossambicensis Peter in Abh. Ges. Wiss. Göttingen, Math.-Phys. Kl., N. Folge **13**: 109 (1928); in Repert. Spec. Nov. Regni Veg. Beih. **40**(1): 123 (1929). Types: Mozambique, between Beira and Manga, 27.ix.1925, *Peter* 31090 (B sheet 1, lectotype here designated, B sheet 2); same locality, 30.viii.1925, *Peter* 30518 (B syntype); Beira, 24.ix.1925, *Peter* 31084 (B 2 syntypes).

Ottelia macrantha (C.H. Wright) Dandy in J. Bot. **72**: 138 (1934).

Annual or perennial aquatic herb, 20–50 cm high, dioecious or exceptionally monoecious. Stem erect, corm-like, with fibrous orange roots to 18 cm long. Leaves from stem base; juvenile leaves submerged, mostly falling early, linear to narrowly lanceolate, up to 25 × 1 cm or more, margins often undulate; adult leaves with distinct petiole, 9–45 cm long, 1–4 mm wide, triangular in cross-section, smooth or with rows of minute prickles on base margins and nerves, base to 2.5 cm wide, sheathing, pale; blades floating, sometimes erect in air, ovate to broadly ovate, more rarely elliptic, 3.5–13.5 × 2.2–7.5 cm, (1.15)1.5–2(2.7) times longer than wide, coriaceous, 5–7 prominent longitudinal veins interspersed with finer ones, distinctly connected by fine cross-veins; margins entire and smooth, base rounded or slightly cordate, not cuneate, apex rounded to obtuse. Male inflorescences with peduncle 6–47 cm long; spathes ovoid, oblong or lanceolate, 36–75 × 10–30 mm, flattened, often with 2 wings c.1.5–2 mm wide, smooth or with minute translucent hairs, inflated when ripe, opening unilaterally with apex torn to several irregular acute lobes. Male flowers 6–40 or more in each spathe, 1–4 appearing simultaneously at anthesis; pedicels exserted c.40 mm above spathe; sepals lanceolate, 8–12(21) × 2–4 mm, with darker veins and tips; petals obovate, 16–27(32) × 12–17 mm, white with cream or yellowish base; stamens 12, filaments flattened, anthers yellow; pistillode globose, 3-lobed, 2–3 mm high, yellow. Female inflorescences with peduncle 16.8–20.5 cm long; spathes narrowly ovoid, 30–60 × 7–10 mm; unwinged, grooved, with c.10 darker longitudinal stripes, smooth or with c.0.15 mm prickles. Female flowers 1 per spathe, hypanthium exserted 6–17 mm above spathe; sepals oblong-lanceolate, 8–16(23) × 2–6 mm, up to 28 × 5 mm in fruit, petals obovate, 16–30(48) × 12–23(30) mm, white with yellowish base; staminodes 3; styles 9–12, bifid, stigmatic branches yellow. Fruit spindle-shaped, 2.5–4 cm long. Seeds numerous, ellipsoid, 3–5 × 1.3–2 mm, apiculate at both ends.

Caprivi Strip. Mentioned from E Caprivi by Clarke & Klaassen (2001). **Zambia**. C: Mpika Dist., S Luangwa Nat. Park (Game Reserve), Mfuwe, ♂ fl. 3.v.1965, *Mitchell* 2778 (K) (intermediate with *O. fischeri*, K). E: Luangwa Valley, Chikwa, ♂ fl. 3.vi.1954, *Robinson* 827 (K). **Zimbabwe**. N: Hurungwe Dist., Marongora, c.4 km E of Nyakasanga R., Madzia pans, ♂ fl. 26.ix.1981, *Pope* 2000 (BR, MO, SRGH, U). E: Chipinge Dist., Chibunje, between Manzwire and Maduku, ♂ fl. 31.i.1975, *Gibbs Russell* 2715 (BR, K, M, SRGH). S: Mwenezi Dist., 5 km from Malipati, ♂ fl. 25.iv.1961, *Drummond & Rutherford-Smith* 7524 (K, SRGH). **Malawi**. S: Lengwe Nat. Park (Game Reserve), ♂ spathe 2.vii.1970, *Hall-Martin* 799 (K, PRE). **Mozambique**. Z: Madal Plantations, 32 km N of Quelimane, ♂ fl.& fr. 25.viii. 1962, *Wild & Pedro* 5897 (BM, K). T: Sisitso Station, Zambezi R., S bank, ♂ fl. 11.vii.1950, *Chase* 2606 (BM). MS: Gorongosa Nat. Park, 2.5 km from Chitengo camp on road no.1, 40 m, ♂ & ♀ fl. 30.iv.1964, *Torre & Paiva* 12162 (intermediate with *O. fischeri*, LISC). GI: Guijá, near Caniçado, Limpopo R., ♂ & ♀ fl. 6.v.1944, *Torre* 6579a (LISC).

Also in Somalia, Kenya, Tanzania, Angola and South Africa. Often in large masses in swamps, pools, perennial pans and dams, wet mud of seasonal pans and in slow-flowing streams; associated with *Chara, Nymphaea, Limnophyton, Sesbania*, etc., water depth 25–45 cm; 10–1000 m (to 1500 m in E Africa).

Conservation notes: Widespread taxon; not threatened.

O. exserta is a rather variable species. There are also intermediates between *O. exserta* and *Ottelia fischeri*, hence the two taxa may prove to be conspecific. Cultivation experiments with plants of diverse origin in waters of different depth

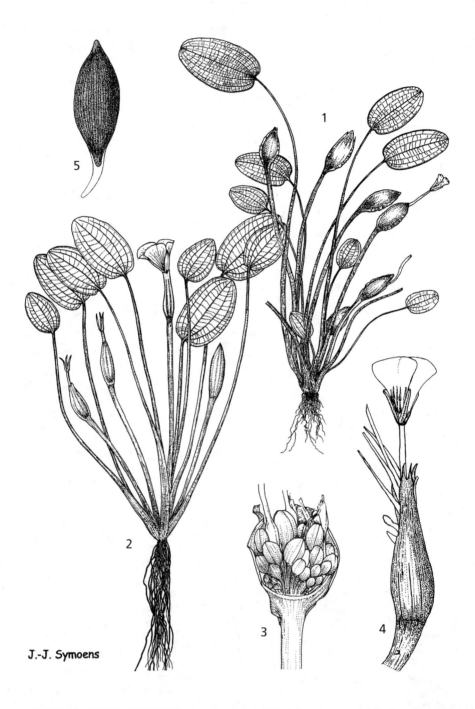

J.-J. Symoens

Fig. 12.2.11. OTTELIA EXSERTA. 1, male plant (× ¹/₃); 2, female plant (× ¹/₃); 3, male spathe forced open (× ³/₄); 4, male spathe and flower (× ¹/₂); 5, seed (× 6). 1 from *Phipps* 2908, 2, 5 from *Peter* 31090, 3 from *Codd* 6136, 4 from *Bidgood, Mwasumbi & Vollesen* 1199. Drawn by J.-J. Symoens.

and current speed and molecular studies are required. Plants from Mozambique identified as *O. macrantha*, with very broad to nearly circular leaves and large flowers (female petals 40–48 mm long, male petals 30 mm long) seem to represent a local form of *O. exserta*. They grow together at the same sites as the more usual small-flowered form of the species.

3. **Ottelia fischeri** (Gürke) Dandy in J. Bot. **72**: 137 (1934). —Simpson in F.T.E.A., Hydrocharitaceae: 20, fig.7 (1989) in part. Types: Tanzania, Usiha, *Fischer* 588 (B† holotype); Ussangu, Msimasi, vi.1899, *Goetze* 1026 (B†, BR neotype here designated).

    Boottia fischeri Gürke in Engler, Pflanzenw. Ost.-Afr. **C**: 95 (1895). —Wright in F.T.A. **7**: 8 (1897). —Gürke in Urban & Graebner, Festschr. Aschers.: 536 (1904).

Annual or perennial aquatic herb, dioecious, to 60(200) cm long. Stem erect, corm-like. Leaves from stem base; juvenile leaves submerged, blades narrowly elliptic, to 11.5 × 4 cm, attenuate at base; adult leaves with distinct petiole to 50 cm or more long; sheaths to 6 cm long; blades submerged or floating, elliptic to narrowly elliptic, more rarely elliptic-ovate or ovate, (8)11–22(42) × 5.6–13.5 cm, 1.5–2.6(3.5) times longer than wide, with 7–9 prominent longitudinal veins, interspersed with 7–9 finer longitudinal veins connected by faint, slightly ascending veins, base usually cuneate (rounded and non-cuneate only where blade longer than 14 cm), apex rounded or obtuse; margins entire, smooth. Male inflorescences with peduncle 5–100 cm long, 0.5–1.7 cm wide, broadening to 22 mm under spathe; spathes ovoid, oblong or lanceolate, (48)63–102 × 18–38 mm, smooth, inflated when ripe. Male flowers up to 60 or more per spathe, up to 6 developing together at anthesis, funnel-shaped; pedicels 60–130 mm long, exserted to 90 mm above spathe at anthesis; sepals lanceolate, 18–25 mm × 4–6.5 mm wide, greenish with darker veins and tip, translucent margins; petals obovate, 30–50 × 22–30 mm, white with a cream or rose base. Stamens 12, unequal, the shortest 8–20 mm long, filaments flattened; anthers 3.2–5 mm long, yellow; staminodes 3, filiform, forked, 15–17 mm long; pistillodium globose, 3-lobed, 2–3 mm high. Female inflorescences with peduncle probably as long as in male plants; spathe oblong-cylindrical, 43–78(95) × 12–25 mm, smooth, with longitudinal veins or ribs and paler furrows, opening apically with 2 larger lobes and several smaller acute teeth. Female flowers 1 per spathe; hypanthium exserted (6)13–19 mm above spathe, up to 8 mm wide in fruit; sepals lanceolate, 22–30 × 4.5–6.5 mm, to 35 × 10 mm in fruit, 7-nerved, green, with translucent margins; petals obovate to nearly circular, 40–50(60) × 20–30 mm, white with yellowish base; staminodes awl-shaped, 10–12 mm long; ovary green, 9–15 carpels, ovoid-oblong; styles 9–12(15), deeply bifid, stigmatic branches 10–23 mm long, yellow. Fruit and seeds not seen.

Zambia. C: Lusaka Dist., near Mbiya school, Lake Chalimbana, ♂ spathes 27.iii.1963, *van Rensburg* KBS 1837 (K). **Malawi**. S: Chiromo, junction of Shire and Ruo rivers, ♂ spathe 13.vii.1958, *Seagrief* CAH 3095 (K, SRGH). **Mozambique**. Z: Between Quelimane and Nicuadala, ♂ fl. 27.vii.1942, *Torre* 4414 (LISC). T: Nhandoa, ♀ fl. 20.iv.1905, *Le Testu* 726 (P). MS: Dondo, Pungué R., ♂ fl. 5.ix.1944, *Mendonça* 2007 (LISC).

Also in Uganda, Kenya and Tanzania. Ponds, pools, swamps, marshes, drainage ditches, on clay and rich silt associated with *Nymphaea*, *Leersia*, *Polygonum*, *Cyperus*, *Typha*, etc.; water depth to 60 cm; 10–1100 m (to 1500 m in Tanzania).

Conservation notes: Moderately widespread taxon within the Flora area; not threatened.

In 1895, Gürke published an incomplete diagnosis of *Boottia fischeri* based on *Fischer* 588 (B), which was probably destroyed. However, a few years later he examined better specimens and proposed a new type specimen (*Goetze* 1026) and gave a detailed description where he apparently combined the characters of both specimens (Urban & Graebner, Festschr. Aschers.: 536, 1904). Both specimens in

Berlin were destroyed during the war, but a duplicate of *Goetze* 1026, consisting of a petiolate leaf and a flowering male spathe, is deposited at BR and is designated here as the neotype.

The specimens from the Kafue, lower Shire, lower Zambezi and lower Pungwe river basins are in the upper size range of the species.

In the Quelimane area of Mozambique the spathes are sold in markets and eaten in curry (*Schäfer* 6755), while in Manica District the corms are eaten (*Pedro* 4247). Large-flowered plants are possibly suitable for cultivation as aquatic ornamentals.

4. **Ottelia luapulana** Symoens sp. nov.[2] Type: Zambia, Luwingu Dist., Goodall Channel, 16.ii.1959, *Watmough* 261 (not 269) (K sheet 1 holotype, SRGH). FIGURE 12.2.**12**.

Submerged, perennial, dioecious herb, 55–65 cm high. Stem erect, corm-like, simple, 1–2.5 cm long. Leaves radical, in dense rosette; petiole 14–34 cm long, sheathing at base, scabrid; blade lanceolate or oblanceolate, (8)13–31 × (1)2–5(6) cm, thin, green to olive-green, sometimes flushed red, long-narrowed into petiole, subacute to obtuse; with 5–9 prominent scabrid longitudinal veins, interspersed with many fine parallel veins, irregularly connected by transverse veins; margins with many fine teeth. Male inflorescence with peduncle 21–57 cm long, simple, exceptionally 2-forked, with many small teeth, some recurved; spathe cylindrical, (39)44–72 × 5–9.5 mm, tapering above, not winged, with many small teeth, sometimes smooth above; sutures of bracts forming 2 longitudinal membranous and paler stripes; apex with 2 lobes 1–5 mm long, ± divergent. Male flowers c.10 per spathe, pedicel exserted 10–35 mm above spathe at anthesis; sepals lanceolate, 13–18 × 1–3.5 mm, 3-veined, darker and obtuse at apex; petals c.25 × 27 mm, very thin, yellow; stamens 12, unequal, 8–16 mm long, the longest sometimes staminodial; filaments broadened, 5–10 mm long; anthers 3.4–5 mm long; pistillodium c.2 mm high, 3-lobed. Female spathes with peduncle 28 cm long, smooth or with small teeth, spathe body narrowly cylindrical, 45–60 × 3.5–5.5 mm, 10-nerved, with small teeth, sometimes smooth above; apex 2-lobed. Female flowers 1 per spathe; sepals lanceolate, 3-veined, 18–19 × 2.5–3.5 mm; petals c.25 mm long, very thin, yellow; ovary cylindrical; hypanthium exserted 10 mm above spathe at anthesis; stigmas orange-yellow. Fruits and seeds not seen.

Zambia. N: Luwingu Dist., Nsombo, Bangweulu Swamps, ♂ fl. 14.x.1947, *Greenway & Brenan* 8205 (BM, K).

Only known from the Bangweulu–Luapula River basin. Also in Congo (Luapula R., Katanga Province). Swamp lagoons and channels, water depth c.60 cm, associated with *Nymphaea, Najas* and Cyperaceae; 1000–1140 m.

Conservation notes: Endemic to N Zambia and adjacent parts of Congo; probably Vulnerable.

---

[2] Herba aquatica submersa, dioica, 55–65 cm alta. Caulis erectus, cormosus, simplex. Folia radicalia in densa rosula, petiolus basim vaginans, muricatus, lamina lanceolata, tenuis, in petiolum longe attenuata, subacuta, vel obtusa, muricata. Spathae masculae, cylindricae, productae, basim paulo incrassatae, haud alatae. Flores masculi c.10 in spatha, pedicelli exserti usque ad 10–35 mm, sepala 13–18 × 1–3.5 mm, petala lutea, stamina 12, inaequalia, 8–16 mm longa, longiora staminodialia. Spathae femineae 1-florae, anguste cylindricae. Flores feminei 1 in spatha, sepala 18–19 × 2.5–3.5 mm, petala c.25 mm longa, lutea, hypanthium usque ad 10 mm exsertum sub anthesi. Fructus et semina non visa. Ab *O. lisowskii* petalis luteis et spathis masculis productioribus differt.

Fig. 12.2.**12**. OTTELIA LUAPULANA. 1, male plant ($\times$ ^{1}/$_3$); 2, leaf apex ($\times$ ^{3}/$_4$); 3, male spathe ($\times$ 1); 4, male spathe split and forced open ($\times$ ^{3}/$_4$); 5, male flower ($\times$ ^{4}/$_5$); 6, short stamen ($\times$ 2); 7, female spathe and flower ($\times$ 4/5). 1, 3, 5–6 from *Greenway & Brenan* 8205, 2, 4, 7 from *de Witte* 13. Drawn by O. Delcourt. Reproduced with permission from Flore d'Afrique Centrale (Meise).

5. **Ottelia lisowskii** Symoens sp. nov.[3] Type: Congo, Haut-Shaba, W of Minga, near Kalemba, Kikole R., 1080 m, 23.iv.1971, *Lisowski* 10589 (POZG holotype, BR, K).

Aquatic perennial herb, submerged or half-floating, 50–90 cm high, dioecious. Stem erect, corm-like, simple, c.2 cm long. Leaves in a dense rosette, 46.5–85 × 2.5 cm, linear to narrowly lanceolate, long attenuate towards base, thin, olive-green, with 5–7 prominent longitudinal veins interspersed with c.6 fine parallel veins, irregularly connected by transverse veins; margins smooth, apex acute. Male inflorescence with peduncle 30–80 × 0.2–0.4 cm; spathe oblong-lanceolate, 45–65 × (7)11–15 mm, 3.5–5 times longer than wide; smooth. Male flowers c.15 per spathe, pedicel exserted 2 cm; petals white; stamens 9, in 3 circles, successively 3.6 mm, 4.5 mm and 5.8 mm long; staminodes 3, threadlike, 6 mm long. Female spathe 1-flowered, oblong-elliptic, 36 mm × 17 mm, compressed, with 7 longitudinal veins on each face, smooth, Fruit ellipsoid, c.27 mm × 13 mm wide, surmounted by 2 cm long hypanthium. Seeds numerous, ellipsoid, c.1 × 0.5 mm, apiculate; testa areolated.

Zambia. W: Ndola Dist., Shibuchinga, fl. 2.iii.1964, *Fanshawe* 8376 (K, NDO).
Also in Congo (Katanga). In rivers; c.1300 m.
Conservation notes: Restricted distribution; probably Vulnerable.

6. **Ottelia cylindrica** (T.C.E. Fr.) Dandy in J. Bot. **72**: 137 (1934). Type: Zambia, Katakwe near Mporokoso, ♂ & ♀ fl., 30.x.1911, *R.E. Fries* 1170 (UPS holotype).

Boottia cylindrica T.C.E. Fr. in R.E. Fries, Wiss. Ergebn. Schwed. Rhod.-Kongo-Exped.: 190, fig.18 (1916).
Boottia stratiotes T.C.E. Fr. in R.E. Fries, Wiss. Ergebn. Schwed. Rhod.-Kongo-Exped.: 189, fig.17 (1916). Type: Zambia, Bangweulu, Kasomo, Chimona R., 20.ix.1911, *R.E. Fries* 688 (UPS holotype).
Ottelia stratiotes (T.C.E. Fr.) Dandy in J. Bot. **72**: 138 (1934). —Astle in Kirkia **7**: 93 (1968).
Ottelia cylindrica var. *stratiotes* (T.C.E. Fr.) Cook, Symoens & Urmi-König in Aquatic Bot. **18**: 271 (1984) in key, invalidly published.

Perennial aquatic herb, up to 120 cm high, Dioecious, leaves, peduncles and spathes smooth or with 0.4–0.9 mm long prickles. Rhizomes creeping, 85–175 × (3)6–8 mm, simple or branched, with numerous roots and contracted stems up to 5 cm long. Leaves completely submerged, grass-like to very narrowly lanceolate, 17–111 cm long by 0.25–12 mm wide, often thin and translucent, pale green or brownish, long attenuate to base into a 0.5–2 mm wide sheathing petiole; apex long acute, ending in a thread-like process, 10–20 mm long, with a few small teeth; up to 9 darker longitudinal nerves; margins smooth or minutely rough. Peduncles ± as long as leaves, 1–4 mm thick, thicker under the spathe, pale green or brownish, smooth or sparsely prickly under spathe, female peduncle becoming spiralized after fertilization. Male spathes oblong-ellipsoid or oblong-lanceolate, slightly inflated, 21–46 × (3.5)7.5–12 mm, smooth or with small prickles. Male flowers up to 25 (or more) per spathe, mature pedicel exserted 17–25 mm above spathe at anthesis; sepals lanceolate, 8–18 × 1.5–4 mm, white or pinkish, apex obtuse and darker; petals obovate, 13–24 mm long, to 14 mm wide, delicate and transparent, white; stamens 9–12, unequal, the longest 8–13 mm, filaments broadened towards base, anthers c.2 mm long; pistillode globose, c.1 mm wide, 3-lobed, yellow or orange. Female spathes narrowly cylindrical, straight or slightly curved, 28–65 ×

---

[3] Herba aquatica, submersa vel seminatans, dioica, perennis, 50–90 cm alta. Caulis erectus, cormosus, simplex. Folia radicalia, in densa rosula, sessilia, linearia ad anguste lanceolata, longe attenuata versus basim, nervis et marginibus laevibus; apex acutus. Spathae masculae oblongo-lanceolatae, anguste (c.1 mm) alatae. Flores masculi: c.15 in spatha; pedicelli exserti usque ad 2 cm; petala alba, sepala 13 × 3 mm; stamina 9, inaequalia, staminodia 3, filiformia. Spathae femineae 1-florae; corpus oblongo-ellipticum, 36 × 13–15 mm. Flores femineae: sepala 16 × 2 mm in fructu. Fructus ellipsoideus, c.27 × 13 mm. Semina multa, c.1 × 0.5 mm, ellipsoidea, apiculata; testa areolata. Ab *O. luapulana* differt foliorum marginibus nervisque laevibus, floribus albis, et spathis masculis minus productis, ab *O. brachyphylla* altitudine distincte majori et floribus albis.

2.5–7 mm, opening at apex in 2 translucent lobes; longitudinal nerves 6(8), darker, smooth or bearing scattered small prickles, sometimes dense (up to 10–15 per cm). Female flowers 1 (exceptionally 2) per spathe, hypanthium exserted 6–36 mm above spathe; sepals linear-lanceolate, (9)14–23 × 1–4 mm, 3-nerved, green with translucent margins near base, tip darker; petals obovate (15)20–25 × 10–20 mm, white, sweet scented; staminodes 3, subulate, 6–7(10) mm long; ovary c.15 mm long, styles 3, bifid, 12–16 mm long, stigmatic branches 7–9 mm long. Fruit oblong, c.3 cm long, 3 mm wide. Seeds ellipsoidal, 1.3 mm long.

Zambia. N: Mporokoso Dist., Kasaka R. on road to Mporokoso, 1350 m, 1.iv.1957, *Richards* 8960 (BM, K).

Also in the Congo (Katanga) and Angola. In fast flowing streams and drainage channels, among peat-bogs; 1100–1800 m (to 1900 m in Angola).

Conservation notes: Localised species, possibly Vulnerable in the Flora area.

The number and degree of development of the prickles is variable and there are intermediates between entirely smooth specimens and those bearing small prickles on leaves, peduncles or spathes; moreover some specimens have both smooth and prickly spathes. Consequently, the recognition of *O. stratiotes* as a taxon distinct from *O. cylindrica* should not be maintained, even at an infraspecific level.

7. **Ottelia muricata** (C.H. Wright) Dandy in J. Bot. **72**: 138 (1934). —Obermeyer in F.S.A. **1**: 111, fig.35 (1966). —Podlech in Merxmüller, Prodr. Fl. SW Afr., fam.142: 3 (1966). —Gibbs Russell in Kirkia **10**: 437 (1977). —Clarke & Klaassen, Water Pl. Namibia: 72 (2001). —Cook, Aq. Wetl. Pl. Sthn. Afr.: 146, fig.147 (2004). Types: Botswana, Ngamiland, Botletle R., viii.1896, *F.D.& E.J. Lugard* 13 (K lectotype) & Chobe R., *McCabe* 32 (K syntype). FIGURE 12.2.13.

Boottia muricata C.H. Wright in F.T.A. **7**: 569 (1897).

Boottia aschersoniana Gürke in Warburg, Kunene-Samb.-Exped. Baum: 173 (1903). — T.C.E. Fries in R.E. Fries, Wiss. Ergebn. Schwed. Rhod.-Kongo Exped.: 189 (1916). Type: Angola, Cuito, 1200 m, 10.iii.1900, *Baum* 772 (M holotype, W).

Perennial or annual aquatic herb, up to 140 cm high, dioecious, with prickles on all green organs. Rhizome oblique, simple or forked, 30–130 × 9–10 mm, with roots up to 27 cm long and several leaf rosettes. Leaves all submerged, without clear differentiation into petiole, strap-like, linear to narrowly oblanceolate, triangular or nearly flat in cross-section, dorsally convex, (12)25–57(200) × 0.5–1.2(1.9) cm, bright green, attenuate towards sheathing base, broader leaves with up to 5 nerves, each bearing a prickle row on the dorsal side; margins with coarse prickles, 2–5 per cm, 0.6–2.5 mm long, spreading or pointing towards apex, hard when dry; apex obtuse in juvenile leaves, otherwise ending in a thread-like process with a few small teeth. Male inflorescence with peduncles to 125 cm or more long, mostly 2.5–5 mm thick, narrowing towards base, mostly bearing coarse prickles (2–3 per cm); male spathes up to 12(or more)-flowered, subcylindrical to ellipsoid-oblong, 25–51 × (5)7–15 mm, 2.3–4 times longer than wide, with 7–10 longitudinal ridges, each with a row of prickles, 1–3 mm long, spreading or pointing upwards, sometimes recurved, the spathe dehiscing distally in 2 lobes. Male flowers expanding singly above water, pedicels exserted 14–40 mm above spathe; sepals lanceolate, 10–15 × 2–3 mm, apex obtuse or subacute, darker; petals broadly obovate, 15–33 × 10–16 mm, 7-veined, white with yellow base and sometimes pink veins; stamens mostly 12, in circles of 3, of which 0–6 in the 2 inner may be staminodial, filaments flattened, from external to inner 5–12 mm long; anthers c.3 mm long; pistillode 1.5 × 1 mm, orange. Female inflorescence 1-flowered, peduncles to 125 cm or more long; spathes narrowly cylindrical, not inflated, 40–70 × (2)3–7 mm, mostly 10–15 times longer than wide, with 6–9 longitudinal ridges, each with a row of spreading or recurved coarse prickles, spathe opening distally in 2 lobes. Female flowers with linear sepals, 13–22 × 1.5–4 mm, with purple streaks, apex subacute; petals obovate, 21–30(40) mm long, white; staminodes 6 (3 long, 3 small and scale-like); ovary with parietal placenta not protruding into ovary cavity; styles 3, bifid, stigmatic branches papillate, c.12 mm long; mature hypanthium exserted 15–23 mm above spathe. Fruits and seeds not seen.

Caprivi Strip. E Caprivi, Linyanti R., ♂ & ♀ fl. 10.x.1970, *Vahrmeijer* 2186 (K, PRE). **Botswana**. N: 13 km S of Shakawe, Okavango R., ♂ fl. 17.x.1969, *Brown* 10009 (K, SRGH). **Zambia**. B: near Senanga, ♂ fl. 30.vii.1952, *Codd* 7234 (BR, COI, K, L, MO, P, PRE, SRGH). N: Shiwa Ngandu, Lake Ishiba Ngandu, off mouth of Manshya R., ♂ & ♀ fl. 14.viii.1936, *Ricardo* 33 (BM). S: near Katombora, Zambezi R. above Livingstone, ♂ fl. 14.vii.1927, *C.E. Moss* 14861 (K). **Zimbabwe**. N: Mentioned by Gibbs Russell (1971).

A readily recognizable species, although variable in leaf width and strength of prickles. A typical species of the Zambezian Region, also in Congo (Luapula R.), Angola and Namibia. Lake edges, dams, swamps and swamp channels, ditches, also in running water, associated with *Echinochloa* and *Phragmites*, water depth 0.75–3.5 m; 1000–1500 m.

Conservation notes: Widespread species; not threatened.

8. **Ottelia kunenensis** (Gürke) Dandy in J. Bot. **72**: 137 (1934). —Obermeyer in F.S.A. **1**: 111, fig.34 (1966). —Podlech in Merxmüller, Prodr. Fl. SW Afr., fam.142: 3 (1966). —Clarke & Klaassen, Water Pl. Namibia: 72 (2001). —Cook, Aq. Wetl. Pl. Sthn. Afr.: 146, fig.147 (2004). Type: Angola, between Kiteve and Humbe, on Cunene R., 3.vi.1900, *Baum* 962 (B† holotype, BM lectotype designated here, COI, G, K, M, W).

> *Boottia kunenensis* Gürke in Warburg, Kunene-Samb.-Exped. Baum: 172 (1903). —Dinter in Repert. Spec. Nov. Regni Veg. **15**: 351 (1918).

Annual or perennial aquatic herb, up to c.90 cm high, dioecious. Leaves all submerged, sessile, long and strap-shaped, linear to narrowly lanceolate, (17)30–67 × 0.5–1.5(2.5) cm, pale-green, often with purple blotches, thin and translucent; margins ± wavy, with minute unicellular prickles (18–28 per cm), rarely smooth, apex obtuse to acuminate; main nerves 5–7, often minutely rough on back. Male inflorescence 5–10-flowered, peduncles triangular in cross-section, to 45 cm or more long; spathes oblong, 30–58 × 5–8 mm, green, sometimes with transverse darker markings, smooth or minutely rough. Male flowers with pedicel up to 5 cm, exserted 10–12 mm above spathe at anthesis; sepals linear-lanceolate, 10–11(18) × 3–4 mm, with translucent margins; petals obovate, 15–16(25) mm long, white with creamy yellow base; stamens 9–15, subsessile anthers c.2 mm long, pistillode globose, 3-lobed. Female inflorescence 1-flowered, peduncles to 90 cm long, coiling after anthesis; spathes cylindrical, 25–45 × 7–15 mm, enlarging in fruit, sometimes with transverse reddish or brownish markings; nerves c.15 on each face, mostly minutely rough; mouth splitting into 3–8 teeth at dehiscence. Female flowers with sepals 13–15 × 2.5–3 mm, margins translucent; petals 26 × 20 mm, white with yellow base; styles 6, 10–15 mm long, deeply bifid, hypanthium exserted to 13 mm above spathe. Fruit ellipsoidal c.22 × 7 mm. Seeds numerous.

Caprivi Strip. Kabuta, Chobe R., ♂ fl. 23.v.1954, *Munro* ML3 (BM). **Botswana**. N: Okavango R., Caprivi border, fl.& fr. 27.iv.1975, *Gibbs Russell* 2824 (K, MO, SRGH). **Zambia**. B: Mongu, ♀ fl. 9.iv.1994, *Bingham* 10044 (K). C: Mumbwa Dist., Shibuyunji, ♂ fl. 14.v.1963, *van Rensburg* 2143 (K). S: Kalomo Dist., Mambora (Mambova), Zambezi R., ♂ fl. 29.v.1954, *Munro* ML18 (BM, K, PRE).

Also in Angola and Namibia. Perennial rivers, floodplains and pools, associated with *Phragmites*, water depth 60–90 cm; 1000–1100 m.

Conservation notes: Moderately widespread in Upper Zambezi area; not threatened.

Ottelia kunenensis is easily recognizable owing to its long, strap-shaped, translucent, finely denticulate leaves.

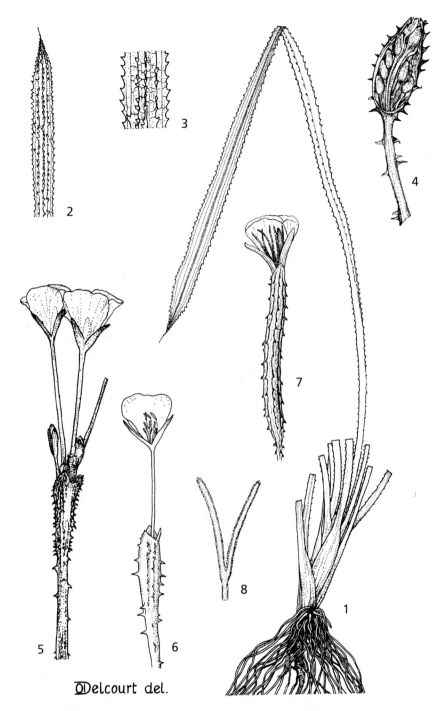

ⅅelcourt del.

Fig. 12.2.**13**. OTTELIA MURICATA. 1, base and leaf ($\times\,^1/_2$); 2, leaf apex, lower surface ($\times\,^1/_2$); 3, leaf part, lower surface ($\times$ 1); 4, male spathe split longitudinally ($\times\,^3/_4$); 5, male spathe with two flowers ($\times\,^3/_4$); 6, top of male spathe with flower, two petals removed ($\times\,^1/_2$); 7, female spathe and flower ($\times\,^3/_4$); 8, top of style and stigmatic branches ($\times\,2^1/_2$). 1–3 from *R.E. Fries* 788, 4 from *Codd* 7234, 5 from *Gibbs Russell* 2844, 6 from *Vahrmeijer* 2186, 7, 8 from *Hess* 52/2138. Drawn by O. Delcourt. Reproduced with permission from Flore d'Afrique Centrale (Meise).

9. **Ottelia verdickii** Gürke in De Wildeman in Ann. Mus. Congo, Bot., sér.4, **1**: 171 (1903). —Simpson in F.T.E.A., Hydrocharitaceae: 19 (1989). Type: Congo (Katanga), Lake Mweru (Moero), ix.1900, *Verdick* s.n. (BR holotype).

Ottelia gigas T.C.E. Fr. in R.E. Fries, Wiss. Ergebn. Schwed. Rhod.-Kongo-Exped.: 191 (1916). Type: Zambia, Kawendimusi, Lake Bangweulu (Bangweolo), 25.ix.1911, *R.E. Fries* 789 (UPS holotype).

Rooted, perennial aquatic herb, 0.6–2.1 m high, submerged or partly floating, monoecious. Roots to 30 cm long, often orange. Stems erect, corm-like, simple, 2.6–4 × 0.6–1.5 cm wide. Submerged leaves linear or narrowly lanceolate,15–100 (or more) × 1.5–5(10) cm, distinctly petiolate or with blade very gradually attenuate into petiole, with 3–7 prominent longitudinal nerves; floating leaves petiolate, petiole 6–44 × 0.3–1.2 cm, with a few minute prickles, blade lanceolate to oblong-elliptic, 15–120 × 5–10 cm, pale to dark green, ± purplish mottled, with 7–11 prominent, longitudinal, scabrid nerves connected by cross-veins; margins ± undulate, smooth or with small prickles, especially near apex; apex obtuse or acute, sometimes hooded. Inflorescence 1-flowered, peduncle 16–200 cm long, 0.35–12(18) mm wide, pinkish at base, dark green in upper parts, rather coarse prickly under spathe, retracting spirally after anthesis; spathe oblong to oblong-lanceolate, sometimes inflated in lower half, ± 3-angled but not winged, 43–110 × 4.5–15(24) mm, several longitudinal ribs with strong conical tubercles or prickles, apex bilobed, often lacerate after dehiscence. Flowers hermaphrodite, falling early; sepals 15–25(34) × 4.2–6.5 mm, up to 12 mm wide after anthesis, green or pale purplish-red, obtuse, 7-veined, margins often translucent; petals obovate, 25–36 × c.30 mm, very delicate, whitish, with 7–9 main veins; stamens 9, filaments 3–5 mm long, narrowly triangular, anthers to c.10 mm long, connective with appendage 1 mm long; ovary with (8)9(11) carpels, narrowly ellipsoid, 57–65 × 10–18 mm, unilocular, placentation laminar-diffuse, with partitions protruding deeply into the ovary; styles (8)9(11), bifid to below middle, stigmatic branches 13–25 mm long, densely papillose, yellowish. Fruit topped by a robust hypanthium up to 3.5 mm wide, sepals up to 35 × 12 mm. Seeds embedded in abundant mucilage, narrowly cylindrical, 3–4(5.5) × 1 mm, with terminal appendage 2–7 mm long.

Zambia. N: Samfya Dist., near Samfya Mission, fl. 26.viii.1952, *White* 3146 (BM, K, WAG). C: between Kabwe (Broken Hill) and Bwana Mukubwa, x.1906, *Allen* 329 (K); Kasanka Nat. Park, Kasaku R., fl. 17.xi.1992, *Breteler* 11841 (WAG).

Also in Angola, Congo and Tanzania. Swamps and rivers, often forming a fringe to papyrus–*Thelypteris* at lake margins, water depth to 2.4 m; 1150–1200 m.

Ottelia verdickii is very variable in size of leaves, spathes and flowers. A small specimen has been described from the Congo as var. *elanga* by De Wildeman (Pl. Bequaert. **1**: 455, 1922), whilst *O. gigas* is the name given by T.C.E. Fries (in R.E. Fries 1916) to the largest specimens from the Bangweulu–Luapula area.

4. BLYXA Thouars

Blyxa Thouars, Gen. Nov. Madag.: 4 (1806). —Cook & Lüönd in Aquatic Bot. **15**: 1–52 (1983).

Freshwater monoecious or dioecious herbs, annual or perennial. Roots adventitious, unbranched. Stems either elongate, erect and floating, or horizontal and stoloniferous, or contracted into a forked or simple rootstock or corm. Leaves submerged, radical or from stem, spirally arranged, without clear differentiation into petiole and blade, midrib ± distinct, with 0–10 secondary and up to 28 tertiary parallel veins, faint cross-veins often present; margins with microscopic unicellular teeth, mainly towards apex. Intravaginal scales usually 2, more at branches, triangular to narrowly triangular, white, margins entire. Spathes comprising 2 united bracts, tubular, narrowly transversely rhombic in cross-section, 2- or 6-nerved, 2-lobed, 1-, 2- or multiflowered. Flowers unisexual or bisexual, emerging at anthesis above water surface, usually one at a time from each spathe, or submerged and cleistogamous, the male borne on elongate pedicels, the female having an elongate hypanthium. Sepals 3, free but remaining tubular during anthesis, linear to linear-lanceolate, persistent; petals 3, free, band-shaped and scarcely wider than sepals, capillary or rudimentary, papillate. Stamens 3 in one whorl in bisexual species, 6 in 2 whorls of 3, or 9 in 3 whorls of 3 in unisexual species. Ovary inferior, sessile within the spathe, 3-carpellate, tubular, elongated into hypanthium, or reduced to a 3-lobed pistillode in male flowers; placentation parietal, ovules numerous, anatropous; stigmas 3, simple, linear, papillate. Fruit an elongate capsule, disintegrating at maturity. Seeds numerous, ellipsoidal or fusiform, with or without ridges, wings and spines.

A genus of 9 species of tropical regions in the Old World and Australia, naturalized in Italy and in USA. Four species in Africa, 3 in the Flora Zambesiaca area.

1. Flowers bisexual; spathes 1 (rarely 2)-flowered; stamens 3; leaves from base, tapering to a fine point at apex · **1.** *aubertii*
 – Flowers unisexual; male spathes many-flowered, stamens 6; female spathes 1 (rarely 2)-flowered; leaf apex rounded or obtuse to subacute · · · · · · · · · · · · 2
2. Stems corm-like or occasionally somewhat elongate; leaves from base, gradually tapering above; leaf apex obtuse to subacute; midrib distinct or faint, often absent near apex; margins near apex with microscopic unicellular teeth · · **2.** *hexandra*
 – Stems creeping, slender and stoloniferous, occasionally thick and rhizomatous; leaves spaced along stolons or clustered into rosettes, linear; leaf apex abruptly rounded, truncate or hooded; midrib distinct, mostly with distinct longitudinal, parallel veins; margins mostly without unicellular teeth · · · · · · · · · **3.** *radicans*

1. **Blyxa aubertii** Rich. in Mém. Cl. Sci. Math. Inst. Natl. France, **1811**(2): 19, 77, t.4 (1814). Type: Madagascar, *du Petit-Thouars* s.n. (P holotype).

Generally an annual herb. Stems erect, simple or forked, rootstock corm-like, 0.5–5 cm long, up to 1 cm wide. Leaves basal, 2.5–60 cm long, 0.2–1.2 cm wide, lanceolate and attenuate to a fine point at apex, or linear and narrowed below middle towards base; midrib distinct. Spathes sessile or with peduncle up to 50 cm long, 1 (rarely 2)-flowered, tubular, laterally flattened, (2)4–7(12) cm long, 6-nerved, persisting in fruit. Flowers 1 (rarely 2) in each spathe, bisexual, sessile, usually emergent, occasionally submerged and cleistogamous; hypanthium (1)5–7(15) cm long. Sepals (4.5)6–8(10) mm long, up to 1 mm wide, green, often striped with purple; petals linear, band-like, often twisted, up to 17 × 0.5 mm, white or reddish, often remaining folded and twisted within calyx. Stamens opposite sepals, filaments up to 4 mm long; anthers 1–1.8 mm long, dehiscing latrorsely before the petals emerge from the calyx; stigmas linear, 10–15 mm long, somewhat ovate in transverse section, the distal part papillose, white or reddish. Fruit cylindrical but narrowed into hypanthium, 20–80 × 5 mm. Seeds ellipsoid, light-brown.

var. **aubertii**. —Cook & Lüönd in Aquatic Bot. **15**: 9 (1983). —Simpson in F.T.E.A., Hydrocharitaceae: 10 (1989). —Cook, Aq. Wetl. Pl. Sthn. Afr.: 141 (2004).

Seeds 1.25–1.8 mm long, smooth or with up to 12 irregular or interrrupted longitudinal ridges or ribs, seed appearing minutely rough; no long spines.

Mozambique. MS: Gorongosa Nat. Park, Cheringoma Plateau, Muaredzi headwaters, fr. v.1972, *Tinley* 2565 (K, LISC, PRE, SRGH).

In Tanzania (Mafia Is.), Mozambique and Madagascar. Still or flowing water to 2 m deep; c.400 m (near sea-level in East Africa).

Conservation notes: A very localised taxon in the Flora area; probably Vulnerable.

Var. *aubertii* differs from var. *echinosperma* (C.B. Clarke) Cook & Lüönd from India, SE Asia, China, Japan and Australasia by having smaller and less spiny seeds.

2. **Blyxa hexandra** Cook & Lüönd in Mitt. Bot. Staatsamml. München **16**: 485, fig.1 (1980). —Cook & Lüönd in Aquatic Bot. **15**: 36, fig.8 (1983). —Simpson in F.T.E.A., Hydrocharitaceae: 10, fig.4 (1989). Type: Angola, Benguela, 30 km from Chicuma in direction of Capala, Cassipera, 1.v.1952, *Hess* 52/1483 (Z holotype, BM, Z, ZT). FIGURE 12.2.**14**.

Submerged annual, dioecious. Stems corm-like, shortly erect. Leaves from base, linear or grass-like, 10–20(40) × 0.1–0.25(0.45) cm, narrowed below middle towards base and gradually attenuate above, minutely obtuse to subacute, light yellowish green, thin and almost translucent; midrib distinct or faint, often absent towards apex, sometimes with 2 very faint secondary veins; margins with microscopic, unicellular teeth, mainly near apex, marginal fibres generally distinct. Male inflorescence with peduncles up to 40 cm long, taller than leaves, spathes cylindrical, 3.5–6.5 cm long. Male flowers up to 20 or more per spathe, 1 or 2 maturing and withering from each spathe each day, pedicels up to 13 cm long; sepals 4–7 × c.1.5 mm, green and somewhat translucent; petals linear, up to 15.6 × 1.5 mm, apically joined at first, later often spreading, white, adaxial surface with longitudinal furrows and coarse pollen-carrying papillae; stamens in 2 whorls of 3, the lower with filaments c.1 mm long and anthers 1.75 mm long, the upper with filaments c.3 mm long and anthers 1.25 mm long; pistillode rudimentary, 3-lobed. Female inflorescence with peduncles up to 25 cm long, spathes narrowly cylindrical, 5.5–8.5 cm long, 6-nerved. Female flowers 1 (rarely 2) per spathe, hypanthium up to c.20 cm long, exserted for 4–11.5 cm above spathe; sepals 6–10.5 × 0.8–1.5 mm, somewhat hooded at apex, pale green, often with dark purple streaks; petals filiform, 20–25 mm long, white, withering as stigmas spread; staminodes 3, rudimentary or absent; stigmas linear, up to 27 mm long, mimicking petals of male flowers, adaxial surface with longitudinal furrow and papillae. Fruit cylindrical, 4–6 cm long. Seeds ellipsoid-fusiform, 1.5–2.5(3) × 1 mm, shortly beaked; testa smooth or weakly tuberculate, straw-coloured to brown.

Zambia. N: Kaputa Dist., Mweru Wantipa road from Bulaya, ♂ fl. 9.iv.1957, *Richards* 9111 (BM, K); Kasama Dist., 61 km W of Isoka, fr. 12.viii.1966, *Gillett* 17421 (BM, EA, K).

Also in Congo (Katanga), Burundi, Tanzania and Angola. Shallow water in lakes, ponds, dambos and temporary pools, also in muddy streams, often associated with papyrus, *Thelypteris*, *Limnophyton*, water depth 10–30 cm; 1000–1700 m.

Conservation notes: Within the Flora area a very limited distribution in NE Zambia; possibly Vulnerable.

3. **Blyxa radicans** Ridl. in J. Linn. Soc., Bot. **22**: 236, t.14 (1886). —Cook & Lüönd in Aquat. Bot. **15**: 39, fig.9 (1983). Type: Angola, Huíla, Lubango, Lopolo stream, v.1860, *Welwitsch* 6471 (BM sheet 2 lectotype, selected by Cook & Lüönd).

Perennial submerged herb, dioecious. Stems creeping, slender and stoloniferous, occasionally thicker and rhizomatous, up to 40 cm long, irregularly branched, rooting at nodes. Leaves solitary and spaced along stolons or clustered in rosettes, linear, (3)15–30(70) × (0.05)0.2–0.35(0.5) cm, somewhat narrowed below the middle and petiole-like above the dilated, sheathing membranous base, apex abruptly rounded; midrib distinct with up to 9 very fine longitudinal, parallel veins on each side. Male inflorescence with peduncles (0.5)10–40(70) cm long, spathes 2–6 cm long. Male flowers up to 20 (or more) per spathe, 1 or 2 maturing and withering from each spathe each day, pedicels 5–9 cm long; sepals 4.5–8.5 mm long, ± translucent, brownish or violet; petals linear, 16–25 mm × 1.5 mm, apically joined at first, later mostly spreading, white, adaxial surface with longitudinal furrows and pollen-carrying papillae; stamens in 2 whorls of 3, the lower with almost sessile anthers 2.5 mm long, the upper with filaments c.1.5 mm long and anthers 1.5 mm long; pistillode rudimentary, 3-lobed. Female inflorescence with peduncles up to 40 cm long or more, spathes cylindrical, 3–5.5 cm long, 1(2)-flowered. Female flowers sessile, hypanthium to c.10 cm long; sepals 5–7 mm long, opaque, brownish; petals filiform, white, as long as stigmas and withering as the latter spread; staminodes 3, rudimentary, linear, or absent; stigmas linear, 13–19 mm long,

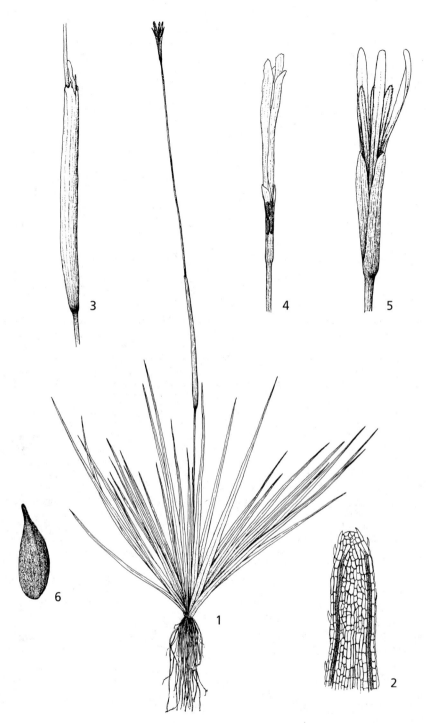

Fig. 12.2.**14**. BLYXA HEXANDRA. 1, female plant (× ³/₄); 2, leaf apex (× 40); 3, male spathe (× 1¹/₂); 4, male flower (× 4); 5, female flower (× 4); 6, seed (× 8). 1–2, 5 from *Milne-Redhead & Taylor* 9165, 3–4 from *Bequaert* 379, 6 from *Gillett* 17421. Drawn by A. Fernandes. Reproduced with permission from Flore d'Afrique Centrale (Meise).

mimicking petals of male flowers, white, the adaxial surface with a longitudinal furrow and papillae. Fruits and seeds not seen.

Zambia. B: Mongu Dist., Namushakende, 24 km S of Mongu, ♂ fl. 26.vi.1955, *King* 41B (K). N: Mbala Dist., near Kambole, Mwami R., ♀ fl. 4.vi.1957, *Richards* 9999 (BM, K). W: Mwinilunga Dist., 28 km N of Mwinilunga, Chitunta R., ♂ fl. 10.vi.1963, *Drummond* 8277 (K, LISC, PRE).

Also in Central African Republic, Congo (Katanga), Angola and Tanzania. Lake margins and ponds, also in streams, water depth 0.15–2 m, associated with *Nymphaea* and *Utricularia*; 900–1700 m (to 2200 m in Angola).

Conservation notes: Moderately widespread in the Flora area; not threatened.

Blyxa radicans is closely related to *B. hexandra*, differing mainly in its stoloniferous or rhizomatous perennial habit and the leaf margins mostly being without unicellular teeth.

5. VALLISNERIA L.

Vallisneria L., Sp. Pl.: 1015 (1753). —Lowden in Aquatic Bot. **13**: 283 (1982).
Physkium Lour., Fl. Cochinch.: 662 (1790).

Perennial, freshwater, submerged herbs (except for flowers at anthesis), dioecious. Roots unbranched, vertical stem axis mostly short and bearing stolons, rarely erect and elongate. Leaves basal or distinctly alternate on elongate stems (*V. caulescens*), sessile, first erect and submerged, but floating when reaching the water surface, linear, strap-shaped, apex obtuse; margins entire or bearing minute unicellular teeth, more dense near apex; main veins 3–5(9), longitudinal, parallel, connected by cross-veins, only midrib reaching the apex, the other longitudinal nerves gradually joining together near apex. Inflorescences in leaf-axils, pedunculate; spathes comprising 2 united bracts. Male spathes shortly pedunculate, ovoid, 2-lobed at apex, containing many flower buds. Male flowers very minute, pedicellate, abscising as buds and rising to the water surface where they open and float freely, slightly irregular; sepals 3, ovate-oblong to broadly ovate, reflexed at anthesis; petal 1, rudimentary; stamens (1)2–3, usually 2 fertile, free or with united filaments, and 1 staminode 1–4-sporangiate. Female spathes tubular with bifid apex, 1-flowered, on a long coiled peduncle bringing flower up to surface at anthesis, contracting helicoidally after fertilization. Female flowers sessile, regular; sepals (2)3, oblong or oblong-ovate; petals (2)3, rudimentary, translucent; staminodes (2)3, rudimentary, sometimes absent; ovary linear, 1-locular, placentation laminar-diffuse, ovules numerous (c.150), oblong, orthotropous, crassinucellar, embedded in a viscous pulp. Styles (2)3, short or highly reduced; stigmas (2)3, linear, 2-lobed or bifid, papillate on adaxial face. Fruit a linear capsule, often very long, opening by decay of the pericarp. Seeds numerous, oblong to fusiform; testa membranous, areolate or striate.

A genus of 3 to 10 species, probably c.6, widely distributed throughout warmer regions of the world. Only 1 species in Africa and the Flora Zambesiaca area.

Vallisneria spiralis L., Sp. Pl.: 1015 (1753). —Ridley in J. Linn. Soc., Bot. **22**: 236 (1886). —Gürke in Engler, Pflanzenw. Ost-Afr. **C**: 95 (1895). —Wright in F.T.A. **7**: 5 (1897). —Rendle in J. Linn. Soc., Bot. **38**: 24 (1907). —Peter in Abh. Königl. Ges. Wiss. Göttingen, Math.-Phys. Kl., Newe Folge **13**(2): 109 (1928); in Repert. Spec. Nov. Regni Veg., Beih. **40**(1): 123 (1929). —Simpson in F.T.E.A., Hydrocharitaceae: 12, fig.5 (1989). Type: Micheli, Nov. Pl. Gen. **3**: 12, t.10/1, 2 (1729).

Rooting perennial; stems creeping and stoloniferous, terete, 0.6–2 mm wide. Leaves narrowly strap-shaped, 1–50(200) × (0.15)0.3–1.2 cm, bright to dark green, with numerous minute red-brown streaks; margins entire to serrate, especially near apex. Male inflorescence solitary in leaf axils, peduncle up to 7 cm long; spathe ovoid, 5–6 × 3–4 mm, with up to 50 flower buds; male

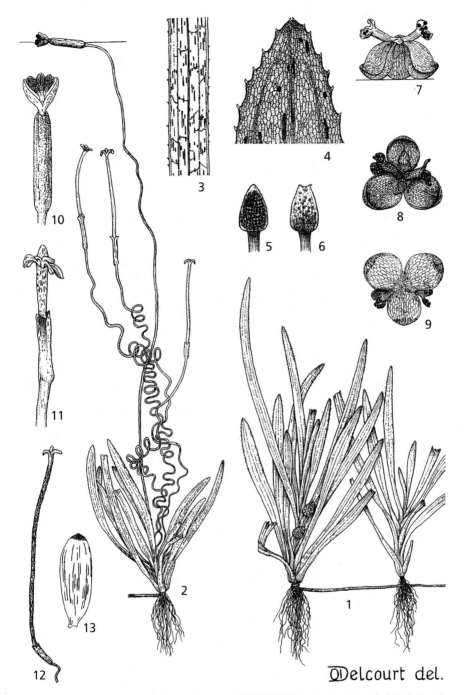

Fig. 12.2.**15**. VALLISNERIA SPIRALIS forma AETHIOPICA. 1, male plant (× ³/₄); 2, female plant (× ³/₄); 3, detail of leaf (× 2); 4, leaf apex (× 10); 5, male spathe (× 2¹/₂); 6, dehiscing male spathe (× 2); 7, male flower side view (× 25); 8, male flower from above (× 25); 9, male flower from underneath (× 25); 10, female spathe with open flower at anthesis (× 2¹/₂); 11, spathe and young fruit with remains of perianth and stigma (× 2¹/₂); 12, spathe with fruit (× ³/₄); 13, young seed (× 25). 1–2, 5–6, 10 from *Symoens* 14973, 3 from *Symoens* 14756, 4, 7–9, 11–13 from *Symoens* 14958. Drawn by O. Delcourt. Adapted from Flora of Ethiopia and Eritrea.

flowers c.1 mm wide; sepals 0.5 mm long, petals 0.3 mm long; 2 free stamens with divergent filaments, c.0.5 mm long, no hairs at base. Female spathes solitary (exceptionally 2) in leaf axils, on peduncles up to 100 cm long; spathe body 8–20 × 1–2 mm, greenish to light brown with dark reddish streaks, apex with 2 obtuse lobes; female flowers with 3 sepals, ovate, to 3 mm long, brownish with dark reddish dots or streaks, petals 3, narrowly ovate, c.0.5 × 0.2 mm; staminodes 3, alternipetalous; ovary 3-carpellate, 10–25 × 1–2 mm, stigmas 3, shallowly or deeply bifid, papillate on adaxial face, 2–3 mm long. Fruits 10–40 × 1–1.5 mm, brownish with dark reddish dots or streaks. Seeds numerous, 1.3–2 × 0.5 mm, brown, with reddish striations.

In tropical and warm regions of the Old World – Europe, Asia (from the Middle East to Japan) and Australia; widespread in continental Africa north and south of the equator, also in Réunion and Mauritius.

Forma **aethiopica** (Fenzl) T. Durand & Schinz, Consp. Fl. Afr. **5**: 2 (1894). —Symoens in Fl. Ethiopia **6**: 5 (1997). —Cook, Aq. Wetl. Pl. Sthn. Afr.: 148 (2004). Type: Sudan, Shiluk, White Nile, Mahabali Is., iv.1837, *Kotschy* 284 (W lectotype, BM, G, K). FIGURE 12.1.**15**.

> *Vallisneria aethiopica* Fenzl in Flora **27**: 311 (1844), invalid name. —Wright in F.T.A. **7**: 5 (1897). —Podlech in Merxmüller, Prodr. Fl. SW Afr., fam.142: 4 (1966). —Hepper in F.W.T.A., ed.2, **3**: 7 (1968). —Gibbs Russell in Kirkia **10**: 437 (1977). —Clarke & Klaassen, Water Pl. Namibia: 70 (2001).

Leaves coarsely serrate near apex and conspicuously denticulate to lower third, and often to base.

Caprivi Strip. Mentioned from W Caprivi by Clarke & Klaassen (2001). **Botswana**. N: Hetsameumo Cliffs, Okavango R., ♀ fl. 24.x.1979, *P.A.Smith* 2874 (K, SRGH); near Xhauga, Okavango R., ♂ fl. 4.x.1973, *P.A.Smith* 705 (K, M, SRGH). **Zambia**. N: Mpulungu, Lake Tanganyika, 20.x.1947, *Greenway & Brenan* 8246 (BM, EA, K). C: near Kafue, Kafue R., ♀ fl.& fr. 23.xi.1959, *Drummond & Cookson* 6760 (COI, LISC, SRGH). S: Livingstone, Victoria Falls, 21.xi.1949, *Wild* 3130 (BM, SRGH). **Zimbabwe**. N: Hurungwe Dist., Sanyati-Chiroti river junction, 21.xi.1953, *Wild* 4227 (K, SRGH, Z). W: Victoria Falls, ♂ fl. 21.xi.1949, *Wild* 3142 (BM, SRGH). C: Chegutu (Hartley), Umfuli R., 26.iv.1964, *Ellis in GHS* 152479 (SRGH, Z). E: Mentioned by Gibbs Russell (1977). **Malawi**. N: Nkhata Bay Dist., Mbuzi Hill, Likoma Is., ♀ fl. 24.viii.1984, *Salubeni, Seyani & Nachamba* 3871 (K, MAL). S: Mangochi Dist., Monkey Bay, ♂ & ♀ fl. 27.viii.1978, *W.N. Grey* 28 (K). **Mozambique**. T: Boroma, Zambezi R., ♀ fl. 20.ix.1942, *Mendonça* 363 (LISC, SRGH).

Widespread in Africa south of the Sahara – Senegal, Mali, Ghana, Niger, Nigeria, Chad, Central African Republic, Sudan, Ethiopia, Congo, Burundi, Uganda, Kenya, Tanzania and Namibia. In still water and watercourses associated with *Ceratophyllum demersum, Hydrilla verticillata, Najas horrida, Potamogeton pectinatus* and *P. schweinfurthii*, water depth 0.2–1.8 m; 200–1200 m.

Conservation notes: Widespread species; not threatened.

Vallisneria spiralis shows a wide variation in plant height, length, width and degree of denticulation of the leaves. The typical form, found in Europe, has entire leaf margins or denticulate only in the upper third. Most populations from Subsaharan Africa have leaves coarsely serrate near the apex and conspicuously denticulate to the lower third, often to the base. These plants were mostly referred to as *Vallisneria aethiopica*. However, given the variability *V. aethiopica* does not deserve to be more than a form of *V. spiralis*. Despite the small number of African specimens examined by Lowden (1982), he placed all African material in *V. spiralis* var. *denseserrulata* Makino. The diagnostic features he used to distinguish this variety were the deeply cleft stigmatic lobes and the conspicuously toothed leaves. Deeply cleft stigmas were occasionally observed on freshly collected plants from Lake Victoria, but it is necessary to check the constancy of this characteristic on fresh material elsewhere. It has not been observed in southern African material (Cook 2004).

6. ENHALUS Rich.

Enhalus Rich. in Mém. Cl. Sci. Math. Inst. Natl. France **1811**(2): 64, 71, 74 (1814). —Hartog, Sea-grasses of World: 214 (1970)

Perennial, submerged marine herbs, dioecious. Roots coarse, simple. Stem rhizomatous, elongate, rarely branched, thick; internodes very short, the segmentation being obscured by the remains of old leaves. Leaves sessile, strap-shaped, with many longitudinal nerves and septate air-channels externally visible as a fine striping, base, sheathing; margins with very coarse vascular bundles remaining after decay as dense, persistent fibres on rhizome; apex rounded or obtuse, often asymmetrical, sometimes slightly serrulate. Male inflorescence: peduncle cylindrical, remaining submerged; male spathe consisting of 2 partially united bracts, the free margins of outer one embracing the inner, containing numerous stalked flower buds breaking off just before anthesis, rising to water surface where they open and float freely. Male flowers: sepals 3, oblong, acute, strongly reflexed; petals 3, ovate, acute, reflexed; stamens 3, alternipetalous, erect, anthers subsessile, erect, opening laterally. Female inflorescence: peduncle long, contracting spirally after fertilisation; spathe consisting of 2 almost free bracts, outer embracing the inner, persistent. Female flowers: 1 per spathe; sepals 3, oblong acute, strongly reflexed; petals 3, erect; ovary narrowly cylindric, 6-carpellate, placentas protruding deeply towards centre, rostrate and densely set with many erect hairs; styles 6, forked from near base, the stigmatic branches subulate, densely papillose, except at apex; ovules several, with 2 teguments, embedded in mucilage. Fruit ovoid to subglobose, acuminate, densely set with erect appendages, opening irregularly by decay of pericarp. Seeds obconical, angular.

A monospecific genus.

Enhalus acoroides (L.f.) Royle, Ill. Bot. Himal. Mtns.: 453 (1840). —Hartog, Sea-grasses of World: 215, fig.60 (1970). —Simpson in F.T.E.A., Hydrocharitaceae: 22, fig.8 (1989). —Kuo & Hartog in Short & Coles, Global Seagrass Res. Meth.: 49 (2001). —Bandeira & Björk in S. Afr. J. Bot. **67**: 421 (2001). —Bandeira & Gell in Green & Short, World Atlas Seagrasses: 93 (2003). Type: Sri Lanka, *J. König* s.n.(BM holotype). FIGURE 12.2.**16**.

  Statiotes acoroides L.f., Suppl. Pl.: 268 (1781).

Coarse marine herb. Rhizome 10–15 mm in diameter, brown. Leaves 30–150 × 0.7–1.7 cm, bright to dark-green; sheaths compressed, up to 15 cm long, whitish, translucent; marginal nerves persistent and blackish after leaf decay. Male inflorescence with peduncle 5–10 cm long, spathe bracts broadly ovate-lanceolate, 5 × 3 cm, slightly keeled, keel and nerves with many long rough hairs, margin curled. Male flowers with thin pedicel 3–12 mm long, breaking off 0.3–0.5 mm below flower; sepals c.2 × 1 mm, white; petals c.1.25–1.75 mm long, white. Female inflorescence with peduncle 40–50 cm, sometimes longer; spathe bracts 4–6 × 1–2 cm, with many long rough hairs, margin not curled. Female flowers with oblong sepals, c.1 × 0.5 cm, with round apex and recurved margins; petals c.4–5 × 0.3–0.4 cm, white with reddish apex; ovary densely set with long fringe-like hairs, stigmatic branches 10–12 mm, densely papillose. Fruit 2.5–7 × 2.2–3 cm, green or dark brown to blackish. Seeds 8–14 in fruit, 10–15 × 12 mm, brown.

Mozambique. N: Pemba, 23.x.1993, *Bandeira* 449 (LMU).
Widely distributed around the Indian Ocean and W Pacific coasts; in Africa from N Kenya to Mozambique, also Madagascar and Seychelles. Along sheltered coasts in shallow water on sandy and muddy bottoms, often on mudbanks adjacent to mangroves, forming pure stands or with other sea-grasses (mostly *Halophila ovalis*); to 6 m deep, but flowering only in places where uncovered briefly during very low tides or where inflorescences can reach the water surface.

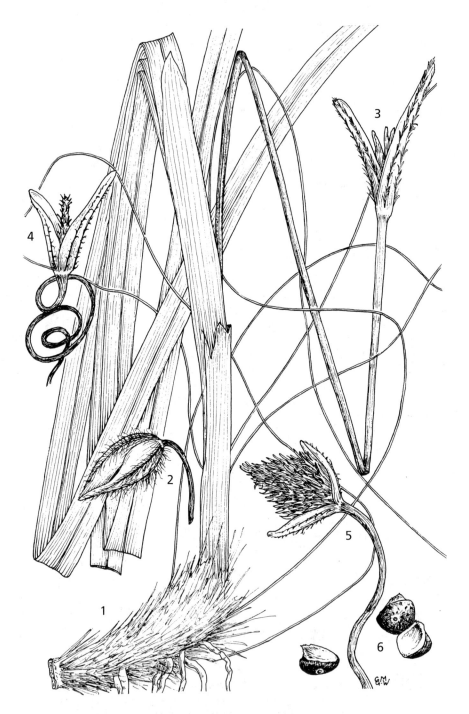

Fig. 12.2.**16.** ENHALUS ACOROIDES. 1, habit (× ²/₃); 2, male spathe (× ²/₃); 3, female flower (× ²/₃); 4, young fruit (× ²/₃); 5, mature fruit (× ²/₃); 6, seeds (× 2). 1, 3, 5 from *Balansa* 3162, 2 from *Parthasanathy* 134, 4 from *Ito* 1454, 6 from *Womersley* NGF 46410. Drawn by Christine Grey-Wilson. From F.T.E.A.

Conservation notes: Local distribution in the Flora area; although widespread and often dominant globally, it is threatened locally by seine fishing. Possibly Vulnerable in the Flora area.

Stands of *Enhalus* provide an important habitat for marine epiphytes and a diverse range of invertebrates; it is also a common food of turtles and dugongs.

7. HALOPHILA Thouars

Halophila Thouars, Gen. Nov. Madag.: 2 (1806). —Hartog, Sea-Grasses of World: 238 (1970).

Lemnopsis Zoll., Syst. Verz. **1**: 74 (1854), non Zipp. (1829).

Perennial or annual marine herbs, totally submerged including the flowers, monoecious or dioecious. Roots unbranched, arising at creeping stem nodes, covered with long fine hairs and sand-binding. Stems elongate, dimorphic, creeping stems with 2 scales at each node, one embracing the creeping stem, the other a short erect stem bearing foliage leaves. Foliage leaves in opposite rows or in pseudo-whorls, sessile or petiolate, ovate, elliptic, oblong or linear, entire or serrulate; midrib distinct, mostly connected by oblique secondary nerves ("cross-veins") to two longitudinal (intramarginal) nerves running along or very near margins and reaching it near the top. Flowers unisexual, male and female spathes similar, solitary in leaf axils, sessile, comprising 2 free bracts, one embracing the other, bracts elliptic, obovate to suborbicular, membranous, acute, rounded to emarginate or indented, keeled, with entire or minutely serrate keel and margins; spathes 1(2)-flowered, rarely with 1 of each sex; 2 scales in axil of each bract. Male flowers pedicellate, remaining attached; tepals 3, imbricate, elliptic or ovate, hooded; stamens 3, alternating with tepals, anthers sessile, erect, oblong, linear, 2- or 4-sporangiate. Female flowers sessile; tepals 3, reduced to lobes at apex of hypanthium; staminodes absent; ovary of 3–5 carpels, ellipsoid or ovoid, beaked, unilocular, ovules few to many, anatropous, styles 3–5, stigmas filiform, entire, papillate on adaxial surface. Fruit ovoid or globose, beaked, with membranous pericarp, opening by dehiscence or decay. Seeds few to numerous, globose or subglobose; testa tuberculate, reticulate or smooth. Embryo consisting of a large hypocotyl and a coiled cotyledon, the plumula lying in a depression of hypocotyl sheathed by expanded base of cotyledon.

About 11–16 species found in tropical and subtropical seas, extending into some warm-temperate waters. Three species in the Flora Zambesiaca area.

The taxonomy of the genus is still uncertain in spite of molecular studies (see Waycott et al. in Bull. Mar. Science **71**: 1299–1308, 2002 & Uchimura et al. in Bot. Mar. **49**: 111–121, 2006).

1. Leaves linear; margins serrulate, at least near apex; stipules 8–17 mm long · **3.** *stipulacea*
– Leaves elliptic to obovate, rarely linear; margins smooth; stipules 0.35–6 mm long · 2
2. Leaves 10–30 mm long, with 11–16 pairs of cross-veins · · · · · · · · · · · **1.** *ovalis*
– Leaves 6–12 mm long, with 3–11 pairs of cross-veins · · · · · · · · · · · · · · **2.** *minor*

1. **Halophila ovalis** (R. Br.) Hook. f., Fl. Tasman. **2**: 45 (1858). —Gürke in Engler, Pflanzenw. Ost-Afr. **C**: 94 (1895). —Cohen in S. Afr. J. Sci. **36**: 246 (1939). — Obermeyer in F.S.A. **1**: 101, fig.30 (1966). —Hartog, Sea-Grasses of World: 240, fig.62 (1970). —Simpson in F.T.E.A., Hydrocharitaceae: 25, fig.10 (1989). —Kuo & Hartog in Short & Coles, Global Seagrass Res. Meth.: 50 (2001). Type: Australia, Queensland, *Brown* 5816 (K lectotype chosen by Simpson (1989), BM).

Caulinia ovalis R. Br., Prodr. Fl. Nov. Holl. **1**: 339 (1810).

Perennial, rarely annual marine herb, dioecious. Rhizome narrowly terete, fragile, 0.2–1.5(2) mm thick, often dichotomously branched, internodes 1–5 cm long; nodes each with 1(2) roots and 2 suborbicular or obovate; keeled scales, 3–6(8) × 1.5–3 mm, greenish or transparent. Erect stems extremely short, mostly not visible, the creeping axis apparently with a pair of petiolate leaves at nodes; petiole (1)1.5–2.5(12) cm long; blade obovate, spathulate, oblong-elliptic, rarely linear, 7–16(40) × (0.8)2.5(20) mm, green, slightly reddish or brown-tinged, surface mostly smooth, sometimes bullate, with 2 longitudinal veins along and near margin, cross-veins (7)10–25(28) at each side of midrib, often forked, joining the intramarginal veins; base rounded, truncate or attenuate, apex rounded, obtuse or subapiculate, margin entire. Spathe bracts ovate, often keeled, membranous, acute, 3–5(10) mm long. Male flowers with pedicel up to 2.5 mm long; tepals 3–4(6) × 2–3 mm, elliptic, spreading or reflexed, obtuse or slightly apiculate; anthers oblong, 1.5–2.5(4) mm long. Female flowers with trilocular ovary, 1–1.5(2.5) mm long, ovules numerous, hypanthium 3–10 mm long, styles 3, stigma 10–25 mm long. Fruit globose, 3–4(6) mm wide, beak persistent, up to 6 mm long, whitish or brown. Seeds (6)20–30, globose, c.1 mm diameter; testa tuberculate or reticulate, light brown.

Widely distributed along the coasts of the Red Sea, Persian Gulf and around the Indian Ocean and the western Pacific extending in some areas beyond the tropics, such as in W Australia and Japan. Common to locally abundant on fine mud and sandy bottoms to hard coral debris and madreporic limestones, from mid-tidal level to 12 m depth. The species occurs over a wide range of salinity and temperature (even being found in association with a submarine hot spring in Japan). and is capable of surviving in turbid or polluted waters. A preferred food of dugong.

H. ovalis shows a great variability in plant size, leaf shape and venation. The above description covers the whole species, including the diverse subspecies as revised by Kuo & Hartog (2001) and Hartog & Kuo (in Larkum, Orth & Duarte, Seagrasses: Biol. Ecol. Conserv.: 1–23, 2006).

Subsp. **ovalis**. —Munday & Forbes in J. S. Afr. Bot. **45**: 3 (1979). —Kalk in Nat. Hist. Inhaca Is., ed.3: 72, fig.3.15 (1995). —Bandeira & Björk in S. Afr. J. Bot. **67**: 421 (2001). —Bandeira & Gell in Green & Short, World Atlas Seagrasses: 99 (2003). —Cook, Aq. Wetl. Pl. Sthn. Afr.: 142, fig.141 (2004). FIGURE 12.2.**17**.

> *Halophila ovata* sensu C.H. Wright in F.T.A. **7**: 10 (1897), non Gaud.
> *Halophila ovalis* 'normal form' of Macnae in Macnae & Kalk, Nat. Hist. Inhaca Is.: 28, fig.8b (1958).

Leaf blades obovate, spathulate, oblong-elliptic to somewhat ovate, 10–40 × 5–20 mm, surface smooth, mostly with more than 12 cross-veins at each side of midrib, ascending at 45–65°, often forked.

Mozambique. N: Quirimba Is., 23.viii.1996, *António* s.n. (LMU). GI: Ponta Tondo, opposite Santa Carolina Is., 11.xi.1958, *Mogg* 29057 (J, LISC). M: Inhaca Is., 29.xii.1956, *Mogg* 27174 (J, K).

Found along the eastern coast of Africa from Egypt to the Cape Provinces of South Africa; also in Madagascar, the Seychelles, Mauritius and Rodriguez. On a range of substrates from coarse rubble to soft mud; from mid-tidal level to c.12 m depth. Flowers in January.

Conservation notes: Widespread taxon; not threatened.

Subsp. **linearis** (Hartog) Hartog, Sea-Grasses of World: 251 (1970). —Munday & Forbes in J. S. Afr. Bot. **45**: 3 (1979). —Kuo & Hartog in Short & Coles, Global Seagrass Res. Meth.: 50 (2001). Type: Mozambique, Maputo, Inhaca Is., W coast, vii.1931, *Moss* 20552 (BM holotype, designated by Hartog 1957); and Inhaca Is., vii.1937, *Cohen* 20652 (BM, J, superfluous type designated by Hartog 1970).

Fig. 12.2.**17**. HALOPHILA OVALIS subsp. OVALIS. 1, habit (× ²/₃); 2, habit (× 2); 3, male spathe and flower (× 2); 4, female spathe and fower (× 1); 5, fruit (× 2). 1–2 from *Faulkner* 3306, 3–5 after Isaac (1968). Drawn by Christine Grey-Wilson. From F.T.E.A.

Halophila linearis Hartog in Act. Bot. Neerl. **6**: 46, fig.1 (1957).
Halophila ovalis 'lanceolate form' of Macnae in Macnae & Kalk, Nat. Hist. Inhaca Is.: 28, fig.8c (1958).

Leaf blades linear or narrowly lanceolate, 15–30 × 0.8–4 mm wide, gradually merging into petiole, 10–16 ascending cross-veins at each side of midrib. Male flowers with tepals 2.5–3 mm. Female flowers with 3 styles, 10–15 mm long.

Mozambique. GI: Inhambane, Mocucuni Is., 26.xi.1958, *Mogg* 29309 (LISC). M: Inhaca Is., ♀ fl.& fr. 27.x.1962, *Mauve & Verdoorn* 24 (BM, PRE).

Conservation notes: Apparently endemic to S Mozambique; probably Vulnerable. Threatened, as with other seagrasses of the island, by trampling and fishing in shallow water, and collection of invertebrate food, and by digging of seagrass meadows at spring low tide.

The ecology of subsp. *linearis* is similar to that of subsp. *ovalis*; often both grow mixed together with many intermediates (Macnae & Kalk 1958). According to Obermeyer (1966), the narrow-leaved plants may represent a variety of *H. ovalis* which developed colonies locally by vegetative multiplication.

2. **Halophila minor** (Zoll.) Hartog in Fl. Males. **1**(5): 410, fig.17b (1957). —Kuo & Hartog in Short & Coles, Global Seagrass Res. Meth.: 50 (2001). —Bandeira & Björk in S. Afr. J. Bot. **67**: 421 (2001). —Bandeira & Gell in Green & Short, World Atlas Seagrasses: 99 (2003). Type: Indonesia, Flores, near Bari, 12.vii.1847, *Zollinger* 3334 (P holotype, BM, L, U).

Lemnopsis minor Zoll., Syst. Verz. **1**: 75 (1854).
Halophila ovata sensu Hartog, Sea-Grasses of World: 251 (1970), in part. —Simpson in F.T.E.A., Hydrocharitaceae: 25 (1989), non Gaud. (1827).

Perennial, dioecious, submerged marine herb. Rhizome filiform or narrowly terete; roots 1 per node. Erect stems hardly developed. Petiole blades ovate, 6–12 × 3.5–6 mm, cross-veins (4)7–12, occasionally branched cross-veins, narrow space between intramarginal vein and blade margin. Male flowers with tepals 1–1.5 mm long, anthers 2.5–3.5 mm long. Female flowers with 3 styles, (6)8–20 mm long. Fruits 2–4 mm wide.

Mozambique. N: mentioned by Bandeira & Gell (2003) from Quirimbas archipelago, Mecúfi–Pemba, Fernão-Veloso and Ilha de Moçambique.

Widely distributed from East Africa to the western Pacific. Reported from Kenya and Mozambique. On sand and mudflats exposed at very low tides, down to 2 m depth.

Conservation notes: Widespread taxon; not threatened.

Related to *H. ovalis* and sometimes difficult to distinguish from small specimens of the latter. However, molecular studies using nuclear ribosomal DNA supports the notion that they are separate species (Waycott *et al.* in Bull. Mar. Sci. **71**: 1306, 2002).

3. **Halophila stipulacea** (Forssk.) Asch. in Sitzungsber. Ges. Naturf. Freunde Berlin **1867**: 3 (1867). —Engler & Prantl, Pflanzenfam. **2**(1): 249, fig.183 (1889). — Hartog, Sea-grasses of World: 258, fig.63 (1970). —Simpson in F.T.E.A., Hydrocharitaceae: 27 (1989). —Kuo & Hartog in Short & Coles, Global Seagrass Res. Meth.: 51 (2001). —Bandeira & Björk in S. Afr. J. Bot. **67**: 421 (2001). — Bandeira & Gell in Green & Short, World Atlas Seagrasses: 99 (2003). Type: Yemen, *Forsskål* (not located).

Zostera stipulacea Forssk., Fl. Aegypt.-Arab.: 158 (1775).

Perennial, dioecious, submerged marine herb. Creeping stems rhizomatous, elongate, 0.5–2 mm thick, unbranched, greenish or whitish, with 1 root at each node; internodes 1–4 cm long; nodal scales elliptic or obovate, 8–17 × 4–10 mm, folded, covering petiole, incised along keel, green or translucent, sometimes tinged with purple, the 2 apical lobes obtuse. Erect stems narrowly cylindrical, 1.5–15 × 0.2–0.5 mm. Leaves in pairs, 1 pair on each erect stem; petiole 0.5–1.5 cm, blade linear to oblong or oblong-elliptic, 1.5–6(8) × 0.2–0.6(0.8) cm, base cuneate, apex rounded or obtuse, surface glabrous, papillose or with minute hairs, occasionally bullate; margins serrulate near apex, cross-veins ascending at a 45–60° angle, joining the marginal leaves. Male flowers with pedicel 7–15 mm long; tepals ovate, U-shaped in cross-section, 3–4 × 2.5 mm, with a prominent central vein; anthers 2–3.5 mm long. Female spathe with ovate keeled bracts, c.10 × 0.7 mm, the keel denticulate, whitish, translucent; ovary with 3 carpels, ovoid to ellipsoid, 2–4 mm long, hypanthium 3–6 mm long; styles 3, extending into filiform stigmas, 20–25 mm long. Fruit ellipsoid, c.5 × 2 mm, beak up to 6 mm long. Seeds 30–40, globular, reticulate, light brown.

Mozambique. N: Mecufi, 21.x.1993, *Bandeira* 341 (LMU).

Common around the Red Sea and present along the East African coast from Somalia to N Mozambique, also in Madagascar, Mauritius and Rodriguez; introduced into the Mediterranean. In shallow water on muddy and sandy substrates, and on sediment-covered coral platforms; often associated with *Halophila ovalis*; from the upper sublittoral down to c.7 m deep.

Conservation notes: Common globally although local in the Flora area; not threatened.

It is a migrant into the eastern Mediterranean Sea, believed to have been transported through the Suez Canal during the 19th century, and has been reported from the western Mediterranean since 1988.

Plants from the Indian Ocean have membranous leaves which never become puckered, and with scales often falling early. Those from the Red Sea and Persian Gulf have cartilaginous leaves, becoming puckered with age, with white, stiff and persistent scales.

8. THALASSIA K.D. König

Thalassia K.D. König, Ann. Bot. **2**: 96 (1805). —Hartog, Sea-Grasses of World: 222 (1970).

Schizotheca Solms in Schweinfurth, Beitr. Fl. Aethiop. **1**: 194, 246 (1867).

Perennial marine herbs, totally submerged (including flowers), dioecious. Roots unbranched, covered with fine hairs and sand-binding. Stems dimorphic, rhizomes horizontal, buried in the substrate, with short erect stems produced at regular intervals. Leaves usually 2–6, in opposite rows on erect stems, sessile, strap-shaped or somewhat falcate, opaque or translucent, with up to 19 longitudinal nerves, numerous fine, longitudinal air channels parallel with nerves, base distinctly sheathing; margins green, entire, apex rounded or obtuse. Male spathes 1–2 in leaf axils, pedunculate, comprising 2 bracts united on one side, bracts oblong or lanceolate, translucent, entire or serrulate, acute to obtuse. Male flowers mostly 1 per spathe, pedicellate, remaining attached; tepals 3, elliptic, hooded, strongly recurved at anthesis; stamens (3)6–9(13), anthers nearly sessile, oblong, erect, latrorsely dehiscent. Female spathes solitary in leaf axils, distinctly pedunculate, comprising 2 membranous bracts united on both sides, 2-lobed at apex, lobes acute to obtuse. Female flowers 1 per spathe, almost sessile; tepals 3, elliptic, recurved at anthesis; staminodes absent; ovary of 6–8 carpels, conical, rough, unilocular or imperfectly 2–3-locular, elongate into hypanthium, styles 6, each with a deeply bifid, papillate stigma, ovules 1 per carpel, with only 1 integument. Fruit globose, densely spinose, the fleshy pericarp bursting open irregularly. Seeds 3–9, conical with thickened basal rim.

Only two species – *T. testudinum* K.D. König from the Atlantic coasts of tropical America and *T. hemprichii* from tropical coasts of the Indian Ocean and the western Pacific.

Thalassia hemprichii (Solms) Asch. in Petermann's Geogr. Mitth. **17**: 242 (1871). —
Gürke in Engler, Pflanzenw. Ost-Afr. **C**: 95 (1895). —Wright in F.T.A. **7**: 9 (1897).
—Macnae in Macnae & Kalk, Nat. Hist. Inhaca Is.: 28, fig.8g (1958). —
Obermeyer in F.S.A. **1**: 101, fig.30 (1966). —Hartog, Sea-Grasses of World: 232,
fig.61 (1970). —Munday & Forbes in J. S. Afr. Bot. **45**: 3 (1979). —Simpson in
F.T.E.A., Hydrocharitaceae: 24, fig.9 (1989). —Kalk, Nat. Hist. Inhaca Is., ed.3:
129 (1995). —Kuo & Hartog in Short & Coles, Global Seagrass Res. Meth.: 49
(2001). —Bandeira & Björk in S. Afr. J. Bot. **67**: 421 (2001). —Bandeira & Gell
in Green & Short, World Atlas Seagrasses: 99 (2003). Type: Eritrea, Massawa,
1820–1826, *Ehrenberg* 170 (B† holotype, BM lectotype, K, L, P). FIGURE 12.2.**18**.
   Schizotheca hemprichii Ehrenb. in Abh. Königl. Akad. Wiss. Berlin 1832: 429 (1832),
invalidly published. —Solms in Schweinfurth, Beitr. Fl. Aethiop.: 194, 246 (1867).
   Thalassia sp. sensu Cohen in S. Afr. J. Sci. **36**: 247 (1939).

Submerged marine herb. Rhizome terete, 2–5 mm wide, greenish to light brown, with many
annular scars; erect stems terete, to 5 mm wide. Leaves 4–25(40) × 0.3–0.7(1.1) cm, bright to
dark olive-green with red-violet spots or streaks; nerves 9–13(17), margins appearing smooth,
apex rounded, finely serrated. Sheaths 3–7 cm long, old ones persisting as torn membranous
remnants. Male spathe with peduncle 3–4(5) cm long, spathe-bracts 1.7–2.5 cm, entire or finely
serrated at apex. Male flowers 1 per spathe, pedicel 2–3 cm long; tepals 7–8 × 3 mm, entire;
stamens (3)6–9(12), anthers 7-11 mm long. Female spathe with peduncle up to 1.5 cm long,
spathe-bracts 1.7–2.5 × 1 cm. Female flowers subsessile, ovary of 6 carpels up to 10 mm long,
hypanthium 2–3 cm long, styles 6, 5–7 mm long, stigmatic branches nearly as long or up to twice
as long as style, becoming recurved at maturity. Fruit 2–2.5 × 1.7–3.2 cm, light green, with a 1–2
mm long beak, splitting into 8–20 irregular valves. Seeds 3–9, c.8 mm long, conical part
greenish, thick basal portion dark brown.

Mozambique. N: Moma Dist., Moma Is., st. 27.x.1965, *Mogg* 32525, 32578 (J, LISC).
GI: Bazaruto Is., Ponta Estone, st. 28.x.1958, *Mogg* 28671 (J, K, LISC, M, PRE). M:
Inhaca Is., W coast, imm.fr. 11.vii.1949, *Mogg* 22939 (K).

Widely distributed throughout tropical regions of the Indian Ocean and western
part of the Pacific, along the East African coast from Egypt to Mozambique, also in
Madagascar and the Seychelles. On sandstone and coral rocks, especially on dead
reef platforms and sublittoral flats consisting of coral-debris or coral-sand, often
luxuriant in shallow eulittoral pools on coral-sand covered with fine mud, where
it forms extensive meadows, reportedly eaten by fish. Often associated with
Halodule wrightii with *Thalassia* occupying small depressions and *Halodule* higher
places exposed at low spring tides; from low water mark to 7 m depth. Flowers in
July and August.

Conservation notes: Widespread species; not threatened except locally by food
collectors on intertidal beds.

Thalassia hemprichii can be readily distinguished from *Thalassodendron ciliatum*
(Forssk.) Hartog (Cymodoceaceae). The latter has the old leaf sheaths falling away,
the erect stems are long, ligneous and bare, not covered with leaf sheaths but with
prominent annular leaf-scars, and the leaf margins are more deeply serrulate near
the apex.

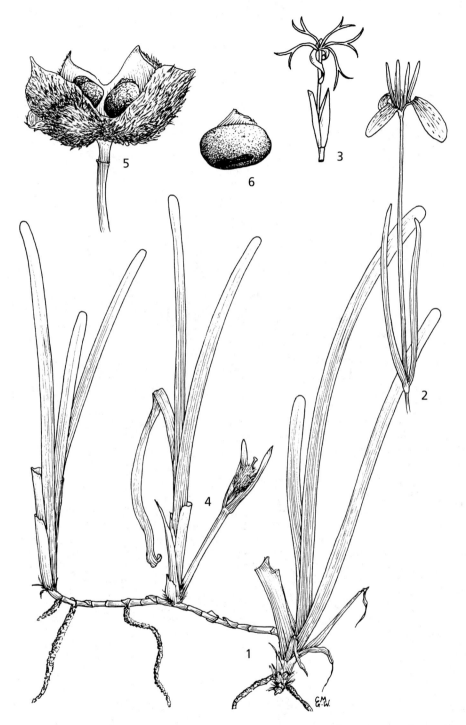

Fig. 12.2.**18**. THALASSIA HEMPRICHII. 1, habit (× ²/₃); 2, male flower (× 2); 3, female flower (× ²/₃); 4, female spathe and young fruit (× ²/₃); 5, dehisced fruit (× 2); 6, seed (× 4). 1, 4 from *Rajeshwari & Lakshmanan* 3, 2 from *Ostenfeld* 1920, 3 from Isaac (1968), 5–6 from *Burkill* 1118. Drawn by Christine Grey-Wilson. From F.T.E.A.

NAJADACEAE

by J.J. Symoens & L. Triest

Annual or rarely perennial, aquatic, obligately submerged herbs of fresh or brackish waters, monoecious or dioecious, glabrous, bottom-rooted, but parts often becoming detached and then free-floating. Roots adventitious, simple, devoid of root caps. Stems slender or robust, usually richly-branched, rooting from the base and lower nodes, internodes sometimes armed with prickles (subgen. *Najas*). Leaves sessile, alternate, but mostly appearing to be subopposite or in pseudowhorls of 3–7, with an open basal sheath and a simple, linear blade; sheath folded, its basal part enclosing 2 tiny hyaline intravaginal scales, its upper part mostly serrulate or prickly-dentate; blade 1-nerved, sometimes dorsally armed with prickles on the midrib, margins bearing (0)4–74(185) teeth on each side, apex acute to acuminate. Flowers submerged, inconspicuous, unisexual, solitary at the base of an axillary shoot or sometimes aggregated above, sessile or shortly pedicellate, naked or surrounded by 1 or 2 membranous envelopes. Male flower consisting of a solitary anther, sessile or subsessile, surrounded by 1 or 2 envelopes; outer envelope ("spathe") present and then toothed or lacerated at apex, or absent; inner envelope ("perianth") adhering to the anther and ending in 2 lips at apex; pedicel at first short, elongating just before anthesis and pushing the anther through the spathe; anther 1- or 4-sporangiate, dehiscing apically. Female flower consisting of a solitary ovary ("gynoecial wall"), subsessile, 1-locular, sometimes ± surrounded by a lobed or toothed spathe; ovule solitary, basal, subsessile, erect, anatropous, with 2 integuments; style short, cylindrical, with 2–3(4) linear, often unequal, sometimes barren stigmatic arms. Fruit 1-seeded, achene-like, but not actually drying, the fleshy pericarp adhering closely to the seed and dehiscing by decay, thin and membranous when dried. Seed elliptic-oblong to ovate, sometimes asymmetrical at apex or U-shaped, without endosperm; testa hard, pitted with areoles irregularly arranged or regularly arranged in longitudinal rows, the end walls often raised; embryo straight with large hypocotyl and radicle, plumule well developed, cotyledon terminal.

A family with only 1 genus, *Najas*, found in temperate and tropical regions worldwide.

Miki (in Bot. Mag. (Tokyo) **51**: 472–480, 1937) emphasized the remarkable similarities in vegetative and reproductive structures of *Najas* and the submerged Hydrocharitaceae and suggested that *Najas* was derived by simplification from the Hydrocharitaceae. From studies of the seed coat structure, Shaffer-Fehre (in Bot. J. Linn. Soc. **107**: 169–188, 189–209, 1991) confirmed this relationship and formally included the genus *Najas*, and consequently its unigeneric family the Najadaceae, within the Hydrocharitaceae. This view is also confirmed by rbcL and matK gene sequence data in Les et al. (in Molec. Phylogen. Evol. **2**: 304–314, 1993; in Syst. Bot. **22**: 443–463, 1997) and by Tanaka et al. (in J. Plant Res. **110**: 329–337, 1997). The inclusion of *Najas* brings the number of genera in the family Hydrocharitaceae to 18 and the number of species to c.115. Although Najadaceae is treated separately here, as it has been in other regional African floras, the arguments supporting its inclusion in Hydrocharitaceae should not be ignored.

1. NAJAS L.

Najas L., Sp. Pl.: 1015 (1753). —Triest in Mém. Acad. Roy. Sci. Outre-Mer, Cl. Sci. Nat. Méd. **21**(4): 19 (1987); **22**(1): 35 (1988).

Characters as for the family.

A genus of c.40 species, widely distributed, with its greatest diversity in tropical and subtropical regions, but absent from very cold areas. Thirteen species in Africa and neighbouring islands, with 4 in the Flora area.

1. Plants always dioecious, mostly relatively robust; spiny prickles usually present on internodes and dorsal side of leaf midrib; seeds asymmetrically ovoid; areoles on seed unequally sized, irregularly arranged · · · · · · · · · · · · · · · · · · · **1.** *marina*
- Plants monoecious, very rarely dioecious, mostly slender; stems and leaf midrib unarmed; seeds ellipsoid oblong to fusiform; areoles on seed equal in size, regularly or irregularly arranged · 2
2. Leaf sheath rounded at apex; areoles regularly arranged · · · · · · · · · **2.** *horrida*
- Leaf sheath truncate to auriculate at apex; areoles irregularly or regularly arranged · 3
3. Leaf sheath with 2–8 spine-cells on each side, apex truncate to slightly auriculate; male and female flowers each in a spathe; less than 23 areoles arranged irregularly in each longitudinal row · · · · · · · · · · · · · · · · · · **3.** *testui*
- Leaf sheath with 3–14 spine-cells on each side, apex deeply auriculate; male and female flowers not in a spathe; more than 23 areoles arranged regularly in each longitudinal row · **4.** *graminea*

Subgen. **NAJAS**

Najas subgen. **Najas**. —Triest in Mém. Acad. Roy. Sci. Outre-Mer, Cl. Sci. Nat. Méd. **21**(4): 22 (1987); **22**(1): 42 (1988).

Najas subgen. *Eunajas* (Asch.) Rendle in Trans. Linn. Soc., Bot. **5**: 389 (1899).

A subgenus with only 1 species widespread under different subspecies; from Europe, Macaronesia and West Africa to eastern Asia, and in Australia, America and Hawaii. Found in waters of relatively high ionic content, brackish or alkaline, with conductivity mostly greater than 500 µS/cm and pH up to 9.

1. **Najas marina** L., Sp. Pl.: 1015 (1753). —Rendle in Trans. Linn. Soc., Bot. **5**: 389, t.39 fig. 1–30 (1899). —Triest in Mém. Acad. Roy. Sci. Outre-Mer, Cl. Sci. Nat. Méd. **21**(4): 23 (1987); **22**(1): 43 (1988); in F.T.E.A., Najadaceae: 2 (1989). Type: Vaillant, Caract. Quatorze Gen. Pl., figs.1–2 (1722) (lectotype).

Submerged aquatic herb, normally rooted, dioecious, mostly robust. Stems up to 150 cm long and 0.5–4 mm in diameter, branched above, mostly with coarse prickle-like teeth. Leaves fleshy, 3.6–65 mm long, 1.1–5.7 mm wide including teeth (0.4–2.5 mm excluding teeth); margins with (0)4–17(40) coarse prickle-like teeth on each side; leaf teeth up to 2 mm long borne on broad triangular excrescences; tooth length:leaf width ratio 0.4–2.5; apex acute; abaxial side of midrib bearing (0)2–8(40) teeth similar to those of stems and leaf margins; leaf sheath 1.5–5 × 1–5.5 mm, length:width ratio 0.4–1.2, rounded, entire or serrulate with 0–3 spine-cells on each side. Flowers axillary, solitary. Male flower enclosed in a spathe, 2–5 (including spathe neck) × 0.8–3 mm; spathe neck 0.5–1 mm long, tapering above and bearing brownish spine-cells at the apex; inner envelope protruding 0.2–0.7 mm above the anther; anther 1.2–4 × 0.7–2.5 mm, 4-sporangiate. Female flower naked, 2–5.7 mm long; ovary 1–3.5 (0.3–1.9) mm; style 0.3–1.7 mm long (including 2–4 stigmatic branches). Fruit 4–5(8) mm long, ellipsoidal, narrowed at the top and bearing the withered style and remains of stigmas; pericarp persistent, thin, membranous. Seed ovoid, slightly asymmetrical, 1.9–7.5 × 0.8–3.3 mm; length:width ratio 1.2–3.5; testa reddish-brown, pitted with areoles irregular in shape, dimensions and arrangement.

A very polymorphic, cosmopolitan species found in Africa, Europe, Asia, Australia, America and Hawaii. Many subspecies and varieties are recognized, some with a relatively restricted distribution. 5 subspecies in Africa and the surrounding islands, mainly in coastal areas and alkaline rift lakes. Only one subspecies in the Flora Zambesiaca area.

Subsp. **armata** (Lindb.f.) Horn in Kew Bull. **7**: 29 (1952). —Robyns & Tournay in Fl.
Sperm. Parc Nat. Alb. **3**: 17 (1955). —Gibbs Russell in Kirkia **10**: 434 (1977) (as
N. marina). —Triest & Symoens in F.A.C., Najadaceae: 4 (1983). —Triest in
Mém. Acad. Roy. Sci. Outre-Mer, Cl. Sci. Nat. Méd. **21**(4): 28, t.1,2, fig.2 (1987);
in **22**(1): 68, t.6, fig.10 (1988); in F.T.E.A., Najadaceae: 3, fig.1 (1989). —Cook,
Aq. Wetl. Pl. Sthn. Africa: 187, fig.192 (2004). Type: Egypt, Fareskour, *Delile* s.n.
(MPU holotype).

    Najas marina L. var. *muricata* (Delile) K. Schum. in Martius, Fl. Bras. **3**: 725 (1894). —
Rendle in Trans. Linn. Soc., Bot. **5**: 397, t.39 fig.12 (1899).

    Najas marina L. subsp. *delilei* (Rouy) Oberm. in F.S.A. **1**: 64, fig.24 (1966), combination
erroneously attributed to Maire.

    Najas pectinata sensu Richards & Morony, Check List Fl. Mbala & Dist.: 252 (1969). —
Triest & Symoens in F.A.C., Najadaceae: 5 (1983) as regards *Kinet* 178, non (Parl.) Magn.

Stems very spiny with 7–13 spines per cm. Leaves 6–20(24) mm long, (1.1)1.5–5.3 mm broad
including teeth on both sides (0.5–2.3 mm broad excluding teeth); each margin with 4–10
teeth. Seeds 2.5–3.5 × 0.8–2.1 mm; length:width ratio 1.3–2.3(3.2).

Botswana. N: Boteti R., main stream, 7.xii.1978, *P.A. Smith* 2572 (PRE, SRGH). SE:
Tsokotsa (Tsokotse) Pan, 7.xii.1978, *P.A. Smith* 2565 (K, PRE, SRGH). **Zambia**. N:
Mbala Dist., Mpulungu, Lake Tanganyika, 20.x.1947, *Greenway & Brenan* 8243 (BM,
K). **Zimbabwe**. S: mentioned as *N. marina* by Gibbs Russell (1977), but probably
absent. **Malawi**. C: Nkhotakota, Lake Malawi, 20.vi.1904, *Cunnington* 9 (BM). S: near
Liwonde, Shire R., fr. 20.viii.1956, *Jackson* 1828 (K). **Mozambique**. GI: Inhambane
Dist., Lake Poelela, iii.1938, *Gomes e Sousa* 2107 (K)

Widespread in Africa, mainly in coastal areas and rift lakes; also in Spain, Crete,
Turkey, Syria, Israel, Iraq, Sri Lanka and Australia. Lakes, pans and rivers; may choke
rivers; waters with a relatively high ionic content, brackish or alkaline (conductivity
mostly above 500 μS/cm); water depth 0.2–1.2 m; sometimes associated with *Najas
horrida*; sea level to 1000 m.

Conservation notes: Widespread species; not threatened.

Subgen. CAULINIA

Najas subgen. **Caulinia** (Willd.) Rendle in Trans. Linn. Soc., Bot. **5**: 398 (1899). —
Triest in Mém. Acad. Roy. Sci. Outre-Mer, Cl. Sci. Nat. Méd. **21**(4): 34 (1987);
22(1): 70 (1988).

    Caulinia Willd. in Mém. Acad. Roy. Sci. Hist. (Berlin) **1798**: 87 (1798).

A subgenus with about 39 species, distributed in Africa, Europe, Asia, Australia,
New Guinea, New Caledonia, Hawaii and America.

2. **Najas horrida** Magnus, Beitr. Kenntn. Najas **7**: 46 (1870). —Rendle in Trans.
Linn. Soc., Bot. **5**: 422, t.42 figs.183–191 (1899) in part. —Bennett in F.T.A. **8**:
228 (1901) for most part. —Süssenguth in Mitt. Bot. Staatssamml. München **8**:
339 (1953). —Triest in Mém. Acad. Roy. Sci. Outre-Mer, Cl. Sci. Nat. Méd. **21**(4):
44–52, t.8–10, fig.5 (1987); **22**(1): 91 (1988); in F.T.E.A., Najadaceae: 5, fig.2
(1989). —Clarke & Klaassen, Water Pl. Namibia: 19, 66 (2001). —Cook, Aq.
Wetl. Pl. Sthn. Africa: 186, 192 (2004). Type: Nigeria, Nupe, 1858, *Barter* 1065 (K
lectotype, BRVU, LE, P). FIGURE 12.2.19.

    Najas interrupta K. Schum. in Engler, Pflanzenw. Ost-Afrikas **C**: 94 (1895). —Bennett in
F.C. **7**: 51 (1897). —Rendle in Trans. Linn. Soc., Bot. **5**: 423 (1899). —Bennett in F.T.A. **8**:
228 (1901). —Peter in Abh. Ges. Wiss. Göttingen, Math.-Phys. Kl., N. Folge **13**(2): 18, 31,

108 (1928); in Feddes Repert. Spec. Nov. Regni Veg. Beih. **40**: 116 (1929). —Horn af Rantzien in Kew Bull. **7**: 38 (1952). Type: Tanzania, Kwimba Dist., Lake Victoria between Magu & Kagehi, Unja, *Fischer* 614 (K lectotype, BM, LE, Z).

    Najas pectinata sensu Durand & Schinz, Consp. Fl. Afr. **5**: 500 (1894) in part. —Horn af Rantzien in Kew Bull. **7**: 38 (1952). —Obermeyer in F.S.A. **1**: 83, fig.24 (1966) for most part. —Podlech in Merxmüller, Prodr. Fl. SW Afr. **146**: 3 (1966). —Binns, Check List Herb. Fl. Malawi: 69 (1968). —Hepper in F.W.T.A. **3**: 20, fig.326 (1968) in part. —Gibbs Russell in Kirkia **10**: 434 (1977). —Triest & Symoens in F.A.C., Najadaceae: 5 (1983) for most part. —Barnes & Turton, List Fl. Pl. Botswana: 37 (1986), non (Parl.) Magnus.

    Najas minor sensu Bennett in F.T.A. **8**: 227 (1901), non All. (1773).

    Najas marina L. var. *angustifolia* sensu R.E. Fries in Svensk Bot. Tidskr. **7**: 255 (1913). — T.C.E. Fries in R.E. Fries, Wiss. Ergebn. Schwed. Rhod.-Kongo-Exped.: 188 (1916). —R.E. Fries, Ergänzungsheft: 62, 64, 65, 72 (1921), non (A.Braun) K.Schum.

    Najas welwitschii sensu Richards & Morony, Check List Fl. Mbala & Dist.: 252 (1969) in part as regards *Richards* 15303, non Rendle (1899).

Submerged aquatic herb, monoecious, 30–100 cm high, forming dense clumps, some upper leaf-tips just exserted above water surface. Stems much branched, terete, thin, 0.4–0.2 mm in diameter, unarmed; lower internodes up to 12 cm long, abbreviated above, shoots appearing bushy. Leaves aggregated in upper axils, spreading to falcate, narrowly linear, flat, acute, 3.2–20(29) mm long, 0.56–2.40(3.22) mm wide including teeth on both sides (0.16–1.6 mm wide excluding teeth); margins serrulate with (2)5–12(16) teeth on each side; leaf teeth 0.15–1.38 mm long, borne on broad triangular excrescences and ending in a dark brown spine; tooth length:leaf width ratio (0.35)0.55–1.57(3.08); midrib prominent, without teeth; transverse septa clearly apparent; leaf sheath 1.2(1.5–2.3)3.9 mm long including spine-cells and (1.1)1.5–2.7(5.8) mm broad, truncately-rounded; length:width ratio 0.5–1.4, margins serrulate or lacerate, with 1–15 spine-cells on each side. Male flower enclosed in a spathe, 1.3–3.5 mm long including spathe neck; spathe neck cylindrical 0.3–0.7 mm long, bearing brown spine-cells at the apex; inner envelope subelliptical, abruptly narrowing above, forming 2 small lip-like gibbosities at top; anther 0.9–2.3 × 0.5–1.2 mm, 4-sporangiate. Female flower naked, (1.3)2–3(4.2) mm long; ovary turbinate, 0.5–2.4 × 0.2–0.9 mm, contracted in a short style with 2 linear-tapering stigmas. Fruit ellipsoid-oblong. Seed 1.6–2.6(3.2) × 0.48–0.84 mm; length:width ratio 3.2–4.2; testa with c.20 regular longitudinal rows of areoles, (18)25–35(40) areoles in each row, squarish, rectangular, or rarely hexagonal.

Caprivi Strip. Mentioned from E Caprivi by Clarke & Klaassen (2001). **Botswana**. N: Okavango Swamps, Txatxanika Lagoon, 1.iii.1972, *Gibbs Russell & Biegel* 1487 (BR, K, LISC, MO, SRGH). **Zambia**. N: Lake Chali, 5.vi.1962, *Symoens* 9554 (BR, BRVU, K). S: Choma Dist., ♀ fl. 15.ii.1959, *Robinson* 3245 (K, M, PRE). **Zimbabwe**. N: Hurungwe Dist., Sanyati-Chiroti river junction, ♀ fl. 21.xi.1953, *Wild* 4229 (K, LISU, MO, SRGH, PRE). W: Hwange Dist., Katsatetsi R., 10.v.1972, *Gibbs Russell* 1953 (K, MO, SRGH). C: Mazowe Dist., Henderson Res. Station, 15.iv.1973, *Gibbs Russell* 2562 (K, MO, SRGH). E: Chipinge Dist., lower Rupembe R., 24.i.1957, *Phipps* 149 (K). S: Mwenezi Dist., near Fishans, Runde (Lundi) R., 28.iv.1962, *Drummond* 7774 (K, LISC, SRGH). **Malawi**. C: Dedza Dist., Ndebvu, Dzalanyama Forest, 29.vi.1967, *Salubeni* 761 (K, MAL, PRE). S: Chikwawa, Elephant Marsh near Namichamba, 14.vii.1980, *Osborne* 14 (MAL). **Mozambique**. T: Tete, Digue, Mazoe R., 21.ix.1948, *Wild* 2583 (K, SRGH). M: Moamba Dist., Ressano Garcia, 22.xii.1897, *Schlechter* 11883 (BM, BR, COI, E, G, L, LE, MO, P, WAG, Z).

Widespread in Africa including Egypt, Senegal, Mali, Niger, Nigeria, Cameroon, Chad, Central African Republic, Sudan, Ethiopia, Congo, Burundi, Uganda, Kenya, Tanzania, Namibia, South Africa (former Transvaal to Cape Provinces) and Madagascar. Lakes, dams, lagoons, ponds, pans and rivers in water with a wide range of ionic content with conductivity 50–1200 μS/cm and pH 7–9.5; water depth 1–4 m; sometimes associated with *Najas marina* var. *armata*, *Lagarosiphon ilicifolius*, *Potamogeton octandrus*, *Nymphaea caerulea* and *Nymphoides indica*; 100–1600 m.

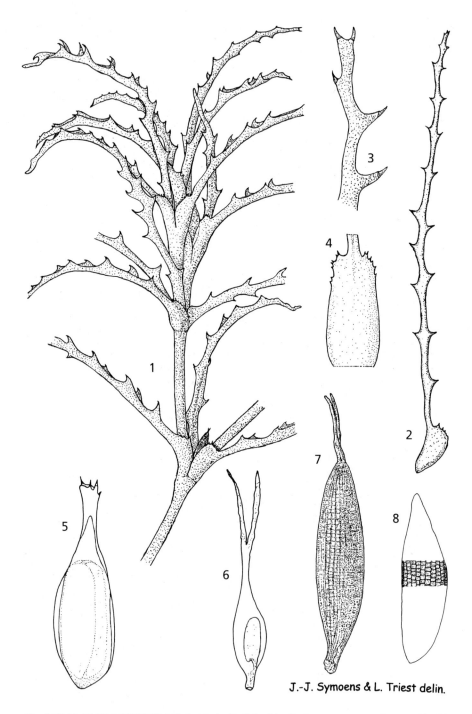

J.-J. Symoens & L. Triest delin.

Fig. 12.2.**19**. NAJAS HORRIDA. 1, habit (× 4); 2, leaf (× 5); 3, upper part and apex of leaf (× 25); 4, leaf sheath (× 12¹/₂); 5, male flower (× 25); 6, female flower (× 25); 7, fruit (× 25); 8, seed (× 16¹/₂). 1, 3, 7 from *Symoens* 6925 (Katanga), 2 from *Drummond* 7774, 4 from *Mortimer* 176, 5 from *Lind, Agnew & Kettle* 5893 (Kenya), 6 from *Vollesen* 2862 (Tanzania), 8 from *Humbert* 9266 (Kenya). Drawn by J.-J. Symoens and L. Triest.

Conservation notes: Widespread species; not threatened.

Although *N. horrida* is generally considered to be monoecious, when examining South African specimens both in the herbarium and in the wild state, Obermeyer (in F.S.A. **1**: 83, 1966) found them to be dioecious. However, Triest (in Mém. Acad. Roy. Sci. Outre-Mer, Cl. Sci. Nat. Méd. **21**: 48, 1987) noticed that he rarely found both sexes together on the same herbarium sheet; even so it was impossible to decide whether the flowers on different branches originated from the same plant as branches easily break off and different individuals may be mounted together.

After the publication of Rendle's revision (Trans. Linn. Soc., Bot. **5**: 399 & 421–424, 1899), the name *N. horrida* Magnus was given to specimens with falcate, often rigid leaves and teeth exceeding the leaf-width, while the name *N. interrupta* K. Schum. was given to related plants with non-falcate and mostly lax leaves and teeth not exceeding the leaf-width. However, *N. horrida* shows much plasticity and there are many intermediates (e.g. *Symoens* 9554, 9596, *Watmough* 268 from Zambia; *Goldsmith* 205/60, *Gibbs Russell* 2562, *Phipps* 761 from Zimbabwe; *Wild* 2583 from Mozambique). Horn af Rantzien (in Kew Bull. **7**: 38–39, 1952) expressed doubts about the distinction between the species and Obermeyer (1966) and Triest (1987) considered the two as synonyms.

After Rendle's revision, and in spite of his reservations about the conspecificity of *N. horrida* and *N. pectinata* (Parl.) Magnus, many authors accepted the synonymy of both species. However, *Najas horrida* differs from *N. pectinata* in its 4-sporangiate anther and naked female flower, while *N. pectinata* has a 1-sporangiate anther and spathaceous female flower and is only known from N Egypt (including Sinai).

3. **Najas testui** Rendle in J. Bot. **75**: 51, fig.A–I (1937). —Horn af Rantzien in Act. Hort. Gotoburg. **18**: 194 (1950); in Kew Bull. **7**: 32 (1952). —Triest in Mém. Acad. Roy. Sci. Outre-Mer, Cl. Sci. Nat. Méd. **21**(4): 52, t.1, fig.6 (1987); **22**(1): 100 (1988); in F.T.E.A., Najadaceae: 7 (1989). Type: Central African Republic, 120 km NE of Yalinga, 22.i.1922, *Le Testu* 3625 (BM holotype, K, L, P).

   Najas meiklei Horn in Kew Bull. **7**: 34, fig.1–2 (1952). —Hepper in F.W.T.A. **3**: 20 (1968) excl. *Hall* 723. Type: Nigeria, Mivna Division, below Gwari hill near Mivna, 3.xii.1949, *Meikle* 710 (K holotype).

   Najas welwitschii sensu Richards & Morony, Check List Fl. Mbala & Dist.: 252 (1969) in part as regards *Richards* 15172, non Rendle (1899).

   Najas graminea sensu Triest & Symoens in F.A.C., Najadaceae: 6 (1983) in part as regards *Liben* 840, non Delile (1813).

Submerged, rooted aquatic herb, monoecious, slender. Stems unarmed, c.10 cm high, spreading and profusely branched above; branches thin. Leaves (6.5)12–25(32) mm long, (0.26)0.42–0.72(1.1) mm wide including teeth (0.24–0.53(0.87) mm wide excluding teeth), linear-lanceolate, flat, acute; margins finely serrulate with 7–21 small teeth on each side; leaf teeth (0.05)0.09–0.14 mm long, spinulous, mostly inconspicuous teeth, supported by 2 ± projecting epidermal cells; tooth length : leaf width ratio 0.13–0.3(0.5); midrib without teeth; transverse septa clearly apparent or absent; apex mostly obtuse or truncate and 2-dentate, sometimes more acute and 1-dentate; leaf sheath truncate to slightly auriculate, (1.3)1.9–2.4(2.9) mm long (including auricle and spine-cells); auricles $^{1}/_{4}$–$^{1}/_{3}$ of total sheath length, obtuse, serrulate or lacerate with 2–8 spine-cells on each side; intravaginal scales narrowly subulate, 0.5–0.88 mm long. Male and female flowers solitary or aggregated, the male and the female mostly on different branches. Male flower enclosed in a spathe, (1.2)1.9–2.5 (including spathe neck) × 0.4–1 mm; spathe neck 0.5–0.6 mm long, tapering at the top and bearing mostly 2 brownish spine-cells at the apex; inner envelope protruding 0.16–0.32 mm above the anther; anther 1.0–2.4 × 0.36–1 mm, 4-sporangiate. Female flower enclosed in a spathe, (1.1)1.5–2 (including spathe neck) × 0.4–0.6 mm; spathe neck reaching about halfway up the style, without spine-cells at apex; ovary 0.5–1.77 × 0.3–0.6 mm; style cylindric, with 2 long protruding stigmatic branches, the whole 0.8–1.3 mm long. Fruit narrowly elliptic with persistent, thin, membranous gynoecial wall and remaining parts of style, stigmas and spathe. Seed (1.85)2.1–2.3(2.6) × (0.41)0.56–0.79 mm, ellipsoid-oblong; length:width ratio 2.3–3.8; testa yellow or dull brown, with 14–20 longitudinal rows of areoles, 10–23 areoles in each row, irregular, mostly ± hexagonal, end walls raised.

Zambia. N: Mbala Dist., Kawimbe, Lumi R., ♀ fl.& fr. 24.iii.1959, *McCallum-Webster* A628 (K).

An exclusively African species; also in Sierra Leone, Ivory Coast, Ghana(?), Nigeria, Cameroon, Central African Republic, Sudan, Congo, Rwanda, Uganda, Kenya and Angola. Ponds, ditches, swamps, streams; 1700 m.

Conservation notes: Very local within the Flora area, probably Lower Risk near threatened, but not threatened globally.

Sterile specimens of *N. testui* Rendle, *N. schweinfurthii* Magnus, *N. minor* All., *N. hagerupii* Horn and *N. baldwinii* Horn cannot be distinguished from each other. Ripe seeds are necessary and, for some determinations, also flowers of both sexes. *N. schweinfurthii*, recorded from near Songea (S Tanzania), and *N. baldwinii*, from on the high plateaux of Katanga (Congo), may yet be found in the Flora area. *N. schweinfurthii* differs from *N. testui* by its 1-sporangiate anthers and shorter seeds (1.1–1.6 mm long) and from *N. baldwinii* by its male and female flowers without spathes and its 1-sporangiate anthers.

4. **Najas graminea** Delile, Descr. Egypte, Hist. Nat. **2**: 282 (1812). —Rendle in Trans. Linn. Soc., Bot. **5**: 424, t.42 figs.192–201 (1899) excl. vars. *minor* & *angustifolia*. — Horn af Rantzien in Meddel. Göteborgs Bot. Trädg. **18**: 192, 193 (1950). — Obermeyer in F.S.A. **1**: 83, fig.24.3 (1966). —Triest & Symoens in F.A.C., Najadaceae: 5 (1983) in part as regards *Troupin* 1836. —Triest in Mém. Acad. Roy. Sci. Outre-Mer, Cl. Sci. Nat. Méd. **21**(4): 69, t.17–18, fig.11 (1987); in **22**(1): 135, t.27, 28, fig.26 (1988); in F.T.E.A., Najadaceae: 8 (1989). —Cook, Aq. Wetl. Pl. Sthn. Africa: 185 (2004). Type: Egypt, Rosetta and Nile Delta, *Delile* s.n. (MPU holotype, H).

Submerged, rooted aquatic herb, normally monoecious, slender, often plumose above. Roots many, long, filiform. Stems 10–60 cm high, terete, unarmed, 0.4–1.5 mm in diameter, sparingly branched. Leaves mostly 7.5–33 mm long and 0.24–1.3 mm wide including teeth (0.19–0.93 mm wide excluding teeth), linear, flat, acute; margins minutely serrulate, with 18–74 teeth on both sides; leaf teeth 0.02–0.07(0.12) mm long, mostly inconspicuous, each consisting mainly of a brown spine-cell; tooth length:leaf width ratio 0.03–0.10(0.22); midrib without teeth; septa absent; leaf sheath mostly deeply auriculate, (1.4)2–3(3.9) mm long including auricles and spine-cells; auricles subulate, up to half of total sheath length, acute, each bearing 3–14 spine-cells. Male and female flowers solitary or in groups 2–4, generally on the same branches, sometimes in pairs consisting of a male and a female flower borne side by side. Male flower about 1 mm long; spathe absent; inner envelope protruding 0.05–0.13 mm above the anther; anthers 0.7–1.3 × 0.3–0.5 mm, not tapering towards the top, 4-sporangiate; pedicel elongating before dehiscence and becoming almost as long as anther. Female flower 1.3–1.6 mm long; spathe absent; ovary 0.7–1.6 × 0.26–0.87 mm; stigmatic arms 2–3. Fruit 1.5 mm long, narrowly oblong, tapering at tip, bearing remaining parts of style; pericarp persistent, thin, membranous. Seed 1.75–2.5 × 0.4–0.6 mm, narrowly oblong; testa pitted, with 25–30(35) longitudinal rows of areoles, 23–25(60) areoles in each row, hexagonal, sometimes squarish or rectangular, end walls slightly raised.

Var. **graminea**. —Triest in Mém. Acad. Roy. Sci. Outre-Mer, Cl. Sci. Nat. Méd. **21**(4): 69, t.17–18, fig.11 (1987); **22**(1): 137, t.27, 28, fig.26 (1988); in F.T.E.A., Najadaceae: 8 (1989).

Plant slender. Stem 0.4–1.5 mm in diameter. Leaves (7.5)14–25(33) mm long and (0.24)0.5–0.9(1.3) mm broad including teeth on both sides; leaf teeth 0.02–0.07(0.12) mm long; leaf teeth length:leaf width ratio 0.03–0.22; leaf sheath 1.4–5 × 0.9–2.1(2.7) mm; auricles with 3–14 spine-cells on each external side and 1–7 on inner edge.

Botswana. N: 40 km NNW of mouth of Nata R., Mompswe (Mumpswe), fr. 21.iv.1957, *Drummond & Seagrief* 5163 (K, LISC).

Widespread in Africa – Algeria, Egypt, Senegal, Gambia, Mali, Cameroon, Chad, Central African Republic, Sudan, Socotra, Congo, Kenya, Tanzania and South Africa (former Transvaal); also in the subtropical and tropical regions of Asia and Australia, in New Guinea and New Caledonia. Found as an adventive in Europe and the USA (California). Lakes, ponds, rice-fields, small rivers, ditches; because of its vigorous growth it is often an important weed in irrigation channels and rice-fields; 900 m.

Conservation notes: Probably more widespread in the Flora area than recorded, but poorly collected; not threatened.

N. graminea is often recognized in the field by its very lax, almost graminoid appearance and the long leaves with numerous minute teeth and distinctly auriculate sheaths. Although both male and female flowers are normally without spathes, De Wilde (in Acta Bot. Neerl. **10**: 166–167, figs.3–5, 1961) has described and illustrated exceptional spathaceous male and female flowers in specimens.

APONOGETONACEAE

by E.S. Martins

Rhizomatous perennial monoecious or dioecious herbs, sometimes apomictic, with tubers, usually growing in fresh water or on wet soil. Leaves all basal, alternate, sessile or long petiolate, with blades elliptic to linear or awl-shaped, entire, submerged or floating. Inflorescence usually a long petiolate single spike or 2 opposite spikes or rarely digitately divided into 3–4(11) spikes, at first enclosed in a membranous caducous spathe, or the inflorescence much abbreviated and resembling a *Ranunculus* flower in the South African *Aponogeton ranunculiflorus*. Flowers bisexual or unisexual. Tepals 1–6 or lacking, petal-like, persisting or not in fruit. Stamens 1–6, rarely more, lacking in female or apomictic plants; filaments free, filiform or flattened; anthers very small, 2–thecous, extrorse, yellowish, brown, violet, purplish or blackish. Ovary superior with 3–8 carpels, free or slightly united near their base, each one 1-locular, with 1–14 erect ovules inserted at the base or along the carpel suture. Fruit of 3–8 free ovoid or bottle-shaped beaked follicles. Seeds 1–14 on each follicle, discoid to fusiform, without endosperm; testa simple or double, the inner brown and close to the embryo, the outer pellucid.

A monogeneric family widely distributed in tropical and subtropical regions from Africa and Madagascar to India, SE Asia and Australia, but absent from the Americas.

APONOGETON L. f.

Aponogeton L.f., Suppl. Pl.: 32 (1781). —Bruggen in Bull. Jard. Bot. Belg. **43**: 193–233 (1973); in Bibl. Bot. **137**: 1–76 (1985).

Description as for the family.

A genus of about 40 species.

Many species have been widely cultivated as aquarium plants. In addition to the species described below, the key includes the South African *A. natalensis*.

1. Inflorescence with 1 spike · 2
– Inflorescence with 2 or more spikes · 4

2. Flowers spirally arranged on rachis; tepals persistent and usually shorter than 3 mm, violet; floating leaves less than 5 times as long as wide; specimens mostly apomictic (female flowers with up to 8 carpels) · · · · · · · · · · · · **3.** *afroviolaceus*
   – Flowers inserted on one side of rachis only; tepals caducous, longer than 3 mm; leaves more than 5 times as long as wide, linear or band-shaped, submerged or floating; specimens not apomictic (bisexual flowers with 3–4 carpels) · · · · · 3
3. Leaves linear, petiolate and floating, rarely subulate and submerged; spathe early caducous; tepals 1-nerved · **2.** *stuhlmannii*
   – Leaves band-shaped, sessile, submerged; spathe persistent with inflorescence curved back into it; tepals with several parallel nerves · · · · · · · **1.** *vallisnerioides*
4. Flowers inserted dorsally on rachis; leaves awl-shaped, 3-edged, rarely gradually expanded into a very narrow lanceolate blade; tepals pink or white · · · **6.** *junceus*
   – Flowers spirally arranged on rachis; leaves with a distinct blade, or if without then with bluish-purple tepals · 5
5. Spikes 35–90 mm long in flower, up to 150 mm long and 15 mm wide in fruit; tepals yellow; seeds more than 3 mm long; specimens not apomictic; bisexual flowers with 3 carpels at the spike apex but with up to 8 at the spike base · · · ·
   · **4.** *desertorum*
   – Spikes up to 50 mm long in flower, little elongated in fruit; tepals white or pink to violet; seeds less than 3 mm long; specimens sometimes apomictic (female flowers with up to 7 carpels) · 6
6. Tepals mauve to violet; secondary leaves with oval to lanceolate floating blade 3–12 cm long, or leaves sessile, linear, without floating blade; seeds with a double testa, the outer one translucent, the inner brown · · · · · · · · · · · · **5.** *abyssinicus*
   – Tepals white to very pale pink; secondary leaves always with a linear-lanceolate to oblong, erect or floating blade up to 17 cm long; seeds with a simple brown testa · 7
7. Secondary leaves always floating; spikes up to 15 mm long; spathe to 10 mm long; tepals 2–3 mm long; fruiting carpels up to 5 × 3 mm with a long terminal beak · **7.** *rehmannii*
   – Secondary leaves mostly submerged or emergent; spikes up to 55 mm long; spathe to 25 mm long; tepals 3–4 mm long; fruiting carpels up to 10 × 4.5 mm with a short terminal beak · *natalensis*

1. **Aponogeton vallisnerioides** Baker in Trans. Linn. Soc., London **29**: 158 (1875). — Durand & Schinz, Consp. Fl. Africa **5**: 493 (1894). —Bennett in F.T.A. **8**: 218 (1901). —Hutchinson & Dalziel, F.W.T.A. **2**: 306 (1936). —Troupin in Bull. Jard. Bot. État **23**: 225 (1953). —Bruggen in Bull. Jard. Bot. Belg. **43**: 197, fig.4/1 (1973); in F.A.C., Aponogetonaceae: 3, pl.1 (1974). —Lye in F.T.E.A., Aponogetonaceae: 2, fig.1 (1989). Type: Uganda, Acholi/Lango Dist., Ukidi, xi.1862, *Speke & Grant* s.n. (K holotype). FIGURE 12.2.**20**.

Tuber globular, usually up to 10 mm (rarely 20 mm) in diameter, with many slender roots on the upper half, the lower half without roots. Leaves submerged, very numerous, sessile, ligulate, 40–200 × 1.5–8 mm, obtuse at apex, with 7–9 scarcely distinct parallel nerves. Inflorescence with a 3–35 cm long peduncle (depending on water depth), slender; spathe up to 6 mm long in our area (to 18 mm outside Flora area), persisting when flowering, but caducous when in fruit. Spike single, up to 20 mm long (50 mm when in fruit) and 10 mm wide, rather dense, usually curved back over spathe. Flowers inserted on only one side of rachis, bisexual. Tepals 2, white or rarely with a pinkish or mauve colouring, 3–4.5 × 1.5–3 mm, oval, with several parallel nerves, caducous. Stamens 6, 1.5–2 mm long; filaments widened towards base; anthers 0.4–0.8 mm long, violet to blackish. Carpels 3, each one with 6–8 ovules; fruiting carpels up to 6 × 2 mm, including a c.1 mm long terminal beak. Seeds 1.5–2.5 × 0.5–0.7 mm, narrowly ellipsoid, with

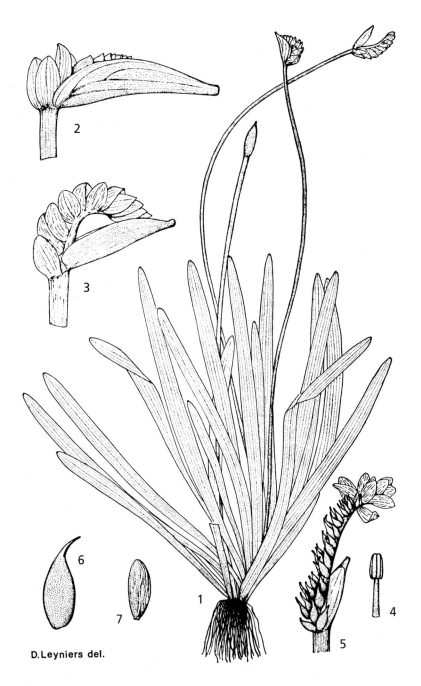

Fig. 12.2.**20**. APONOGETON VALLISNERIOIDES. 1, habit (× 1); 2 & 3, inflorescences (× 3); 4, stamen (× 10); 5, fruiting spike (× 3); 6, fruiting carpel (× 10); 7, seed (× 10). 1–3 from *Troupin* 1585, 4 from *César & Ménaut* 167, 5–7 from *Boutique* 112. Drawn by D. Leyniers. Reproduced with permission from Flore d'Afrique Centrale (Meise).

D. Leyniers del.

double testa, the outer one lax, pellucid and reticulate, the inner one dark brown and closely fitting in the embryo; plumule absent.

Zambia. N: Kasama Dist., Kilolo village, Mpanda R., Chilumbelo rocks, 1200 m, fl.& fr. 13.xii.1964, *Richards* 19371 (K). W: Ndola rural, fl.& fr. 30.iii.1963, *Fanshawe* 7753 (K, SRGH).

Also known from Senegal, Sierra Leone, Ivory Coast, Nigeria, Burkina Faso, Cameroon, Central African Republic, Sudan, Congo, Rwanda, Uganda and Kenya. In shallow pools and marshes; 1200–1220 m.

Conservation notes: Widespread species, although more restricted within the Flora area; not threatened.

2. **Aponogeton stuhlmannii** Engl. in Notizbl. Königl. Bot. Gart. Berlin **1**: 26 (1895); Pflanzenw. Ost-Afrikas **C**: 94 (1895). —Bennett in F.T.A. **8**: 218 (1901). — Obermeyer in F.S.A. **1**: 90, fig.27 (1966). —Podlech in Merxmüller, Prodr. Fl. SW Afrika, fam.143: 3 (1966). —Bruggen in Bull. Jard. Bot. Belg. **43**: 201, fig.4 (1973); in F.A.C., Aponogetonaceae: 4 (1974). —Gibbs Russell in Kirkia **10**: 435 (1977). —Lye in F.T.E.A., Aponogetonaceae: 2 (1989). —Clarke & Klaassen, Water Pl. Namibia: 68 (2001). —Cook, Aq. Wetl. Pl. Sthn. Africa: 60 (2004). Type: Tanzania, Mwanza Dist., Uzinza (Usinja), Bugando, *Stuhlmann* 3541 (B holotype).

Aponogeton gracilis Schinz [in Durand & Schinz, Consp. Fl. Afr. **5**: 492 (1895), nomen] ex A. Bennett in F.C. **7**: 43 (1897); in Bull. Herb. Boiss., sér.2, **1**: 763 (1901). Type: South Africa, Transvaal, Houtbosch, *Rehmann* 5761 (Z holotype, K).

Aponogeton gramineus Lye in Norw. J. Bot. **18**: 190, fig.2 (1971); in F.T.E.A Aponogetonaceae: 4 (1989) in note. Type: Zambia, Mporokoso Dist., Mweru Wantipa road to Bulayo, 1050 m, 12.iv.1957, *Richards* 9178 (K holotype).

Tuber globular, up to 10 mm in diameter with many slender whitish roots. Leaves up to 10, the primary ones (when present) submerged, without distinguishable petiole and blade, subulate to linear, usually up to 20 mm long; secondary leaves floating, with petiole 5–20 cm long (depending on water depth), filiform, blade 20–50 × 4–10 mm, lanceolate to narrowly elliptic, cuneate to obtuse or rounded at base, obtuse at apex, with 5–9 longitudinal parallel nerves. Peduncle up to 15 cm long, very slender, not thickened towards apex; spathe 6–10 mm long, greenish, early caducous. Spike single, 6–13 mm long in flower, 14–23 mm in fruit, lax to rather dense. Flowers bisexual, 7–15, inserted on one side of rachis, or on two sides but the spike seeming cylindrical early on. Tepals 2, white but often tinted lilac-mauve, 3–4 × 1–2 mm, 1-nerved, caducous. Stamens 6, 2–2.5 mm long; filaments widened towards base, lilac but whitish lower down; anthers c.0.5 mm long, brownish-yellow, but turning black when withering. Carpels 3–4, green, each with 2–8 ovules. Fruiting carpels 4–6 mm long (including a 1–2 mm long curved beak) and 1.25–2 mm wide. Seeds 1.5–2.5 × c.0.5 mm, narrowly ellipsoid; testa double, the outer one somewhat lax, brown, reticulate, the inner one closely fitting the embryo, dark brown; plumule absent.

Botswana. SE: Gaborone, Aedume Park, 24°42'S, 25°54'E, 1050 m, fl.& fr. 23.xii.1977, *Hansen* 3313 (K, SRGH). **Zambia.** B: Mongu, fl. 26.i.1966, *Robinson* 6812 (K). N: Mporokoso Dist., Kawambwa, 1200 m, fl. 7.i.1960, *Richards* 12081 (K). W: Mwinilunga Dist., Zambezi rapids, 6.4 km from Kalene Mission, 1200 m, fl. 13.xi.1962, *Richards* 17197 (K). S: Choma Dist., Choma Chimwani Hills, fl. 27.xi.1962, *Astle* 1718 (SRGH). **Zimbabwe.** W: Matopos Nat. Park, c.50 m from Rhodes' tomb, 1350 m, fl.& fr. 24.ii.1981, *Philcox & Leppard* 8827 (K). C: Gweru Dist., Mlezu School Farm, 18 km SSE of Kwekwe (Que Que), 1280 m, fl.& fr. 19.ii.1966, *Biegel* 922 (K, SRGH). E: Nyanga Dist., c.1 km NE of Bonda Mission, 2100 m, fl. 26.xii.1972, *Biegel* 4132 (K, LISC, SRGH). S: Masvingo (Victoria), fl., *Munro* 2231 (K).

Also known from Kenya, Tanzania, Namibia and South Africa. In small swamps and pools, sometimes in depressions on granite outcrops, in shallow water; 1050–2100 m.

Conservation notes: Widespread species; not threatened although said to be Data

Deficient for Zambia in Sabonet Red Data List (Golding 2002).

Some specimens from Zambia growing in mud in shallow rockpools, were previously considered as *A. gramineus* Lye. They have primary linear leaves very rarely up to 60 mm long and 3 mm wide, similar to those of *A. vallisnerioides* but usually smaller. The spike shows the flowers on two sides of the rachis, but this character is sometimes difficult to see. Floral characters are very similar to those of *A. stuhlmannii* sensu stricto, hence it is difficult to separate the taxa. Van Bruggen (1973 & 1974) considers these characters of little taxonomic significance and accepts no segregation, while Lye (1989) considers that *A. gramineus* may only be an infraspecific variant of *A. stuhlmannii*. The small amount of material seen does not allow for a definitive conclusion. It seems preferable to follow Van Bruggen and consider *A. stuhlmannii* as a single species until more field studies are done.

3. **Aponogeton afroviolaceus** Lye in Bot. Not. **129**: 68 (1976); in F.T.E.A. Aponogetonaceae: 4 (1989). —Gibbs Russell in Kirkia **10**: 435 (1977). —Cook, Aq. Wetl. Pl. Sthn. Africa: 57 (2004). Type: Kenya, Kiambu Dist., Thika Town, 1450 m, 1.vii.1971, *Lye, Katende & Faden* 6348 (MHU holotype, K).

Aponogeton violaceus Lye in Norw. J. Bot. **18**: 187, fig.1 (1971). —Bruggen in Bull. Jard. Bot. Belg. **43**: 205, fig.4/4 (1973); in F.A.C., Aponogetonaceae: 2 (1974). Type as for *A. afroviolaceus*.

Tuber oval or cylindrical, up to 25 mm long and 15 mm diameter, with numerous slender whitish roots on the upper part. Leaves usually up to 15, the primary ones submerged, with petiole up to 8 cm long and blade up to 40 × 5 mm, narrowly lanceolate, cuneate at base, acute at apex but with the tip blunt, with 3 or 5 longitudinal nerves; secondary leaves floating, with petiole up to 20 cm long and blade 20–80 × 3–16 mm, oval to linear-oblong, cuneate to truncate at base, acute at apex with the tip blunt, with 5 or 7 longitudinal nerves. Peduncle up to 30 cm long, slender, not thickening to apex, greenish, but often white to mauve below, dull maroon to brown above; spathe early caducous. Spike single, up to 30 mm long and 5–10 mm wide, usually rather dense, elongated up to 40 mm when fruiting. Flowers 10–30, spirally arranged, bisexual or apparently female in apomictic specimens. Tepals 2, persistent, purple or mauve, 2–6 mm long, obovate or spathulate, 1-nerved. Stamens 6 (0 in apomictic specimens), up to 3.5 mm long; filaments thickened towards base, purple; anthers 0.3–0.7 mm long, yellow to dark brown. Carpels 3 (up to 8 in apomictic specimens), each with 2–5 ovules (to 9 in apomictic specimens). Fruiting carpels up to 7 mm long, including a c.1.5 mm long beak. Seeds 1–2 mm long, narrowly ellipsoid to cylindrical and slightly curved; testa double, the outer one transparent and reticulated, the inner one closely fitting the embryo, brown; plumule absent.

Var. **afroviolaceus**

Plant with 5–15 leaves; leaf-blade 8–16 mm wide, ovate to oblong-elliptic sometimes scarcely distinguishable from the petiole. Flowers 10–30. Tepals 2–4 mm long. Anthers 0.5–0.7 mm long. Fruiting carpels 3.5–4.5 mm long.

Zambia. B: Machili, fl. 25.xi.1960, *Fanshawe* 5919 (K, SRGH). N: Mbala Dist., Lumi River marsh, Kawimbe, 1740 m, fl. 27.iii.1959, *Richards* 12274 (K). S: Kabulamwanda Dam, 129 km N of Choma, 1030 m, fl. 24.iv.1954, *Robinson* 724 (K). **Zimbabwe.** N: Makonde Dist., Rukuti Farm, 1220 m, fl. 12.i.1969, *Jacobsen* 3632 (PRE, SRGH). C: Chegutu Dist., Poole Farm, fl. 23.xii.1953, *R.M. Hornby* 3337 (K, SRGH).

Also known from Kenya and Tanzania. In water pans, marshes, small temporary ponds and shallow irrigation channels, in shallow water; 1000–1750 m.

Conservation notes: Widespread species; not threatened.

A. afroviolaceus var. *angustifolius* Lye, endemic to S Kenya, usually has only 3–5 leaves with leaf-blades 3–6 mm wide, always distinguishable from the petiole.

4. **Aponogeton desertorum** A. Spreng., Tent. Suppl.: 14 (1828). —Wild in Kirkia **2**: 36 (1961); Plant. Aq. Nuis. Afr. Madagascar: 33, pl.13B (1964). —Obermeyer in F.S.A. **1**: 90, fig.27/1 (1966). —Podlech in Merxmüller, Prodr. Fl. SW Afrika, fam. 143: 2 (1966). —Bruggen in Bull. Jard. Bot. Belg. **43**: 213, fig.6/9 (1973); in F.A.C., Aponogetonaceae: 8 (1974). —Gibbs Russell in Kirkia **10**: 435 (1977). —Clarke & Klaassen, Water Pl. Namibia: 68 (2001). —Cook, Aq. Wetl. Pl. Sthn. Africa: 58 (2004). Type: South Africa, Eastern Cape, Uitenhage, *Zeyher* 191 (B holotype).

> *Aponogeton kraussianum* C. Krauss in Flora **28**: 343 (1845). —Bennett in F.T.A. **8**: 217 (1902). Type: South Africa, Eastern Cape, Uitenhage, foot of Winterhoek Mts., *Krauss* 1604 (B† holotype, K).
>
> *Aponogeton holubii* Oliv. in Hooker's Icon. Pl. **15**: t.1470 (1884). —Durand & Schinz, Consp. Fl. Afr. **5**: 492 (1895). —Bennett in F.C. **7**: 44 (1897); in F.T.A. **8**: 217 (1902). Type: Botswana (Bechuana Country), Hendrick's pan, "Eastern Bamanguato Territory", 30.iii.1876, *Holub* 1036–1039 (K holotype).
>
> *Aponogeton dinteri* Engl. & K. Krause in Bot. Jahrb. Syst. **38**: 92 (1905). Types: Namibia, Otjimbingue, *Fisher* 165 (B† syntype), and *Dinter*, cultivated at Botanic Garden, Berlin (B† syntype).
>
> *Aponogeton eylesii* Rendle in J. Bot. **70**: 76 (1932). Type: Zimbabwe, Harare (Salisbury), Rua R., 1460 m, x.1917, *Eyles* 852 (BM holotype, K).

Tuber globose or oboval, up to 3 cm in diameter. Leaves usually all floating; petiole up to 100 cm long; leaf-blade 5–21 × 1.7–6 cm, ovate or lanceolate to oblong-elliptic, cuneate to almost cordate or very rarely cordate at base, obtuse or rounded at apex, with (5)7–9(13) longitudinal nerves and many parallel transverse veins between them. Peduncle up to 100 cm long, strong, usually thickened towards apex; spathe up to 2 × 1 cm, hooded and apiculate, caducous. Spikes 2, 35–90 × c.7 mm, densely or sometimes laxly flowered, usually very dense, elongated up to 150 mm and widened up to 15 mm in fruiting. Flowers spirally arranged, bisexual. Tepals 2(3), yellow, 2–3.25 × 0.5–1.75 mm, 1-nerved, persistent. Stamens 6; filaments up to 4 mm long, widened towards base; anthers yellow. Carpels 3–8, 1.25–3 mm long and 0.5–1 mm wide, each with 4–10 ovules. Fruiting carpels up to 9 × 4 mm, with a long terminal beak toped by the dark stigmatic ridge. Seeds 3–5 × 1–2 mm, ellipsoid or oblong-ellipsoid; testa double, the outer one lax, winged and reticulated veined, the inner one closely fitting the embryo, brown; plumule absent.

Botswana. N: Hendrick's Pan, "Eastern Bamangwato Territory", fl.& fr. 30.iii.1876, *Holub* 1036–1039 (K). SE: 22.5 km (14 miles) N of Mahalapye, fl. 2.iv.1937, *van der Merwe* 1325 (K). **Zambia.** B: Sesheke, Zambezi R., fl. 1911, *Gairdner* 228 (K) (mixed with *Nymphaea* leaves). N: Mpika, fl.& fr. 30.viii.1966, *Fanshawe* 9781 (K). W: Kasempa Dist., Lunga R., Kafue Nat. Park, fl.& fr. 17.ix.1962, *Uys* 39/62 (COI, LISC, SRGH). C: Serenje Dist., Kundalila Falls near Kanona, Kundalila Reserve, 1200 m, fl. 24.viii.1963, *Richards* 18206 (K). **Zimbabwe.** N: Makonde Dist., Hunyani R., Gomberra Ranch, 990 m, fl.& fr. viii.1959, *Jacobsen* 1431 (PRE). W: Bulawayo, 1350 m, fl.& fr. v.1961, *Miller* 7938 (SRGH). C: Chegutu Dist., Kalembo Dam, fl.& fr. 26.ii.1969, *Mavi* 1021 (K, PRE, LISC, SRGH). E: Chimanimani Dist., Haroni area, Haroni R., 360 m, fl.& fr. 5.ix.1969, *T. Wild & Bisset* 60 (K, SRGH). S: Masvingo Dist., Mushandike Nat. Park, S entrance gate, fl.& fr. 10.i.1975, *Lightfoot* 45 (SRGH). **Mozambique.** MS: Barúê, Catandica (Vila Gouveia) towards Chimoio (Vila Pery), fl.& fr. 16.ix.1942, *Mendonça* 260 (BR, EA, K, LISC, LMU, MO, PRE, SRGH, WAG); Mossurize, Gogói, fl.& fr. 13.vi.1942, *Torre* 4322 (COI, LISC, LISU, LMU, M, UPS).

Also known from Congo (Katanga), Angola, Namibia and South Africa. In rocky rapids and very deep slowly-flowing water; 360–1450 m.

Conservation notes: Widespread species; not threatened.

The tubers are edible.

5. **Aponogeton abyssinicus** A. Rich., Tent. Fl. Abyss. **2**: 351 (1851). —Bennett in F.T.A. **8**: 218 (1902). —Bruggen in Bull. Jard. Bot. Belg. **43**: 221, fig.7/12 (1973); in F.A.C., Aponogetonaceae: 9, photo (1974). —Lye in Lidia **1**: 67, figs.1–5 (1986); in F.T.E.A., Aponogetonaceae: 6, fig.2 (1989). Type: Ethiopia, Tigray, near Axum, 25.ii.1840, *Schimper* 1483 (P holotype, BR, GOET, K, M).

    Aponogeton boehmii Engl. in Notizbl. Königl. Bot. Gart. Berlin **1**: 26 (1895). —Bennett in F.T.A. **8**: 218 (1902). Type: Tanzania, Tabora Dist., Unyamwezi, Wala R., *Böhm* 98 (B holotype, Z).

    Aponogeton braunii K. Krause in Bot. Jahrb. Syst. **57**: 240 (1921). Type: Tanzania, Bukoba Dist., Msera, fl. vi.1913, *Braun* in *Herb. Amani* 5494 (EA holotype, B).

    Aponogeton oblongus Peter in Abh. Königl. Ges. Wiss. Göttingen, Nath.-Phys. Kl., Newe Folge **13**: 41, fig.13 (1928) in key; in Repert. Spec. Nov. Regni Veg. Beih. **40**: 116 (1938) in key. Types: Tanzania, Tabora Dist., Ngulu, Malongwe railway bridge, *Peters* 34581 & 39208 (both B† syntypes).

Perennial herb with tuber shortly cylindrical, ovoid or irregularly shaped, up to 20 mm in diameter with numerous whitish roots above. Leaves few to many, the primary ones submerged, usually linear or ligulate, the secondary ones with a floating blade or, in var. *graminifolius*, all leaves sessile, linear without floating blade. Petiole up to 60 cm long and 3 mm thick, greenish; blade up to 12 × 5 cm, oval to narrowly elliptic or lanceolate, cuneate to truncate (very rarely cordate) at base, acute to obtuse at apex, with 2–4 parallel nerves on each side of the prominent midrib. Peduncle up to 45 cm long, not thickened towards apex, terete or somewhat flattened; spathe 5–20 mm long, pale green with darker longitudinal lines. Spikes 2, up to 50 mm long. Flowers bisexual or female arranged all around rachis. Tepals 2, persistent or caducous, pinkish, mauve, violet or purplish, 1–5 mm long, linear to obovate, 1-nerved. Stamens 6 (0 in apomictic plants); filaments 2–3 mm long, violet or white, thickened towards the base; anthers violet or blackish with yellow pollen. Carpels 3 (up to 7 in apomictic plants), each with 4–14 ovules. Fruiting carpels up to 7 × 2.5 mm, including the 1–2 mm long terminal beak. Seeds up to 1.5 × 0.4 mm, ellipsoid; testa double, the outer one transparent, reticulated nerved and longitudinally ridged but not winged, the inner one closely fitting the embryo, brown; plumule absent.

Lye (Lidia **1**: 67–80, 1986) considers there to be 5 varieties – var. *abyssinicus* occurring in Ethiopia, Rwanda, Uganda, Kenya, Tanzania (T1, 3–5) and Congo; var. *albiflorus* occurring in Kenya and Tanzania (T6, 8); var. *cordatus*, from Kenya and Somalia; var. *glanduliferus* endemic to W Tanzania, Tabora Dist. (T4), all of them with floating leaf-blades; and var. *graminifolius* with linear leaves without a floating blade occurring in SW Tanzania, Ufipa, Kasisiwue plain (T4, Sumbawanga Dist.).

A poor specimen collected in C Malawi, Lisasadzi R. near bridge on main Kasungu road in riverine marsh, 1075 m, fl. 15.i.1959, *Robson & Jackson* 1197 (K), has leaf-blades up to 80 × 2 mm, spathe 9 mm long, spikes 9 mm long and 4 mm wide, and purple tepals c.2 × 1 mm. This I refer to *A. abyssinicus* var. *graminifolius* Lye.

Bruggen (1973: 225) refers to this collection "…the only collection from Malawi I have seen, I could not identify with certainty, because of the incompleteness of the specimen. Because of the lilac tepals I arranged it under *A. abyssinicus*, though it might belong to *A. junceus* subsp. *rehmannii*, which I consider, however, less probable." The relative nearness of the collecting localities of this and other varieties in Tanzania persuaded me to insert the species in this account.

6. **Aponogeton junceus** Schltdl. in Linnaea **10**, Litteratur-Bericht: 76 (1836) as '*junceum*'. —Obermeyer in F.S.A. **1**: 88 in part as regards syn. *A. spathaceus* var. *junceus*, fig.26 (1966). —Jacot Guillarmod, Fl. Lesotho: 100 (1971). —Gibbs Russell in Kirkia **10**: 435 (1977). —Clarke & Klaassen, Water Pl. Namibia: 68 (2001). —Cook, Aq. Wetl. Pl. Sthn. Africa: 59 (2004). Type: Hort. Hamburg. ex South Africa, "Caffraria", *Ecklon* s.n. (HBG holotype).

Aponogeton spathaceus E. Mey. [ex Drège in Flora **26**(2), Beigabe: 54 (1843), nomen; ex Schltdl. in Linnaea **20**: 215 (1847), nomen] ex Hook. f. in Bot. Mag.: sub t.6399 (1878) as '*spathaceum*'. —Durand & Schinz, Consp. Fl. Afric. **5**: 493 (1894). —Bennett in F.C. **7**: 44 (1897) as '*spathaceum*'. —Rendle, Cat. Afr. Pl. Welw. **2**: 94 (1899) as '*spathacea*'. —Bennett in F.T.A. **8**: 216 (1902). —Podlech in Merxmüller, Prodr. Fl. SW Afrika, fam. 143: 2 (1966). Type: South Africa, Eastern Cape, Uitenhage, 1840, *Drège* s.n.

    Aponogeton junceus subsp. *junceus*. —Bruggen in Bull. Jard. Bot. Belg. **43**: 225, fig.7/13 (1973); in F.A.C., Aponogetonaceae: 10 (1974).

Perennial herb with tuber globose or oval up to 3 cm in diameter. Leaves submerged or emerged, apparently rarely floating, 4–18(43) cm long and 4–5 mm wide, usually without a distinct blade, needle-shaped and somewhat three-edged or band-shaped, sometimes with a narrow blade up to 9 × 0.5(1) cm, narrowed to base and apex, with 7–11 longitudinal parallel nerves. Peduncle up to 30(44) cm long, very slender, not thickened towards apex. Spathe up to 20 mm long, caducous. Spikes 2, 10–40 mm long, laxly flowered. Flowers bisexual or female, dorsally placed on the rachis. Tepals 2, pink or white, rarely with purplish marks, 2–4 × 0.7–2 mm, 1-nerved or rarely 3-nerved. Stamens 6 (0 in apomictic plants); filaments 1.5–2.5(3.5) mm long, widened towards base; anthers 0.4–1 × 0.2–0.5 mm, yellow, mauve, violet or purplish; pollen yellow. Carpels 3, or up to 6 in apomictic plants, each with 2–6 ovules. Fruiting carpels up to 7 mm long, usually less, brown, with a long terminal beak. Seeds up to 4 × 1 mm, ellipsoid, slightly curved; testa simple; plumule absent.

Zambia. B: Machili, fl. 20.ii.1961, *Fanshawe* 6293 (K, SRGH). C: Kabwe Dist., 20 km SW from Kabwe, 2 km SE from Great North road, 14°35'S, 28°17'E, 1200 m, fl. 31.i.1973, *Kornaś* 2997 (K). S: Namwala, 3.2 km from Namwala on Kafue Nat. Park road, fl. 13.xii.1962, *van Rensburg* KBS 1083 (K, SRGH). **Zimbabwe**. N: Makonde (Lomagundi) Dist., Rukuti farm, Doma area, 1220 m, fl.& fr. 27.i.1963, *Jacobsen* 2107 (PRE). C: Harare Dist., Mukuvisi (Makabusi), fr. 8.i.1948, *Wild* 2273 (BM, SRGH). S: Masvingo Dist., Mutirikwe Recr. Park (Kyle Nat. Park), imm.fr. 16.ii.1974, *Lightfoot* in GHS 231509 (SRGH).

Also known from Congo, Angola, Namibia, Lesotho and South Africa. In shallow water in ephemeral pools, swamps and flood plains, on muddy soil; up to 1260 m in the Flora area, but reported from 2990 m in Lesotho and Bruggen (1973: 232) mentions 3300 m in Central Africa.

Conservation notes: Widespread species; not threatened.

7. **Aponogeton rehmannii** Oliv. in Hooker's Icon. Pl. **15**: t.1471b (1884). —Bennett in F.C. **7**: 44 (1897); in F.T.A. **8**: 217 (1902). —Engler & Krause in Engler, Pflanzenr. **24**: 15 (1906). —Podlech in Merxmüller, Prodr. Fl. SW Afrika, fam. 143: 2 (1966). —Bruggen in Bibl. Bot. **137**: 63 (1985). —Lye in F.T.E.A., Aponogetonaceae: 9 (1989). —Clarke & Klaassen, Water Pl. Namibia: 68 (2001). —Cook, Aq. Wetl. Pl. Sthn. Africa: 60 (2004). Type: South Africa, Northwest Province (Transvaal), Boshveld, between Kleinsmit and Kameelpoort, 1875, *Rehmann* 4835 (K holotype, Z).

    Aponogeton hereroensis Schinz in Bull. Herb. Boiss., sér.2 **1**: 764 (1901). Types: Namibia, pool E of Windhoek, Siedelungsfarm, ii.1899, *Dinter* 589 (Z syntype) & N of Waterberg, 10.iv.1899, *Dinter* s.n. (Z syntype).

    Aponogeton junceus Schltdl. subsp. *rehmannii* (Oliv.) Oberm. in Fl. Pl. Africa **37**: t.1449 (1965). —Bruggen in Bull. Jard. Bot. Belg. **43**: 229, fig.7/15 (1973). —Gibbs Russell in Kirkia **10**: 435 (1977).

    Aponogeton junceus sensu Obermeyer in F.S.A. **1**: 88, fig.26 (1966) in part as regards syn. *A. rehmannii* & fig.26/a.

Perennial herb with tuber elongate or ovoid, up to 5 cm long and 0.8–2.5 cm in diameter, usually densely set with whitish roots in upper part. Leaves few to many, the primary ones submerged, usually linear or sword-shaped, sometimes with a spathulate or oblanceolate blade, sessile, the secondary ones with a floating blade; petiole 3–15(30) cm long and up to 3 mm thick; blade 2–10 cm long and 4–18 mm wide, narrowly ovate or linear-lanceolate, cuneate, obtuse, rounded or truncate at base, subacute or obtuse and apiculate at apex, with 2–3 weak longitudinal parallel nerves on each side of the prominent midrib. Peduncle 5–15(30) cm long and up to 2 mm thick, not thickened towards apex; spathe 6–10 mm long, early caducous. Spikes 2, very densely flowered, 4–15 mm long, up to 50 mm long in fruit. Flowers bisexual or female, arranged all around rachis. Tepals 1–2, white or rarely pinkish, 1.5–3 × 0.8–2 mm, ovate to obovate, 1-nerved. Stamens 6 (0 in apomictic plants); filaments up to 3 mm long, slightly widened towards base; anthers mauve or purple with yellow pollen. Carpels 3 (up to 6 in apomictic plants), each with 2–6 ovules. Fruiting carpels up to 5 mm long and 3 mm wide with a whitish beak. Seeds usually 2–4 on each carpel, 2–3 mm long, narrowly ovoid or cylindrical, slightly curved; testa simple, brown; plumule absent.

Caprivi Strip. Mentioned from E Caprivi by Clarke & Klaassen (2001). **Botswana.** N: Pan just past Samadupi Drift, 20°05'S, 23°31'E, 1000 m, fl.& fr. 23.i.1972, *Gibbs Russell & Biegel* 1382 (K, LISC, SRGH). SE: Content Farm North, 24°33'S, 25°57'E, 1050 m, fl. 27.i.1978, *Hansen* 3340 (BM, K). **Zimbabwe.** N: Gokwe South Dist., Sengwa Res. Station, fl. 9.iv.1969, *N. Jacobsen* 580 (SRGH). W: Nyamandhlovu Pasture Res. Station, fr. 10.vi.1955, *Plowes* 1857 (K, SRGH). C: Chegutu Dist., Chegutu (Hartley), Ben Bank house dam, fl. 26.ii.1969, *Mavi* 1014 (K, PRE, SRGH).

Reported also from Kenya, Namibia and South Africa. In seasonal pools and small pans, flooded plains, in muddy or sandy wet soils; 900–1160 m (to 1700 m outside the Flora area).

Conservation notes: Widespread species; not threatened.

Bruggen (1973: 229) cites *Junod* 485 (BR, Z) from Mozambique, "Delagoa Bay, Hangwane". I have not seen the specimens cited and I am unsure as to this locality. Until the beginning of the 20th century "Delagoa Bay" was a name frequently applied to a wide region around the Baía de Lourenço Marques or Baía do Espírito Santo, now Maputo Bay, including areas now in Swaziland and South Africa (Mpumalanga).

JUNCAGINACEAE

by E.S. Martins

Annual or perennial aquatic herbs with bulbs or rhizomes; stolons rarely present. Leaves basal, linear, flat to terete, sheathing at base and sometimes ligulate. Inflorescence a spike or raceme. Flowers regular, 3-merous, very rarely 2-merous (*Tetroncium*), usually bisexual. Perianth of 4–6 tepals in 2 whorls; tepals free or connate below, scarious, usually green, deciduous. Stamens usually 6, rarely 4 in 2 series opposite the tepals; filaments short or absent; anthers 2-thecous, extrorse, dehiscing lengthwise. Gynoecium superior; carpels usually 6, 3 of them sometimes abortive, free or axially connate below, unilocular; ovule 1, basal, erect, anatropous, rarely apical, pendulous, atropous; styles usually short or undifferentiated; stigmas plumose or papillose. Fruit a schizocarp of 3–6 free or axially connate indehiscent mericarps. Seeds without endosperm.

A small family of 3 genera found in saline or freshwater habitats, from Australia (*Maundia*), the Straits of Magellan area in South America (*Tetroncium*), and one cosmopolitan (*Triglochin*).

TRIGLOCHIN L.

Triglochin L., Sp. Pl.: 338 (1753); Gen. Pl. ed.5: 157 (1754).

Annual or perennial glabrous small herbs with bulbous organs at the base or at end of stolons. Leaves in basal tufts, graminaceous, sheathing at base, with the blade flat, semiterete or terete, linear to filiform, and often ligulate at the junction of sheath and blade. Inflorescence a single ebracteate raceme or spike, compact when young, elongating in fruit. Flowers regular, bisexual, sometimes partially aborted. Tepals 6, or 3 by reduction, free, concave. Stamens 6 in 2 series, the 3 of the inner series sometimes reduced; anthers sessile, dorsifixed, kidney-shaped. Carpels 6, alternately fertile and sterile, joined when young, later on free; stigmas 3, sessile or subsessile, plumose; mericarps 3, attached near apices to a carpophore. Seeds erect.

Genus of about 25 species, occurring mainly in the temperate regions of the northern and southern hemispheres, particularly in Australia. Only 5 species are known in Africa, 2 of them in the Flora Zambesiaca area.

Fruits narrowly ellipsoid or narrowly ovoid, 5–14 mm long with 3 fertile carpels plus 3 vestigial ones; stolons absent; found in inland wetlands · · · · · · · · · **1.** *bulbosa*
Fruits globulose or ellipsoid, c.2 mm long with 3 fertile carpels plus 3 well developed sterile ones; stolons usually present; found in saline wetlands · · · · · · · **2.** *striata*

1. **Triglochin bulbosa** L., Mant. Pl. Alt.: 226 (1771). —Bennet in F.T.A. **8**: 215 (1901) as '*bulbosum*'. —van der Veken in Bull. Soc. Roy. Bot. Belg. **91**: 99 (1958). — Obermeyer in F.S.A. **1**: 93, fig. 28.1 (1966). —Malaisse in Symoens, Expl. Hydrobiol. Bassin L. Bangweolo & Luapula **18**(2): 15 (1975). Type: "Habitat ad Cap. B. Spei" (LINN 466.3 holotype). FIGURE 12.2.**21**.

Subsp. **bulbosa** —Lisowski, Malaisse & Symoens in F.A.C., Juncaginaceae: 2, pl.1a–f (1982).

   Triglochin milnei Horn in Svensk. Bot. Tidskr. **55**: 85 (1961). —Napper in F.T.E.A., Juncaginaceae: 1, fig.1 (1971). Type: Zambia, Mwinilunga Dist., c.1 km S of Matonchi Farm, 30.x.1937, *Milne-Redhead* 3012 (K holotype).

Slender perennial erect herb up to 50 cm high, glabrous; bulbous conical rhizome often with 1–4 whitish bulbous shoots. Rhizome surrounded by brownish or blackish dense remnant layers of fibrous leaf-bases. Outer leaves usually scale-like, reduced to 5–10(15) mm wide sheaths and sometimes a small ligule and minuscule acute limb, often reddish; inner leaves with sheaths up to 70 × 6 mm, with membranous margins; ligule up to 0.5 mm long, membranous; blade 5–350 × 1–3 mm, linear, semi-terete or somewhat concave on upper surface, obtuse to acute at apex. Inflorescence an erect, few to many-flowered raceme, spike-like when young, longer than the leaves; peduncle terete or more or less angular; pedicels erect, spreading at a 30–45° angle, 1–3 mm long at flowering, elongating up to 10 mm on fruit. Flowers 4–25(44), green, tips of tepals and stigmas tinged with purple; outer tepals 2–3 × 1.4–1.5 mm, ovate or elliptic, conchiform, subacute, entire; inner tepals shorter, obtuse, usually toothed or eroded at apex. Outer anthers 1.5–3 mm long, the inner ones 1–1.5 mm long. Ovary 2–3 mm long, lengthening quickly after fertilization; stigmas 3, free, plumose at tip. Fruit 7–14 × 1.8–3.7 mm, narrowly ovoid or ellipsoid, a schizocarp of 3 fertile mericarps alternating with 3 vestigial ones; fertile mericarps narrowly ovoid or sub-cylindrical, separating from the central axis at maturity. Seeds 5–7 mm long, narrowly ellipsoid, compressed.

Zambia. N: Kasama Dist., Chishimba Falls, fl. 15.x.1960, *Robinson* 3969 (K). W: Kitwe, Baluba, fl.& fr. 2.xii.1963, *Fanshawe* 8160 (K, SRGH). C: Fiwila, Mkushi, 1220 m, fr. 11.i.1958, *Robinson* 2730 (K, SRGH). E: Chadiza, 850 m, fl. 25.xi.1958, *Robson* 686 (BM, K, LISC, PRE, SRGH). S: Muckle Neuk, 19.3 km N of Choma, 1280 m, fl.& fr. 27.xi.1954, *Robinson* 969 (K, SRGH). **Zimbabwe**. C: Harare Dist., Ruwa R., 1490 m,

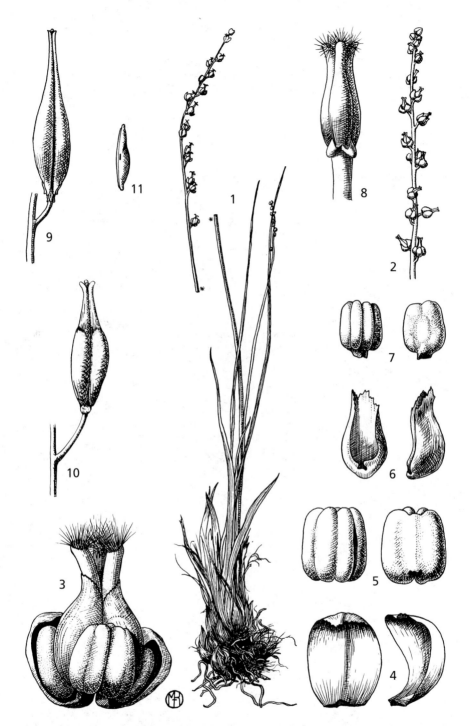

Fig. 12.2.**21**. TRIGLOCHIN BULBOSA. 1, habit (× ²⁄₃); 2, raceme (× 1); 3, flower (× 10); 4, outer tepal (× 10); 5, outer anther (× 10); 6, inner tepal (× 10); 7, inner anther (× 10); 8, gynoecium (× 10); 9, fruit (× 3); 10, other fruit showing variation in shape (× 3); 11, seed (× 3). 1–8 from *Milne-Redhead* 3012, 9, 11 from *Milne-Redhead* 3693, 10 from *Bullock* 2364. Drawn by Olive Milne-Redhead. From F.T.E.A.

fl. 7.xii.1947, *Wild* 2260 (K, SRGH). **Malawi**. N/C: Kasungu Dist., Kasungu Nat. Park, 1030 m, fl. 1.xii.1970, *Hall-Martin* 1014 (SRGH). C: Dedza Dist., Chongoni Forest, fl.& fr. 4.xii.1968, *Salubeni* 1241 (K, LISC, MAL, PRE, SRGH).

Also from Tanzania, Congo, Angola and South Africa. In shallow pools, wet or swampy dambos, secondary woodlands and open places, on mud, dry sandy or peaty moist soils; 850–1500 m.

Conservation notes: Widespread species; not threatened.

2. **Triglochin striata** Ruiz & Pav., Fl. Peruv. **3**: 72 (1802). —Bennet in F.T.A. **8**: 216 (1901). —Buchenau in Engler, Pflanzenr. **16**: 10 (1903). —Horn in Svensk. Bot. Tidskr. **55**: 112 (1961) as '*striatum*'. —Obermeyer in F.S.A. **1**: 94, fig.28/2 (1966). —Ross, Fl. Natal: 56 (1973). —Lisowski, Malaisse & Symoens in F.A.C. Juncaginaceae: 5, pl.1g–i (1982). —Clarke & Klaassen, Water Pl. Namibia: 66 (2001). —Cook, Aq. Wetl. Pl. Sthn. Africa: 156 (2004). Type from South America.

Perennial erect herb 5–50 cm high forming long stolons. Rhizome woody, loosely covered with greyish, creamy or brownish fibres. Outer and inner leaves identically developed; sheaths broad up to 10 cm long, with membranaceous margins; ligule well distinct, 0.5–2.5 mm long, membranaceous; blade 5–30 cm long, semi-terete, succulent, obtuse at apex, longitudinally striate. Inflorescence an axillary raceme, lateral to the leaf tuft and overtopping leaves, flowers in groups, spirally or irregularly arranged on angular rachis; flowers few to many, with pedicels decurrent, very short at first, arcuate and elongate up to 3 mm in fruit. Outer tepals 1.3–1.5 mm long, elliptic, obtuse to acute, inner ones 1–1.2 mm long, conchiform, acute and somewhat hood-shaped. Outer anthers 0.9–1.2 mm long, inner ones smaller, apiculate. Carpels 6, the 3 inner fertile ones alternate with the 3 outer sterile; stigmas 3, sessile. Fruit c.2 mm in diameter, a globulose-ribbed schizocarp of 3 fertile mericarps alternating with 3 vestigial ones; fertile mericarps 1.8–2.3 × 1 × 0.6 mm, cuneate, dorsally 3-ribbed or crested, minutely verruculose, falling off at maturity; sterile mericarps c.0.9 × 0.5 mm, membranaceous except for dorsal nerve, persistent for a short time after fertile ones have fallen.

Mozambique. GI: Inhambane, salt marshes 11.2 km S of airport, 5 m, fl.& fr. 20.vii.1963, *Mogg* 30343 (K, PRE). M: Inhaca Is., salt mash near Portinho, fl.& fr. 4.iii.1958, *Mogg* 31420 (K, SRGH).

Also from Angola, Congo and South Africa (Cape Province, KwaZulu-Natal), from temperate regions of the Americas, Australia and New Zealand, and from Portugal (naturalized). In saline marshes and lagoons, usually near the sea but sometimes inland on saline soils; 0–10 m.

Conservation notes: Widespread species; not threatened.

POTAMOGETONACEAE

by J.J. Symoens

Aquatic herbs, with submerged and/or floating leaves, perennial or rarely annual, glabrous, rooted in the substrate. Vegetative parts comprise horizontal shoots (here called rhizomes) which are mostly stoloniferous and without chlorophyll, and vertical shoots (here called stems) which are flexible, erect or sometimes ± floating, unbranched or ± densely branched, with chlorophyll; tubers or turions (specialized perennating buds) sometimes present on the horizontal and/or erect shoots. Leaves alternate or subopposite, rarely ternate, sessile or petiolate, simple; blade entire, denticulate or serrate; stipules present throughout or only on young shoots (*Groenlandia*), often forming tubular sheaths around the stem or axillary bud, either free or fused with the leaf base; intravaginal scales 2–several, situated in leaf axils, mostly linear to subulate, rarely ovate, entire and acute. Inflorescences pedunculate, with 2 opposite flowers (*Groenlandia*) or more than 2 flowers in a dense or interrupted spike; peduncle rigid and mostly slightly elevated above the water, sometimes flexuous with flowers resting on the water surface, rarely submerged, although the fruits are often withdrawn below the surface; bracts abortive or absent. Flowers hermaphrodite, small, regular, hypogynous, tetramerous. Perianth a single whorl of 4-valvate shortly clawed tepals. Stamens 4, opposite the tepals and attached basally to the claw; anthers sessile, bilocular, extrorse, opening by longitudinal slits. Gynoecium superior; carpels (1)4(8), sessile, free or shortly joined at the base, alternating with stamens; styles usually short; stigma unicellular-papillate or smooth, ± secretory; ovule solitary in the carpel and attached to its ventral margin, orthotropous at first, but becoming campylotropous at maturity, pendant and filling the locule, with 2 integuments. Fruits apocarpous consisting of distinct, sessile fruitlets, ± drupaceous but opening with a dorsal lid, with a rather fleshy parenchymatous mesocarp and a sclerified endocarp, the latter multi-layered (*Potamogeton*) or 1-layered (*Groenlandia*). Seed without endosperm; embryo hook-shaped or spiral (i.e. coiled more than 1 complete turn), with a large hypocotyl and a single obliquely terminal cotyledon that encloses the plumule.

A cosmopolitan family of 3 genera with its highest diversity in the northern hemisphere. Two genera in the Flora Zambesiaca region – *Potamogeton*, cosmopolitan with many species and with its highest diversity in W Europe, Japan and E North America, and *Stuckenia* Börner. The latter was created by the elevation of *Potamogeton* subgen. *Coleophylli* Koch to generic level by Börner (in Abh. Naturwiss. Ver. Bremen **21**: 258, 1912), and the name was reactivated by Holub (in Fol. Geobot. Phytotax. **19**: 215, 1984). Later, *Potamogeton* subgen. *Coleogeton* (Rchb.) Raunk., covering the same species group, was elevated to generic level by Les & Haynes (Novon **6**: 389, 1996) under the superfluous generic name *Coleogeton*. This segregation was not immediately accepted, but is now justified by the results of an increasing number of molecular studies (see Lindqvist et al. in Cladistics **22**: 568–588, 2006 and Wang et al. in Pl. Syst. Evol. **267**: 65–78, 2007).

The identification of species is complicated by the great variation in leaf form with age or from environmental factors such as current speed, water depth, light intensity and nutrient supply. The existence of numerous hybrids may also be troublesome. A binocular microscope is required for examination of leaf veins and lacunae and for fruitlet characters. The pattern of stem anatomy can also be used to assist identification, but requires the preparation of thin cross-sections of the stem, preferably of the internode of the upper part of the flowering stem. The main characters are the shape and size of the stele, the shape of endodermis cells (mostly of O-type with cell walls equally thickened on all faces, or of U-type with the outer face of the cell wall thinner than the inner and lateral faces), the number and size of interlacunar and of subepidermal bundles, and the presence or absence of pseudohypodermis (see Figure 12.2.**22**).

Many species produce modified buds (often termed turions, winter-buds or hibernacula) which serve as means of vegetative spread or of perennation during unfavourable periods. Different types have been recognized and provide valuable taxonomic criteria. These are produced from the rhizomes or are found on the stems and branches, either at the apex or in the leaf axils. Although leaves rather than the axis are their major component, in some species swollen internodes of the rhizome form tubers, which are abundantly filled with starch and topped by a bud which will become the future erect stem. These structures, common in temperate regions, are less well known in African populations. More information is needed on their presence and their role in the life cycle.

Stipules fused for most of their length with leaf, forming a winged leaf base ending in
 a free ligule; floating leaves always absent; leaf margins entire · · · · **1. Stuckenia**
Stipules free from leaf; floating leaves present or absent; leaf margins entire or
 denticulate · **2. Potamogeton**

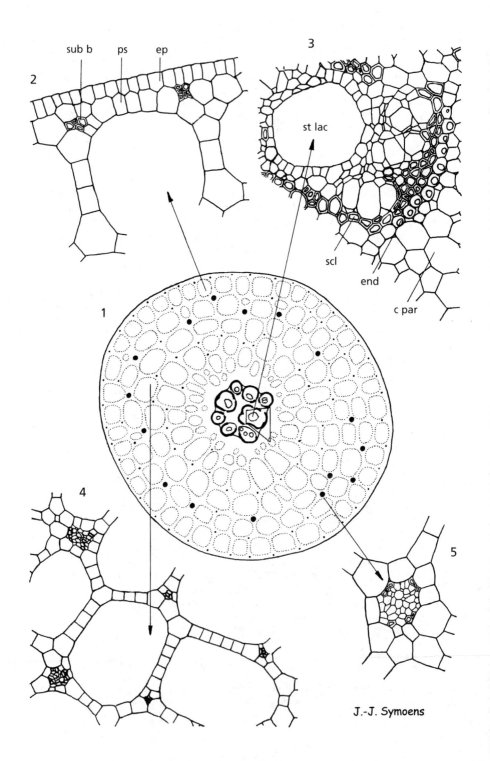

1. STUCKENIA Börner

Stuckenia Börner in Abh. Naturwiss. Ver. Bremen **21**: 258 (1912).
Potamogeton sect. *Coleophylli* Koch, Syn. Fl. Germ. Helv., ed.1 (1837).
Potamogeton subgen. *Coleogeton* (Rcbh.) Raunk., Dan. Blomsterpl. Naturh. **1**/1: 108 (1896).
    Coleogeton (Rchb.) Les & Haynes in Novon **6**: 389 (1996).

Submerged aquatic herbs. Perennial rhizome present, slender to very robust, terete, mostly
with apical tubers, rarely without. Stems unbranched to much-branched, annual to perennial.
Leaves submerged, alternate, filiform to linear, sometimes relatively robust, grooved, with air
chambers bordering on midrib, veins 1–5; apex obtuse, acute or acuminate, margins entire;
floating leaves absent. Stipules forming a sheath, closed or open (i.e. with one edge rolled
inward and enveloped by opposite edge) around stem and fused for most of its length with leaf
base. Spikes long, cylindrical, interrupted after flowering; peduncles flexible. Flowers in whorls;
carpels 4(6), stigmas stalked or sessile with long papillae. Fruits with or without a distinct beak;
dorsal keel indistinct. Embryo curved.

About 10 species, mostly in Asia, and 3 recognised hybrids. Only 1 cosmopolitan
species in tropical Africa, from the north to South Africa.

Stuckenia pectinata (L.) Börner, Fl. Deutsch. Volk: 713 (1912). Type: Austria, *Burser*
s.n. (UPS, Burser Herb. X:124, lectotype chosen by Haynes in Taxon **35**: 569,
1986). FIGURE 12.2.**23**.
    Potamogeton pectinatus L., Sp. Pl.: 127 (1753). —Garcke in Peters, Reise Mossamb., Bot.:
    511 (1864). —Schumann in Engler, Pflanzenw. Ost-Afr. **C**: 93 (1895). —Bennett in F.C. **7**:
    49 (1897); in F.T.A. **8**: 223 (1901). —Peter in Abh. Konigl. Ges. Wiss. Göttingen, Math.-
    Phys. Kl. **13**: 108 (1928). —Dandy in J. Linn. Soc., Bot. **50**: 513 (1937). —Podlech in
    Merxmüller, Prodr. Fl. SW Afr., fam.144: 3 (1966). —Obermeyer in F.S.A. **1**: 61 (1966). —
    Lisowski et al. in F.A.C., Potamogetonaceae: 3 (1978). —Wiegleb & Kaplan in Folia
    Geobot. **33**: 305 (1998). —Clarke & Klaassen, Water Pl. Namibia: 62 (2001). —Cook, Aq.
    Wetl. Pl. Sthn. Africa: 236 (2004).
    Potamogeton livingstonei A. Benn. in F.T.A. **8**: 223 (1901). —Graebner in Engler,
    Pflanzenr. **4**(11): 128 (1907). Types: Malawi, Livingstonia, Lake Malawi, Kambwe Lagoon,
    1877, *Laws* 3 (BM lectotype, K); Likoma Is., 1887, *Bellingham* s.n. (BM syntype); Lake
    Malawi, SW bay, 1861, *Kirk* s.n. (K syntype).
    Potamogeton filiformis sensu Bennett in F.T.A. **8**: 223 (1901). —Peter in Abh. Konigl. Ges.
    Wiss. Göttingen, Math.-Phys. Kl. **13**: 107 (1928), non Pers.
    Coleogeton pectinatus (L.) Les & Haynes in Novon **6**: 390 (1996).

Rooted submerged aquatic with adventitious roots at rhizome nodes; rhizomes perennial,
slender to very robust, terete, sometimes developing starch-filled tubers (generally absent in
tropical plants) surrounded by a scaly leaf at end of the growing season. Stems annual to

---

Fig. 12.2.**22**. POTAMOGETON. Stem anatomy of *Potamogeton richardii* in transverse section. 1,
transverse section of internode, diagrammatic (× 25); small dots indicate subepidermal fibrous
bundles and their remnants now deeper in the lacunar cortex; large dots indicate interlacunar
vascular bundles; the stele (or central cylinder) is of the trio-type with 10 vascular bundles,
irregularly surrounded by sclerenchyma (thicker lines). 2, detail of peripheral tissues (× 200):
ep, epidermis; ps, pseudohypodermis (here 1-layered); sub b, subepidermal bundle. 3, detail of
contact zone of stele and cortex (× 200), from position indicated in 1: st lac, stelar xylem lacuna;
scl, sclerenchyma; end, endosperm (here cells of U-type); c par, cortical parenchyma (compact
in innermost layers). 4, cortical lacunar system (× 100). 5, detail of interlacunar system (× 200).
All figures are orientated as in 1. All from *Bamps, Symoens & Vanden Berghen* 526. Drawn by J.-J.
Symoens. From F.T.E.A.

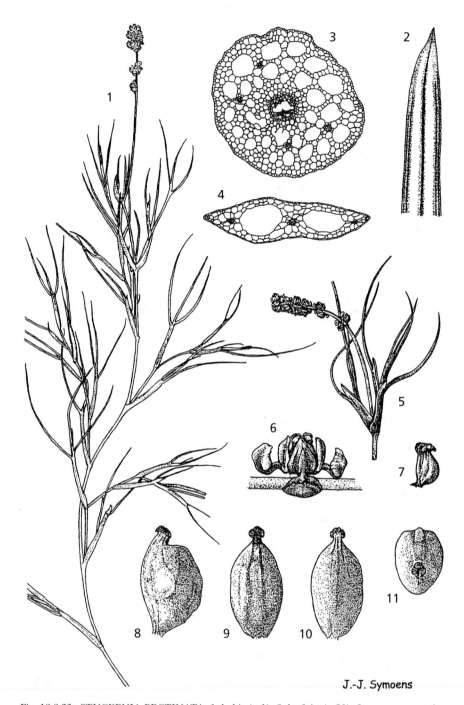

J.-J. Symoens

Fig. 12.2.**23**. STUCKENIA PECTINATA. 1, habit (× 1); 2, leaf tip (× 20); 3, transverse section of stem (× 40); 4, transverse section of upper leaf part (× 40); 5, node and inflorescence (× 1¹/₄); 6, open flower with tepals spread (× 8); 7, carpel (× 8); 8–11, respectively, fruitlet, lateral view, dorsal view, ventral view, apical view (× 8). 1 from *LaBarbera* 16 (Uganda), 2–4 from *Ross* 1469 (Tanzania), 5–7 from *Symoens* 5340 (Rwanda), 8–11 from *Lye* 5329 (Uganda). Drawn by J.-J. Symoens. From F.T.E.A.

perennial, to 4 m long and 0.5–2 mm in diameter, filiform to relatively robust, terete, sometimes pink, usually much-branched. Submerged leaves sessile, (20)30–125(300) × 0.2–4 mm, 24–160(200) times as long as wide, filiform to linear, sometimes robust, bright to olive green, straight and grooved at base, entire at margins, the narrower leaves acute to finely acuminate at apex, sometimes the broader leaves rounded or obtuse and mucronate; midrib bordered on each side by one to several air channels, lateral veins 1–2 on each side, inconspicuous. Stipules persistent, 10–70 mm long, forming a convolute sheath, fused with leaf base for 8–65 mm and ending in a free ligule 3–15 mm long, rounded, obtuse or truncate, later disintegrating into fibres. Intravaginal scales 0.4–0.6 mm long. Floating leaves always absent. Peduncles 20–100(450) mm long, 1.5–5(10) times as long as fruiting spike, as thick as the stem, flexuous; spikes (4)8–14-flowered, with 2–7 pairs or whorls of flowers, contiguous at first, later distant, 13–35(60) mm long in fruit. Tepals orbicular to elliptical, 1–3 mm long. Anthers 0.8–1.3 mm long. Carpels 4; stigmas borne on a short but distinct style. Fruitlets 3.3–4.7(5.1) × 2–3.6 mm, asymmetrically obovoid, ventrally nearly straight, dorsally very convex, hardly 3-keeled; beak 0.2–0.6 mm long, ventral, rarely subventral. Stem anatomy: stele of four bundles, mostly oblong; endodermis of U-type; interlacunar bundles present (in ± complete ring), subepidermal bundles few or absent; pseudohypodermis present, 1(2)-layered.

Caprivi Strip. E Caprivi, Kongola, Kwando pass, 29.ix.1970, *Vahrmeyer* 2139 (PRE). **Zambia**. N: Mpulungu, Lake Tanganyika, fl. 20.x.1947, *Greenway & Brenan* 8242 (BM, PRE). **Malawi**. N: Livingstonia, Lake Malawi (Nyasa), Kampori (Kambore) lagoon, 1877, *Laws* 3 (BM, K). C: Lake Malawi, SW bay, fl. ix.1861, *Kirk* s.n. (K). S: Mangochi Dist., Monkey Bay, shore of Tumbi Is., 1.ix.1968, *Eccles* 190, 503 (PRE, SRGH). **Mozambique**. MS: Chupanga (Shupanga), Rios de Sena, *Peters* s.n. (see Garcke 1895) GI: Inharrime Dist., Lagoa Poelela, 26.i.1954, *Schelpe* 4519 (BM). M: Lagoa Satine, near Zitundo, fl.& fr. 15.vii.1971, *Correia & Marques* 2180 (BM, COI, LISC, LMU).

Cosmopolitan, widely distributed across Africa from North Africa, Mauritania, tropical NE and E Africa, to Congo, Angola, Namibia, South Africa, Madagascar and the Mascarenes. Lakes, rivers, irrigation canals, 50–300 cm deep, also in brackish and polluted water; often forming extensive dominant patches and becoming a serious water weed, blocking water flow in irrigation and drainage canals; sea level–1800 m, up to 2850 m in Ethiopia.

Conservation notes: Widely distributed; not threatened. An important food source for many waterfowl species.

Stuckenia pectinata is an extremely polymorphic species with many regional and local forms. Many subspecific taxa have been described, mostly at the variety level, but there are many transitional forms. Transplantation experiments show that their distinguishing characters may be environmentally induced. Very robust specimens from the African Great Rift lakes with 2–4 mm wide leaves (e.g. *Laws* 3, *Bellingham* s.n. (1887), *Cunnington* 12), were originally described as *Potamogeton livingstonei*. *S. pectinata* differs from *S. filiformis* (Pers.) Börner in having open stipular sheaths, larger fruits and a short but distinct style.

2. POTAMOGETON L.

Potamogeton L., Sp. Pl.: 126 (1753). —Dandy in J. Linn. Soc., Bot. **50**: 507–540 (1937). —Wiegleb & Kaplan in Folia Geobot. **33**: 241–316 (1998).

Aquatic herbs. Rhizome present or absent, if present well differentiated from the stem, filiform, slender or robust, terete or ± compressed; stems terete or compressed, unbranched or ± densely branched; turions present or absent. Leaves alternate, but mostly subopposite toward the inflorescence apex (involucral leaves), sessile or petiolate, 1- to many-nerved, all submerged and with filiform, grass-like or an expanded blade, or dimorphic, the submerged ones thin and ± translucent, sometimes reduced to phyllodes, the floating ones ± coriaceous, opaque, petiolate, and usually broader; stipules always present, although sometimes disintegrating or falling early, sheathing the stem and/or axilllary bud, closed or convolute, mostly free from leaf, rarely fused for less than half its length with the leaf base. Spikes cylindrical or subglobose, 3 to many-flowered, dense or sometimes interrupted, mostly borne above the water and wind-pollinated, but sometimes submerged and water-pollinated. Tepals rounded at apex. Fruitlets rounded on back, with soft mesocarp and multi-layered sclerified endocarp, with or without a distinct beak. Embryo hooked-shaped or spiral.

A cosmopolitan genus with about 60 species and 46 hybrids.

Key to species

1. Floating leaves absent · 2
 – Floating leaves present, different from the submerged leaves · · · · · · · · · · · · 7
2. Leaves less than 4 mm broad, narrowly linear to filiform · · · · · · · · · · · · · · 3
 – Leaves more than 5 mm broad · 5
3. Carpels 1–2 in each flower; leaves usually 0.3–1 mm broad, lateral veins absent or indistinct; lacunae absent or restricted to a very narrow band at the leaf base, rarely extending to the apex; fruitlet dorsal edge often rough · · · · **2.** *trichoides*
 – Carpels 3–7 in each flower; leaves usually 1–2 mm broad, with or without a row of lacunae bordering the midrib; dorsal edge of fruitlets smooth and without keel, or with keel mostly distinct and crenulate · 4
4. Leaves with 1–2 distinct veins on each side of midrib; lacunar system bordering the midrib absent or narrow; fruitlets smooth, without dorsal keel · · · · · · **1.** *pusillus*
 – Leaves with 1 inconspicuous vein on each side of midrib; lacunar system bordering the midrib very broad; fruitlets with dorsal keel mostly distinct and crenulate · **4.** *octandrus*
5. Leaves linear to linear-oblong, mostly 6–12 mm wide, 3–7-veined, finely serrate and usually markedly undulate at margins; spikes 5–16 mm long in fruit; flowers 3–8; beak of fruitlet very prominent, nearly as long as fruitlet body · · · **3.** *crispus*
 – Leaves narrowly lanceolate to elliptic, mostly 7–28 mm wide, 7–21-veined, entire or bearing minute translucent teeth on margins; spikes 14–90 mm long in fruit; flowers numerous; beak of fruitlet less than $^1/_4$ of fruit body length · · · · · · · · 6
6. Submerged leaves sessile or very shortly petiolate (petiole mostly less than 30 mm), finely acute or mucronate; cells of stem endodermis of U-type, interlacunar bundles present, 1-layered pseudohypodermis present · · · · · · · **5.** *schweinfurthii*
 – Submerged leaves distinctly petiolate (petiole 30–150 mm long), narrowly obtuse to subacute, never mucronate; stem endodermis cells of O-type, interlacunar bundles absent, pseudohypodermis absent or present as a discontinuous layer · **6.** *nodosus*
7. Submerged leaves less than 3 mm wide; floating leaves with petiole 3–25(34) mm long, lamina 5–30(38) mm long; spikes 5–16 mm long in fruit · · · **4.** *octandrus*
 – Submerged leaves, if present, more than 3 mm wide, the lower ones often reduced to phyllodes; floating leaves with petiole (14)30–210 mm long, lamina 30–150 mm long; spikes 25–90 mm long in fruit · 8
8. Submerged leaves disappearing early, not present in mature plant; petiole of floating leaves often with discoloured section at junction of lamina; stem endodermis cells of U type, interlacunar bundles present in 2–3 circles, 1-layered pseudohypodermis present · **7.** *richardii*
 – Submerged leaves present and persistent; petiole of floating leaves without discoloured section at junction of lamina · 9
9. Submerged leaves sessile or very shortly petiolate (petiole less than 30 mm, up to 82 mm in intermediate leaves), persistent, 4–21 times as long as broad, apex finely

acute to sharply mucronate; spikes 30-90 mm long in fruit; stem endodermis cells of U-type, interlacunar bundles and 1-layered pseudohypodermis present · **5.** *schweinfurthii*
- Submerged leaves petiolate (petiole 20–150 mm long), persistent but sometimes decaying after fruiting, 5–9 times as long as broad, apex narrowly obtuse to subacute, never mucronate; spikes 14–50 mm long in fruit; stem endodermis cells of O-type, interlacunar bundles absent, pseudohypodermis absent or present as a discontinuous layer · **6.** *nodosus*

1. **Potamogeton pusillus** L., Sp. Pl.: 127 (1753). —Bennett in F.C. **7**: 49 (1897); in F.T.A. **8**: 222 (1901). —Podlech in Merxmüller, Prodr. Fl. SW Afr., fam.144: 3 (1966). —Obermeyer in F.S.A. **1**: 63 (1966). —Wiegleb & Kaplan in Folia Geobot. **33**: 292 (1998). —Clarke & Klaassen, Water Pl. Namibia: 62 (2001). —Cook, Aq. Wetl. Pl. Sthn. Africa: 237 (2004). Type from Europe (LINN 175.15, left hand specimen chosen as lectotype by Dandy & Taylor in J. Bot. **76**: 92, 1938).

  Potamogeton panormitanus Guss., Fl. Sic. Syn. **1**: 207 (1843). —Dandy in J. Linn. Soc., Bot. **50**: 523 (1937). —Lisowski et al. in F.A.C., Potamogetonaceae: 4 (1978). Type from Italy (Sicily).

  Potamogeton friesii sensu Bennett in F.C. **7**: 48 (1897), non Rupr.

  Potamogeton preussii A. Benn. in F.T.A. **8**: 222 (1901) in part for syntypes from Eritrea and Ethiopia.

  Potamogeton pusillus L. var. *africanus* A. Benn. in Ann. Conserv. Jard. Genève **9**: 102 (1905). —Graebn. in Engler, Pflanzenr. **31**: 115 (1907). Syntypes from South Africa.

  Potamogeton subjavanicus Hagstr. in Kongl. Svensk. Vet.-Akad. Handl., n.s. **55**(5): 129 (1916). Types: South Africa, Transvaal, *Wilms* 1656, 1657 (G syntype); Natal, Durban (Port Natal), between Tyger Berg and Blue Berg, *Drège* 1206 (G syntype, BM, K, S) & Umlazi R., *Drège* 4458 (G syntype); Natal, between Umzinto and Ifafa, *Wood* 3055 (BOL syntype, G, K, NH).

Submerged aquatic herb. Rhizomes absent or only present later in growing season, annual to biennial, filiform, terete, with short internodes. Stems annual to perennial, up to 1 m long, 0.3–0.7(1) mm wide, terete or compressed, sparingly to much branched; turions when present mostly sessile and axillary, but sometimes terminal on axillary branches, rigid, narrowly cylindrical or fusiform with a few erect, spreading or recurved free leaves. Submerged leaves sessile, linear, (9)20–85(110) × (0.3)0.8–2(2.5) mm, (15)20–60(90) times as long as wide, flaccid or firm, translucent, bright green to olive green or dark green, sometimes with a brownish tinge, narrowly cuneate at base, tapering or rather abruptly narrowed to an acute or acuminate apex; margins entire, bordered by a narrow marginal vein; midrib occupying 15–35% of leaf width at base, not bordered by lacunae or lacunae poorly developed and restricted to the lower leaf half; lateral veins 1(2) on each side, distinct, joining the midrib 1.5–4 leaf widths below leaf apex. True floating leaves absent but rarely the uppermost leaves with lamina floating at the water surface, subsessile, linear-oblanceolate, 18–38 ×1.3–3.1 mm, 7–20 times as long as wide, bright green, narrowly cuneate at base, acute to narrowly obtuse at apex, 3–5 veined, with broad rows of lacunae bordering the midrib. Stipules axillary, joined, forming a tubular sheath for most of their length when young, but splitting with age, 4–18(32) mm long, translucent, persistent or decaying. Peduncles filiform to slightly club-shaped, ± flexuous, (6)10–30(55) mm long, 1–6 times as long as fruiting spike, as thick as stem, slightly or distinctly compressed. Spikes cylindrical, with 2–7 flowers, in 1–4 whorls, ± contiguous, sometimes the lowest flower remote, 4–15 mm long in fruit. Tepals 0.8–1.8 mm long, mostly persistent; carpels (3)4(7). Anthers 0.7–0.9 mm long. Fruitlets obovoid, ± compressed,1.8–2.7(3.3) mm long, 1–1.5 mm broad, green to pale olive, without dorsal keel; beak centrally placed, straight or somewhat oblique, 0.2–0.4 mm long. Stem anatomy: stele of circular type, endodermis of O-type; interlacunar bundles absent, subepidermal bundles present; pseudo-hypodermis mostly absent, if present 1-layered.

Caprivi Strip. Cited from W Caprivi by Clarke & Klaassen (2001). **Zambia**. S: Mapanza, R. Munyeke, fr. x.1953, *Robinson* 334 (BM, K). **Zimbabwe**. N: Mazowe, Henderson Res. Station Fisheries, fl.& imm.fr. 15.iv.1973, *Gibbs Russell* 2561 (BR, K, SRGH, WAG). W: Matobo Dist., Matopo Hills, fl.& fr. ix.1905, *Gibbs* 190 (BM). C: Harare (Salisbury), fl.& fr. 28.xi.1946, *Wild* 1601 (BM, SRGH). E: Nyanga Dist., near Nyanga, fl.& fr. 7.xi.1930, *Fries, Norlindh & Weimarck* 2756 (BM, LD). **Malawi**. S: Zomba, fl. ix.1895, *Whyte* s.n. (K). **Mozambique**. MS: Inyamadzi Valley, 8.ix.1907, *Swynnerton* 958 (BM).

Across Africa from Azores and Canary Is., North Africa and tropical NE and E Africa to Congo, Namibia and South Africa; also common in Europe and temperate parts of Asia and North America, rarer in the American and Asian tropics, including New Guinea. Mostly in disturbed places (ditches, ponds, reservoirs) and slow-flowing waters from 30–60 cm deep, sometimes in lightly brackish water; 600–1750 m, up to 2900 m in Ethiopia.

Conservation notes: Widely distributed; not threatened.

P. pusillus is a very polymorphic species, varying greatly in leaf length, the number of leaf nerves and rows of lacunae along the midrib, shape of leaf apex, and number of flower whorls in the inflorescence. Owing to the lack of a consistently correlated set of morphological characters, no discrimination of infraspecific taxa based on these criteria seems possible. The united stipules forming a tubular sheath, at least when young, differentiate *P. pusillus* from *P. berchtoldii* Fieber from northern temperate regions, which has stipules open throughout their length. Isozyme studies have confirmed the separate identity of *P. pusillus* and *P. berchtoldii* (Hettiarachchi & Triest in Opera Bot. Belg. **4**: 87–114, 1991; Kaplan & Stepánek in Plant Syst. Evol. **239**: 95–112, 2003). For differences between *P. pusillus* and *P. trichoides*, see note under the latter.

Specimens such as the type of *P. subjavanicus* Hagstr., with the uppermost leaves ± floating and blade oblanceolate, form the transition to *Potamogeton* subsect. *Javanici*. The hybrid with *P. octandrus* has been described as *P.* × *apertus* Miki. More research is necessary to establish the real taxonomic position of such plants.

2. **Potamogeton trichoides** Cham. & Schltdl. in Linnaea **2**: 175, t.4 (1827). — Bennett in Ann. Conserv. Jard. Bot. Genève **9**: 101 (1905). —Peter in Abh. Königl. Ges. Wiss. Göttingen, Math.-Phys. Kl. **13**: 108 (1928). —Dandy in J. Linn. Soc., Bot. **50**: 520 (1937). —Obermeyer in F.S.A. **1**: 65 (1966). —Gibbs Russell in Kirkia **10**: 434 (1977). —Wiegleb & Kaplan in Folia Geobot. **33**: 297 (1998). —Cook, Aq. Wetl. Pl. Sthn. Africa: 238 (2004). Type from Europe (many syntypes; lectotypification needed).

Submerged aquatic herb. Rhizomes absent or only present later in growing season, filiform, terete. Stems annual to perennial, up to 1.5 m long, sparingly to much branched, filiform, terete to slightly compressed, fragile and breaking easily at nodes; turions apical or axillary, not differing greatly from ordinary buds but formed of more crowded darker leaves, more rigid, detaching from stem of parent plant and readily dispersed by water. Submerged leaves sessile, linear, 14–80(130) × 0.3–1(1.8) mm, (30)40–80(110) times as long as wide, bright green to dark green, often with a brownish tinge, narrowly cuneate at base, acuminate at apex, entire, 3-veined, midrib not bordered by rows of lacunae; lateral veins sometimes inconspicuous, without additional sclerenchyma strands, not bordered by a marginal vein. Stipules axillary, convolute, 5–27 mm long, translucent, often with a greenish tinge, persistent. Floating leaves absent. Peduncles 10–75 mm long, (2)3–9 times as long as fruiting spike, as thick as stem; spikes shortly cylindrical, 3–9 mm long in fruit, with flowers contiguous to ± closely spaced. Flowers 3–5, carpels 1(2). Fruitlets 2.5–3.2 mm long; dorsal keel mostly distinct, often muriculate; beak 0.3–0.5 mm long, straight. Stem anatomy: stele of circular type, endodermis of O-type; interlacunar bundles absent, subepidermal bundles present; pseudohypodermis absent.

Botswana. N: Qogai, Thaoge R., fl.& fr. 4.x.1974, *P.A. Smith* 1133 (K, SRGH). **Zimbabwe**. N: Makonde Dist., no locality, vii.1921, *Eyles* 3141 in part (BM sheet 2, SAM 26532). W & E: cited in Gibbs Russell (1977). C: Harare Dist., Cleveland Dam, fl. 30.vi.1927, *Blenkison* in *Moss* 14860 (BM).

Also in North Africa, tropical East Africa, Congo, South Africa (apparently rare). Also in Europe and W Asia. Dams, ponds, seasonal pools and watercourses at 30–50 cm depth; sometimes associated with *P. pusillus*; 1000–1500 m, to 2440 m in East Africa.

Conservation notes: Widely distributed; not threatened.

As mentioned by Wiegleb & Kaplan (1998, p.298), this species is closely related to *P. pusillus*, from which it is sometimes difficult to distinguish in the vegetative stage from morphotypes with extremely narrow leaves. *P. trichoides* differs from *P. pusillus* by having open and convolute stipules and a leaf midrib occupying 30–70% of leaf width at the base. When in flower or in fruit, *P. trichoides* may be distinguished by its gynoecium reduced to 1 or 2 carpels and its fruitlets with a distinct dorsal keel.

3. **Potamogeton crispus** L., Sp. Pl.: 126 (1753). —Bennett in J. Bot. **33**: 137 (1895); in F.C. **7**: 47 (1897); in F.T.A. **8**: 221 (1901). —Graebner in Engler, Pflanzenr. **31**: 99 (1907). —Gomes e Sousa, Pl. Menyarth.: 210 (1936). —Dandy in J. Linn. Soc., Bot. **50**: 537 (1937). —Podlech in Merxmüller, Prodr. Fl. SW Afr., fam.144: 2 (1966). —Obermeyer in F.S.A. **1**: 66 (1966). —Gibbs Russell in Kirkia **10**: 434 (1977). —Wiegleb & Kaplan in Folia Geobot. **33**: 284 (1998). —Clarke & Klaassen, Water Pl. Namibia: 62 (2001). —Cook, Aq. Wetl. Pl. Sthn. Africa: 235 (2004). Type: Europe, LINN 175.6 (lectotype, designated by Haynes in Taxon **35**: 567, 1986).

Potamogeton crispus var. *najadoides* Graebn. in Engler, Pflanzenr. **31**: 100 (1907). Type: South Africa, Mpumalanga, Standerton, *Wilms* 1658 (B†).

Submerged aquatic herb. Rhizomes annual to biennial or perennating, short, strongly branched, compressed, rooting at nodes. Stems annual or partly perennial, up to 1.5 m long, 0.5–2.5 mm in greater width, unbranched or sparingly branched, filiform to robust, compressed and somewhat 4-angled with the broader sides grooved; turions of various shapes, mostly formed by axillary, stiff lateral branches with buds of reduced and horny leaves that develop as they are formed, whilst the supporting shoot grows no further. Submerged leaves sessile, linear to linear-oblong, 25–90(132) × (4)6–12(18) mm, 5–9(13) times as long as wide, bright green to dark green or occasionally slightly reddish, especially along the midrib, slightly clasping at base, rounded or obtuse at apex, margins finely serrate and usually markedly undulate; longitudinal veins 3–7, the laterals close to margins, with 5–7 pairs of fine transverse or ascending veins; 1–2(3) narrow to broad rows of lacunae bordering the midrib. Floating leaves always absent. Stipules axillary, convolute to shortly united at base, (3)5–12(17) mm long, subtriangular, delicate and translucent, truncate or emarginate at apex, decaying early into fibres. Peduncles 14–70(125) mm long, 3–8 times as long as fruiting spike, as thick as the stem; spikes cylindrical, (4)10–18 mm long in fruit, with flowers contiguous to shortly distant. Flowers (3)5–8 in 2–3(5) whorls. Tepals broadly ovate, 1.2–2.1 mm long, green; carpels (2)4. Anthers 0.7–1.3 mm long. Fruitlets united at base, ovoid, 2–3.6 × 2–2.5 mm (excluding beak), dark-olive or brownish; dorsal keel distinct, ± denticulate and with a basal tooth, lateral keels distinct, obtuse; beak very prominent, 1.5–2.4 mm, 0.5–0.8 times as long as the fruitlet body, ± falcate, tapering from a broad base to a slender apex. Stem anatomy: stele oblong with but 1 central bundle and 1 lateral bundle on each side, endodermis of O-type; interlacunar bundles absent; subepidermal bundles absent, pseudohypodermis 1-layered, sometimes absent.

Botswana. SE: c.30 km N of Lobatse, Mogobane Dam, 6.x.1955, *McConnell* in SRGH 68900 (K, SRGH). **Zambia**. S: Mazabuka Dist., c.30 km N of Kafue R. pontoon crossing, 3.ix.1947, *Greenway & Brenan* 8042 (K, PRE). **Zimbabwe**. N: Hurungwe Dist., Sanyati-Chiroti river junction, 21.xi.1953, *Wild* 4228 (BM, BR, K, PRE, SRGH). W: cited in Gibbs Russell (1977). C: Kwekwe Dist., Sable Park, fl. 13.iii.1978, *Chipunga* 178 (BR, K, SRGH). S: Chiredzi Dist., Chitove Camp, Runde (Lundi) R., 16.viii.1984, *Ngoni* 562 (PRE). **Malawi**. N: Dwangwa (Roangwa) R., W shore of Lake Malawi (Nyasa), imm.fl. ix.1861, *Kirk* s.n. (K). **Mozambique**. Z: near Mambusha, Rio Cuácua (Quaqua), 24.viii.1892, *Menyarth* s.n. (Z). M: Lake Mandjene (Mhandlene), imm.fr., viii.1919, *Junod* TVM 21468 (PRE).

Also in North Africa, Sudan, Ethiopia, South Africa, and in Europe, Asia, Australia; introduced in America (from Canada to Argentina) and in New Zealand. In stagnant and slowly flowing water, showing a preference for eutrophic conditions and often gregarious, growing with *Potamogeton octandrus, P. richardii, Ottelia, Nymphaea*, etc.; sea level–1200 m.

Conservation notes: Widely distributed; not threatened.

4. **Potamogeton octandrus** Poir. in Lamark, Encycl. Méth. Bot., suppl. **4**: 534 (1816). —Dandy in J. Linn. Soc., Bot. **50**: 517 (1937). —Podlech in Merxmüller, Prodr. Fl. SW Afr., fam.144: 2 (1966). —Obermeyer in F.S.A. **1**: 65 (1966). —Gibbs Russell in Kirkia **10**: 434 (1977). —Lisowski et al. in F.A.C., Potamogetonaceae: 5 (1978). —Wiegleb & Kaplan in Folia Geobot. **33**: 298 (1998). —Clarke & Klaassen, Water Pl. Namibia: 62 (2001). —Cook, Aq. Wetl. Pl. Sthn. Africa: 235 (2004). Type: Vietnam ('Cochinchina'), *Loureiro* s.n. (BM holotype).

> *Hydrogeton heterophyllus* Lour., Fl. Cochinch. **1**: 244 (1790). Type as above.
>
> *Potamogeton javanicus* Hassk. in Acta Soc. Regiae Sci. Indo-Neerl. **1**: 26 (1856). —Durand & Schinz, Consp. Fl. Afr. **5**: 496 (1894). —K. Schumann in Engler, Pflanzenw. Ost-Afr. **C**: 93 (1895). Bennett in J. Bot. **33**: 138 (1895); in F.C. **7**: 47 (1897); in F.T.A. **8**: 220 (1901). —Graebner in Engler, Pflanzenr. **31**: 46 (1907). —Peter in Abh. Ges. Nov. Regni Veg., Beih. **40**(1): 113 (1929). —Bremekamp & Obermeyer in Ann. Transv. Mus. **16** (1935). — Pedro in Bol. Soc. Estudos Moçamb. **24**: 18 (1954). Type: Indonesia, W Java, near Tjisarupan, foot of Mt. Papandayan, *Hasskarl* s.n. (not located).
>
> *Potamogeton preussii* A. Benn. in F.T.A. **8**: 222 (1901) in part for syntypes from Cameroon: Kumba, Barombi-ba-Mbu, Elephant Lake, 1890, *Preuss* 451 (BM, COI syntypes); Kumba Mountains [Johann-Albrechtshöhe], bank of Elephant Lake, 1895, *Staudt* 462 (BM syntype).

Aquatic herb with submerged and floating leaves. Rhizome system sometimes absent, annual or perennial, poorly developed, filiform, terete. Stems annual, 20–100 cm long, sparingly or much branched, filiform, terete with grooves, sometimes developing axillary tubers. Submerged leaves sessile, linear, 25–55(75) × 0.5–1.2(2) mm, 30–75 times as long as wide, bright green to brown-green, entire at margins, acute to acuminate, 3-veined; midrib with broad rows of lacunae on border; lateral veins sometimes inconspicuous. Intermediate leaves often present, petiolate, oblong to lanceolate. Floating leaves present or absent, petiolate; petiole 3–25(34) mm long, 0.2–1.1 times as long as lamina; lamina narrowly lanceolate to elliptical or oblong, (5)10–30(40) × (2)3–11 mm, 1.3–6(8.5) times as long as wide, opaque, coriaceous to subcoriaceous, bright green to dark green, sometimes with a brownish tinge, 5–7 veined, cuneate at base, apex acute; margins entire. Stipules axillary, convolute, 4–13 mm long, translucent, decaying early. Inflorescences developing mostly in axils of floating leaves; peduncles 9–20(40) mm long, 0.7–2.4 times as long as fruiting spike, as thick or slightly thicker than the stem; spikes cylindrical, 5–16 mm long in fruit, with 7–9 flowers in 3–7 contiguous or shortly distant whorls. Tepals broadly ovate, ± 1 mm long, green; carpels (3)4(7). Fruitlets 1.5–2.4(3.5) mm long, green; beak short to fairly long, hooked; dorsal keel mostly distinct, crenulate. Stem anatomy: stele of oblong or circular type, endodermis of O-type; interlacunar bundles mostly absent, subepidermal bundles present; pseudohypodermis absent or present, 1-layered.

Caprivi Strip. Okavango R., bank opposite Rundu (Runtu), 2.ii.1956, *De Winter &
Marais* 4477 (K). **Botswana**. N: Savuti R., near Zibadinaja lagoon, fl. 22.x.1972, *Gibbs
Russell* 2183 (BR, SRGH). **Zambia**. B: Mongu Dist., c.16 km NE of Mongu, fl.
18.xi.1959, *Drummond & Cookson* 6594 (K, SRGH). N: Mansa Dist., 6 km from
Chisunka, Musonda Falls, fl.& imm.fr. 11.iv.1963, *Symoens* 10214 (BR, BRVU, K). W:
Copperbelt, Luanshya, fl. 4.x.1955, *Fanshawe* 2490 (BM, K). C: Kabwe Dist., 19 km N
of Kabwe, Nasuleka R., fl. 23.ix.1947, *Brenan & Greenway* 7936 (EA, FHO). S: Choma
Dist., 20 km N of Choma, Muckle Neuk, fl.& fr. 11.x.1954, *Robinson* 924 (K).
Zimbabwe. N: Hurungwe Dist., Zvipani (Zwipani), fl. 12.x.1957, *Phipps* 785 (BR, K,
SRGH). W: Hwange Dist., Victoria Falls, Zambezi R., fl. 28.xi.1949, *Wild* 3219 (BM,
LISC). C: Chegutu Dist., Kalembo dam, fl.bud 27.ii.1969, *Mavi* 1025 (BRVU, K, L,
LISC, SRGH). E: Mutare Dist., Vumba Hotel, fr. 27.ii.1956, *Chase* 5981 (BM, K, LISC).
S: Chivi Dist., Razi dam, fr. 12.v.1970, *Biegel & Pope* 3314 (K, SRGH, WAG). **Malawi**.
S: Machinga Dist., Machinga East, Lifune R., fl. 3.v.1982, *Patel* 914 (BR, K).
Mozambique. N: Niassa, Sanga Dist., Unango, R. Malagui, fl.& fr. 2.iii.1964, *Torre &
Paiva* 10955 (LISC). T: Cabora Bassa, Rio Zambeze, fl. 30.xi.1973, *Correia, Marques &
Pires Monteiro* 4000 (LMU, WAG). MS: Cheringoma Dist., R. Nhamissembe
(Inhamissembe) flats, fr. 14.vii.1972, *Ward* 7951 (K). M: Maputo (Delagoa Bay),
Cherinda, fl. ix.1892, *Junod* 473 (BR, G, P).

Also in West Africa (Senegal to Central African Republic), tropical NE and E Africa,
Congo, Angola, Namibia, South Africa (Gauteng, KwaZulu-Natal), Madagascar, S & E
Asia, New Guinea and Australia. Lakes, ponds, backwaters, swamps, rice fields and
flowing water, to 0.6–2 m deep. Associated with *Najas horrida, Potamogeton richardii,
Ludwigia stolonifera, Utricularia inflexa*, etc.; sea level–1800 m.

Conservation notes: Widely distributed; not threatened.

5. **Potamogeton schweinfurthii** A. Benn. in F.T.A. **8**: 220 (1901) nom. conserv. as
regards *Schimper* 1359. —Graebner in Engler, Pflanzenr. **31**: 79 (1907). —Dandy
in J. Linn. Soc., Bot. **50**: 526 (1937). —Podlech in Merxmüller, Prodr. Fl. SW Afr.,
fam.144: 3 (1966). —Obermeyer in F.S.A. **1**: 66 (1966). —Gibbs Russell in Kirkia
10: 434 (1977). —Lisowski et al. in F.A.C., Potamogetonaceae: 6 (1978). —
Wiegleb & Kaplan in Folia Geobot. **33**: 273 (1998). —Clarke & Klaassen, Water
Pl. Namibia: 62 (2001). —Cook, Aq. Wetl. Pl. Sthn. Africa: 237 (2004). —Kaplan
& Symoens in Bot. J. Linn. Soc. **148**: 346 (2005). Type: Ethiopia, Lake Tana,
9.xi.1863, *Schimper* 1359 (K lectotype, BM, CGE, E, LD).

Potamogeton lucens sensu K. Schum. in Engler, Pflanzenw. Ost-Afr. **C**: 93 (1895) as *P. lucers*
in error. —Bennett in F.T.A. **8**: 221 (1901). —Peter in Rep. Spec. Nov. Regni Veg., Beih.
40(1): 113 (1929), non L.

Potamogeton fluitans sensu Bennett in J. Bot. **33**: 138 (1895); in F.C. **7**: 46 (1897). —sensu
Peter in Rep. Spec. Nov. Regni Veg., Beih. **40**(1): 112 (1929), non Roth.

Potamogeton americanus Cham. & Schltdl. var. *thunbergii* (Cham. & Schltdl.) A. Benn. in
F.C. **7**: 46 (1897) in part for each of Drège, Gueinzius & Zeyher.

Potamogeton lucens L. var. *fluitans* sensu Bennett in F.C. **7**: 48 (1897); in F.T.A. **8**: 221
(1901). —Schinz & Junod in Bull. Herb. Boiss., sér.2, **3**: 654 (1903). —Pedro in Bol. Soc.
Estudos Moçamb. **24**: 19 (1954). —Binns, First Check List Herb. Fl. Malawi: 89 (1968), non
Coss. & Germ.

Potamogeton longifolius sensu Burkill in Johnston, Brit. Cent. Afr.: 275 (1897). —Bennett
in J. Bot. **33**: 138 (1905), non Gay.

?*Potamogeton repens* Hagstr. in Fries, Wiss. Ergebn. Schwed. Rhod.-Kongo-Exped.: 185
(1916); in Kungl. Svensk. Vet.-Akad. Handl., n.s. **55**(5): 170, fig.87,88 (1916). Type:
Zambia, Lake Bangweulu, near Kasoma, 19.ix.1991, *R.E. Fries* 655 (UPS holotype).

Potamogeton capensis Hagstr. in Kungl. Svensk. Vet.-Akad., Handl., n.s. **55**(5): 203 (1916).
Types: South Africa, E Cape, Uitenhage, Swartkops R., i.1830, *Zeyher* 919 (S lectotype

chosen by Kaplan & Symoens 2005, BREM, LD); same locality, xii.1829, *Drège* 8799 (S syntype, G, P); KwaZulu-Natal, Umshlangwe R., 19.viii.1893, *Schlechter* 3120 (S syntype, BM, G, K, LD, LE, LY, PRC, W, Z); Mozambique, 'Prov. Delagoa', Mathibis Kom, *Bolus* in *Herb. Norm. Austro-Afr.* 1393 (UPS syntype, BM, BOL, G, K, P, W, Z).

Rooted aquatic herb with submerged and floating leaves. Rhizomes perennial, slender, terete, white, with apical winter buds. Stems annual sparingly to much branched, slender, terete, to 3.5 m long, ± spongy; specialized dormant turions not developing. Submerged leaves almost always present, sessile to shortly petiolate, often encrusted with calcium carbonate; petiole 0–30(82) mm long, less than 0.15(0.35) times as long as lamina, rarely petiole of uppermost leaves as long as lamina; lamina narrowly lanceolate to narrowly elliptical, sometimes those of lower leaves reduced to phyllodes, 45–190(260) × (3)7–28 mm, (4)6–30(70) times as long as wide, membranous and translucent, cuneate, apex acute to mucronate, young leaves yellow green or brownish, older ones bright green or dark olive-green, often with a reddish or brownish tinge, 5–11(13)-veined, with or without narrow rows of lacunae bordering the midrib, margins often slightly undulate, entire or bearing very minute translucent teeth. Intermediate leaves sometimes present. Floating leaves sometimes present, petiolate; petiole (14)30–70(200) mm long, 0.1–0.6 times as long as lamina, never discoloured at junction with lamina; lamina narrowly oblong to elliptical or ovate, 43–130 mm long, 12–30 mm wide, 2–6 times as long as wide, opaque, subcoriaceous to coriaceous, yellow-green to dark green with a reddish tinge, 11–21-veined, entire, mostly narrowly cuneate at base, obtuse to acute. Stipules persistent, axillary, convolute, 20–62 mm long, translucent, acute. Peduncles mostly terminal or lateral in axils of submerged leaves, sometimes in axils of floating leaves, 35–250 mm long, 2–3 times as long as fruiting spike, thicker than the stem. Spikes emergent, cylindrical, 30–90 mm long in fruit, contiguous. Flowers numerous, sepals pale green-brown; carpels 4; stigmas whitish becoming mauve brown. Anthers yellow green. Fruitlets 2.9–3.9(4.1) mm long, pale grey- to yellow-green, nearly straight ventrally; dorsal keel distinct, lateral keels sometimes present; beak short, obtuse. Stem anatomy: stele of trio type, rarely proto or oblong type, endodermis of U-type, interlacunar bundles present in 1 circle (rarely 2, one being incomplete), subepidemal bundles absent or scattered ones present; pseudohypodermis present, 1-layered.

Botswana. N: ?Boteti R., 3.xii.1978, *P.A. Smith* 2547 (K, SRGH). **Zambia**. N: Mporokoso Dist., Kalungwishi R. ferry, 18.i.1960, *Richards* 12414 (K). W: Copperbelt, Chingola, 18.iv.1954, *Fanshawe* 1114 (K). C: Ndola Dist., Walamba, fr. 22.v.1954, *Fanshawe* 1228 (K). S: Choma Dist., Mapanza, Munyeke R., fl. 5.ix.1953, *Robinson* 308 (BM, BR, K). **Zimbabwe**. N: Hurungwe Dist., Sanyati R., near junction of Fulechi R., 11.x.1957, *Phipps* 760 (BR, K, SRGH). W: Hwange Nat. Park, 40 km NE of Robin's Camp, 24.x.1968, *Rushworth* 1234 (K, LISC, SRGH). C: Harare Dist., Mukuvisi (Makabusi), 21.iv.1947, *Wild* 1908 (BM, SRGH). E: Chipinge Dist., E Sabi, lower Rupembi, fl. 24.i.1957, *Phipps* 148 (COI, K, LISC, SRGH). **Malawi**. N: Nkhata Bay Dist., Lake Malawi (Nyassa), Likoma Is., fl. iii.1988, *Kasselmann* 136 (B, M). C: Nkhota Kota Dist., Lake Malawi (Nyasa), off Nkhota Kota, fl. vi.1904, *Cunnington* 8 (BM). S: Machinga Dist., Shire R., 5 km S of ferry, fl. 8.vii.1975, *Seyani* 272 (BR, SRGH). **Mozambique**. MS: Chibabava Dist, R. Mababa, fl. 13.viii.1967, *M.F. Carvalho* 912 (BR). GI: Massinga Dist., Rio das Pedras, imm.fr. 8.vii.1981, *de Koning & Hiemstra* 8932 (BR, K, LMU). M: Maputo (Delagoa), 'Mathibis Kom', viii.1886, *Bolus* 1393 (BM, BOL, G, K, P, UPS, Z).

Also in West Africa (Senegal to Central African Republic), tropical NE and E Africa, Congo, Namibia, South Africa and Madagascar. In its northern range the species is more extensive than was formerly realized, stretching from the Azores to the Mediterranean islands, Libya, Egypt and possibly SW Asia. Lakes, ponds, backwaters, pools, and running waters; water depth 20–200 cm. The leaves are sometimes encrusted with calcium carbonate. Sometimes invasive and considered as a pest in ditches and reservoirs. Associated with *Vallisneria*, *Najas*, *Ceratophyllum*, *Nymphaea*, *Ludwigia*, etc.; sea level–1500 m, to 4250 m in Kenya.

Conservation notes: Widely distributed; not threatened.

P. schweinfurthii, which is a very variable species, has frequently been confused with other broad-leaved species. When Bennett published the name in 1901, he cited three syntypes of which two, including one later chosen as the lectotype by Dandy (1937), proved to belong to *P. nodosus*. However, the third specimen (*Schimper* 1359) corresponds largely to the species largely known under the former name, which has consequently been accepted for conservation (see Kaplan & Symoens in Taxon **53**: 837–838, 2004 and in Bot. J. Linn. Soc. **148**: 346, 2005). *P. schweinfurthii* differs from *P. nodosus* in its sessile or very shortly petiolate submerged leaves and particularly by characteristics of its stem anatomy.

Plants with only submerged leaves closely resemble *P. lucens*, which explains why they have long been identified under this name. *P. lucens* may be distinguished by its more elliptical leaves, 25–65 mm broad and mostly 3–6 times as long as wide, often with an embossed surface, and the broadly acute to rounded and always distinctly mucronate apex. However, some plants (e.g. *Lye* 5245 from Lake Mutanda, Uganda) are intermediate between both species and difficult to assign to either.

Denny & Lye (in Kew Bull. **28**: 120, 1973) suggest the existence of two heterophyllous taxa: one fertile, *P. schweinfurthii*, and one sterile, being the putative hybrid *P. schweinfurthii* × *P. richardii* (*P. thunbergii* sensu Oberm.), which they described as *Potamogeton* × *bunyonyiensis* Denny & Lye based on a collection from Uganda (*Lye* 5216, K holotype, MHU). This hybrid could be distinguished from both *P. schweinfurthii* and *P. richardii* by being virtually sterile and having some characters of each parent: submerged leaves, like *P. schweinfurthii*, and predominantly coriaceous leaves, although more lanceolate than those of *P. richardii*. More recently, Lye (in Fl. Somalia **4**: 14, 1995 and in Fl. Ethiopia **6**: 23, 1997) considered *P. schweinfurthii* itself to be of hybrid origin, with *P. lucens* and *P. richardii* (*P. thunbergii* sensu Oberm.) or *P. nodosus* being the parents. Molecular studies and more transplantation experiments are necessary to establish the relationships of these taxa.

6. **Potamogeton nodosus** Poir. in Lamarck, Encycl. Méthod. Bot., suppl. **4**: 535 (1816). —Dandy in J. Linn. Soc., Bot. **50**: 531 (1937) in part. —Cook, Aq. Wetl. Pl. Sthn. Africa: 235 (2004). —Kaplan & Symoens in Bot. J. Linn. Soc. **148**: 332–340 (2005). Type: Canary Islands, *Broussonet* s.n. (P lectotype, FI-W).

Potamogeton natans sensu Thunberg, Prodr.: 32 (1794). —Gibbs in J. Linn. Soc., Bot. **37**: 474 (1906) as regard *Gibbs* 146. —Eyles in Trans. Roy. Soc. S. Afr. **5**: 293 (1916) as regard *Gibbs* 146, non L.

Potamogeton thunbergii Cham. & Schltdl. in Linnaea **2**: 221, t.6 (1827). —Podlech in Merxmüller, Prodr. Fl. SW Afr., fam.144: 3 (1966). —Obermeyer in F.S.A. **1**: 67 (1966) in part. —Gibbs Russell in Kirkia **10**: 434 (1977) in part. —Clarke & Klaassen, Water Pl. Namibia: 62 (2001). Type: South Africa, W Cape, Swellendam, Hartebeeskraal near Brak R., i.1819, *Mundt & Maire* s.n. (B† holotype, HAL lectotype selected by Kaplan & Symoens 2005, LE).

Potamogeton natans L. var. *capensis* T. Durand & Schinz, Consp. Fl. Afr. **5**: 494 (1894) (in misunderstanding the title of a section in Chamisso & Schlechtendal's paper, Durand & Schinz created this combination and attributed it to the former authors).

Potamogeton americanus Cham. & Schltdl. var. *thunbergii* (Cham. & Schltdl.) A. Benn. in F.C. **7**: 46 (1897) in part for *Thunberg* and for *Mundt & Maire*.

Potamogeton nodosus Poir. var. *billotii* (F.W. Schultz) Hagstr. forma *angustissimus* Hagstr. in Fries, Wiss. Ergebn. Schwed. Rhod.-Kongo-Exped.: 186 (1916); in Kungl. Svensk. Vet.-Akad. Handl., n.s. **55**(5): 188 (1916). Type: Zimbabwe, Hwange Dist., Zambezi R. above Victoria Falls, 1911, *R.E. Fries* 137 (UPS holotype).

Potamogeton stagnorum Hagstr. in Kungl. Svensk. Vet.-Akad. Handl., n.s. **55**(5): 159 (1916); in Fries, Wiss. Ergebn. Schwed. Rhod.-Kongo-Exped.: 187 (1916). —Fries, Wiss.

Ergebn. Schwed. Rhod.-Kongo-Exped., Ergänzung.: 61, 72 (1921). Type: Zambia, Chimona R. at Lake Bangweulu, 20.ix.1911, *R.E. Fries* 691 (UPS holotype).
Potamogeton richardii sensu Wild in Clark, Victoria Falls Handb.: 138 (1952), non Solms.

Aquatic herb with submerged and floating leaves. Rhizomes robust, long-creeping, biennial or perennial, much-branched, producing apical fusiform turions. Stem short-lived or annual, slender to robust, terete, to 2.5 m long, unbranched or sparingly branched. Submerged leaves petiolate, lanceolate to oblong-lanceolate, often absent after fruiting; petiole (20)30–150 mm long; lamina 50–280 mm long, 10–38(50) mm wide, translucent, pale green when fresh and green or brownish green when dried, 11–21-veined, finely denticulate, at least when young, gradually tapering to a cuneate base and an acute apex. Floating leaves petiolate; petiole (30)100–210 mm, mostly longer than lamina, ± brown-reddish; lamina lanceolate-elliptical or oblong-elliptical to broadly elliptical or obovate, (40)50–150 mm long, 20–50(60) mm wide, 11–23-veined, opaque but sometimes only slightly leathery, bright green, cuneate or ± rounded at base, not discoloured and without folds at base, apex acute to slightly obtuse; stipules axillary, open, convolute and enfolding the stem, 20–60(125) mm, with 2 veins more prominent than the others and forming dorsal ridges. Peduncles 40–130 mm, not or slightly thickened above. Spikes cylindrical, 14–50 mm long, 4–10 mm wide, with up to 15 contiguous flower whorls. Flowers numerous, tepals 0.2–0.3 mm long, suborbicular; carpels (3)4. Fruitlets 2.9–3.9(4) mm long, brown or reddish brown, nearly straight ventrally, dorsal keel sharp, the lateral rather prominent, beak 0.3–0.5(0.8) mm long, straight or slightly recurved. Stem anatomy: stele of trio or proto type, endodermis of O-type, interlacunar bundles and subepidemal bundles absent; peudohypodermis absent.

Caprivi Strip. E Caprivi, Singalamwe, 1.i.1959, *Killick & Leistner* 3229 (K, M, PRE). **Botswana**. N: Upstream from Txatxanika Camp, Khwai R., fl.& fr. 1.iii.1972, *Gibbs Russell & Biegel* 1479 (BR, K, SRGH). **Zambia**. B: Kaoma Dist., Luena R., 20.xi.1959, *Drummond & Cookson* 6639 (BM, SRGH). N: Lake Bangweulu, Mboyalumbambe Is., fl.& fr. 6.iii.1996, *Renvoize* 5752 (K). C: Mumbwa Dist., Mumbwa dam, fl.& fr. 19.ix.1947, *Brenan & Greenway* 7884 (EA, FHO, K). **Zimbabwe**. N: Binga Dist., Sinamwenda R., E bank, Elephant Is., fl.& fr. 28.ix.1970, *Mitchell* 1272 (K, SRGH). W: Hwange Dist., Kazuma Range, Katsatetsi R., 10.v.1972, *Gibbs Russell* 1950 (SRGH, WAG). C: Marondera Dist., Chikokorana Pan, fl. 29.iv.1972, *Gibbs Russell* 1985 (K, M, SRGH). E: Mrs. Strickland's Charity Farm, 1934, *Gilliland* 527b (BM label partly burned, K).

Widely distributed in Africa from North Africa (including Saharan oases), Senegal to Central African Republic, Sudan, Angola, Namibia, South Africa and Madagascar, but extremely rare in Ethiopia and East Africa; also found in Azores, Madeira, Canary Is., Socotra, Seychelles, Mascarene Is., Europe, SW and C Asia, tropical Asia to Indonesia, Philippines and New Guinea, N and C America. Shallow parts (1.5–2 m deep) of lakes, ponds, pans, swamps and rivers, often in muddy soil; sometimes in extensive colonies or associated with *Nymphaea*; 1000–1700 m.

Conservation notes: Widely distributed; not threatened in the Flora area, but said to be endangered in East Africa and Ethiopia.

Although widespread from North Africa to the Cape in South Africa, *P. nodosus* has long been overlooked on the African continent. Due to the great morphological variation in *Potamogeton*, it is sometimes difficult to distinguish *P. nodosus* from two other broad-leaved species, *P. schweinfurthii* and *P. richardii*. Moreover, inadequate synonymy created much confusion between them. *P. nodosus* can mostly be recognized by its long petiolate, well-developed submerged leaves with a narrowly obtuse to subacute but never sharply mucronate apex. Identification based on stem anatomy is more certain. If the endodermis is of the O-type and there are no interlacunar bundles in the cortex (rarely one or few), it is *P. nodosus*. If the submerged leaves are completely sessile or subsessile, or the endodermis has U-cells and at least one circle of interlacunar bundles, *P. nodosus* can be excluded.

7. **Potamogeton richardii** Solms in Schweinfurth, Beitr. Fl. Aethiop. **1**: 194 (1867). —Durand & Schinz, Consp. Fl. Afr. **5**: 496 (1894). —Bennett in F.C. **7**: 47 (1897); in F.T.A. **8**: 219 (1901). —Graebner in Engler, Pflanzenr. **4**(11): 56 (1907). —Kaplan & Symoens in Bot. J. Linn. Soc. **148**: 340 (2005). Type: Ethiopia, Adwa (Adoa), 11.vi.1837, *Schimper* 135 (K lectotype, BM, BR, L, LE, M, P, ZT). FIGURE 12.2.**24**.

Potamogeton natans sensu Bennett in F.C. **7**: 46 (1897). —Gibbs in J. Linn. Soc., Bot. **37**: 474 (1906) for *Gibbs* 94. —Eyles in Trans. Roy. Soc. S. Afr. **5**: 293 (1916) in part for *Gibbs* 94 and *Rand* 540. —Peter in Abh. Ges. Wiss. Göttingen **13**: 18, 108 (1928) for "Urundi"; in Rep. Spec. Nov. Regni Veg., Beih. **40**: 113 (1929), non L.

Potamogeton nodosus sensu Dandy in J. Linn. Soc., Bot. **50**: 531 (1937) in part.

Potamogeton thunbergii sensu Obermeyer in F.S.A. **1**: 67 (1966) in part. —Gibbs Russell in Kirkia **10**: 434 (1977) in part. —Lisowski et al. in F.A.C., Potamogetonaceae: 9 (1978). —Symoens et al. in Bull. Soc. Roy. Bot. Belg. **112**: 79 (1979). —Wiegleb & Kaplan in Folia Geobot. **33**: 264 (1998). —Cook, Aq. Wetl. Pl. Sthn. Africa: 238 (2004), non Cham. & Schltdl.

Rooted aquatic herb, with clearly visible floating leaves. Rhizomes slender to robust, terete, pink to orange brown, perennial, abundantly branched, internodes to 8 cm long, with apical scaly turions. Stem terete, up to 1 m long, mostly unbranched, rooting at nodes, annual. Submerged leaves decaying early, generally not present on adult plants, rarely 1 or 2 at the fruiting stage; petiole (10)45–180 mm long, 0.2–1.8 times as long as lamina; lamina lanceolate to oblong, sometimes almost reduced to phyllodes, 80–200 mm long, 5–27 mm wide, 5–13 times as long as wide, bright green to dark green, 5–9(15) veined, with narrow rows of lacunae bordering the midrib, entire at margins, narrowly cuneate at base, gradually narrowed towards an obtuse apex, never mucronate. Intermediate leaves often present. Floating leaves always present on adult plants; petiole 18–110(200) mm long, (0.3)0.6–3 times as long as lamina, often with a discoloured section at junction with lamina, but only on some leaves; lamina elliptical to oblong-ovate, 30–80(124) mm long, (11)20–40(48) mm wide, 1.6–3.5(5) times as long as wide, opaque, leathery, green, brownish green to dark green, or ± pinkish, 11–25-veined, broadly cuneate to rounded at base, broadly acute at apex, entire at margins. Stipules axillary, robust, convolute, 25–40(60) mm long, translucent, persistent as grey fibres after decay. Peduncles inserted in axils of floating leaves, 50–100 mm long, 2–3.5 times as long as fruiting spike, ± as thick as the stem, tinged pale brown. Spikes emergent, cylindrical, sometimes crooked, 30–50 mm long in fruit, contiguous. Flowers numerous; tepals pale green; carpels 4. Anthers white to yellowish. Fruitlets (3.2)3.9–5.2(5.5) × 2.1–3.3 mm, green to pale brown, straight ventrally, dorsally convex, distinctly 3-keeled. Stem anatomy: stele of trio type, endodermis of U-type, interlacunar bundles present in (1)2(3) circles, subepidermal bundles mostly present, pseudohypodermis present, 1-layered.

Zambia. N: Mbala Dist., Mbala (Abercorn), fl.& fr. 5.ii.1964, *Richards* 18927 (BRVU, K). **Zimbabwe**. W: Matobo Dist., Matopos Hills, 1905, *Gibbs* 94 (BM). C: Harare Dist., Cleveland Dam, fr., ii.1917, *Eyles* 659 (BM, K). E: Nyanga Dist., 7 km NNE of Nyanga (Inyanga), Troutbeck, fl. 31.xii.1973, *Bamps, Symoens & Vanden Berghen* 481 (BR, WAG). **Malawi**. N: Mzimba Dist., Mzuzu, Marymount, fr. 24.x.1973, *Pawek* 7442 (BR, LG, MO, PRE, WAG). S: Zomba Plateau, Chagwa dam, fl. 14.iv.1989, *Iversen & Martinsson* 89235 (UPS).

Also in Cameroon, tropical NE and E Africa, Congo, South Africa and Madagascar. Lakes, ponds, ditches and streams, often in acidic water, 30–150 cm deep; sometimes associated with *Potamogeton crispus* and *P. octandrus*; 900–2000 m.

Conservation notes: Widely distributed; not threatened.

P. richardii has often been confused with *P. nodosus*. *P. richardii* may be distinguished by its promptly decaying submerged leaves, its floating leaves often with discoloured petiole at the junction with the lamina, which is mostly rounded at the base, denser mature spikes and by its larger fruits (3.2–4.8 × 2.1–3.3 mm broad in *P. richardii*, 2.7–4

Fig. 12.2.**24**. POTAMOGETON RICHARDII. 1, habit (× ½); 2, flowering spike (× 2); 3, flower (× 10); 4, flower, longitudinal section (× 10); 5, stamen and tepal (× 10); 6, fruiting spike (× 1); 7, fruitlet (× 4); 8, seed (× 8). All from *Lebrun* 7909. Drawn by A. Cleuter. Reproduced with permission from Flore d'Afrique Centrale (Meise).

× 1.7–2.7 mm in *P. nodosus*). The best diagnostic characters are in the stem anatomy – *P. richardii* has endodermis cells with U-thickenings and interlacunar bundles are present, while in *P. nodosus* the endodermis cells have O-thickenings and there are no interlacunar bundles.

ZOSTERACEAE

by S.O. Bandeira

Perennial marine herbaceous plants adapted to live in shallow sea water, monoecious or dioecious, rhizomes creeping or tuberous, nodes with intervaginal scales. Leaves distichous, linear or filiform; leaf sheath open or closed, often with stipuloid flanges, with an auriculate ligule at junction with the blade; blade parallel-veined, sometimes with a midrib. Inflorescence of 1 to several spadices, each enclosed in a spathe formed by sheath of next leaf; flowers small, sessile in two rows on a flattened axis; petals absent, stamen 1, sessile, dorsifixed, anther with longitudinal slits, pollen filamentous, pollination hydrophilous, pollen transported usually in aggregate 'search vehicles' either at or under the water surface. Pistillate flower protogynous, gynoecium 1, 1-locular with basally united styles and 2 stigmatic arms; ovule single, pendulous, orthotropous, bitegmic. Fruit irregularly dehiscing, a small irregular drupe; endosperm absent.

A family of three genera (*Heterozostera*, *Phylospadix* and *Zostera*) with about 17 species, in warm and cool coastal waters.

Zosteraceae here is treated as a separate family but was formerly considered part of Najadaceae (as in Flora of Tropical Africa and Flora Capensis) or Potamogetonaceae (in Flore de Madagascar and Sea-Grasses of the World).

ZOSTERA L.

Zostera L., Sp. Pl.: 968 (1753). —Setchell in Proc. Nat. Acad. Sci. **19**: 810 (1933). —Den Hartog, Sea-grasses of World: 42 (1970). —Les, Moody, Jacobs & Bayer in Syst. Bot. **27**: 468–484 (2002).

Nanozostera Toml. & Posl. in Taxon **50**: 429–437 (2001).

Submerged marine plants; rhizome creeping, not looping, monopodial, dense, rooting at the nodes, with extended internodes; erect stems annual, dense, glabrous, sterile stems short and arising from axils of rhizomatous leaves; flowering stems terminal or lateral, sympodial, longer than sterile stems, branched, each branch with an inflorescence. Leaves distichous, linear; leaf-sheath compressed, embracing the stem, membranaceous, auriculate and ligulate, persisting longer than the leaf-blade; leaf-blade linear, with 3–5 longitudinal veins, margins entire or somewhat denticulate distally. Inflorescence terminal or lateral, forming erect annual branches, each usually terminating in a fertile leaf with the sheath containing a sessile, flattened, linear spadix with or without bract-like lobes (retinacula) that fold over one or more flowers. Flowers in two lines of alternating staminate or pistillate flowers on one surface of the spadix; anthers single, sessile, of 1–3 locules joined by a ridge-like connective; ovary uniovulate with persistent style, stigmas protruding from sheath at anthesis. Fruit membranaceous, cylindrical to ovoid, 1-seeded, indehiscent or irregularly dehiscent; pericarp scarious; seed ellipsoid or ovoid, smooth or ribbed; embryo grooved.

Mostly found in temperate seas; 11 species worldwide, 1 along the East African coast.

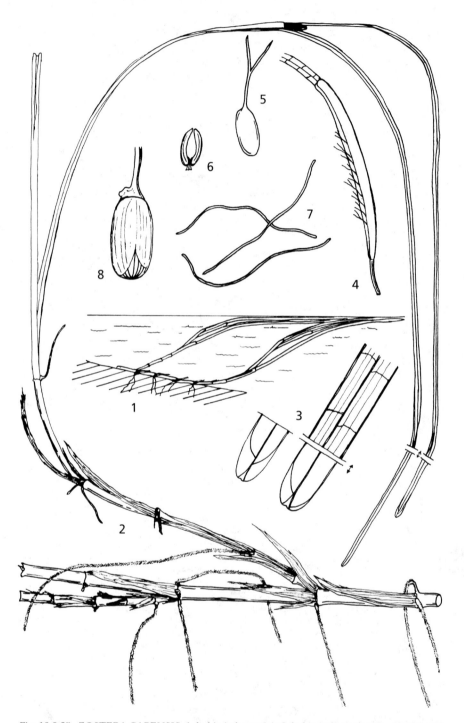

Fig. 12.2.**25**. ZOSTERA CAPENSIS. 1, habit (schematic); 2, habit (× 3); 3, details of leaf (× 15); 4, inflorescence (× 4); 5, carpel (× 16); 6, anther (× 16); 7, pollen (× 40); 8, fruit with base of style (× 10). 2–3 from *Greenway & Rawlins* 8902, 4–5, 8 modified from Flora of Southern Africa, 6–7 from *Zostera marina*. Drawn by H. Beentje. From F.T.E.A.

Zostera capensis Setch. in Proc. Nat. Acad. Sci. Washington **19**: 815 (1933). —Adamson in Fl. Cap. Penins.: 37 (1950). —Obermeyer in F.S.A. **1**: 58, fig.14 (1966). —Bandeira in Guide Seashores E Africa: 64 (1997). —Beentje in F.T.E.A., Zosteraceae: 1 (1999). Type: South Africa, Cape, Knysna lagoon, i.1933, *Duthie* s.n. in DAV 501307 (UC holotype, L). FIGURE 12.2.**25**.

Zostera nana sensu A. Bennett in F.C. **7**: 50 (1897); in F.T.A. **8**: 225 (1901), non Roth.
Zostera marina var. *angustifolia* sensu A. Bennett in F.C. **7**: 50 (1897), non Hornem.
Nanozostera capensis (Setch.) Toml. & Posl. in Taxon **50**: 432 (2001).

Perennial marine herb; rhizome slender, yellowish-red, 0.5–2 mm in diameter with 1–2(3) roots at each node, internodes 3–35 mm; stems of leafy branches arising from axils of rhizome leaves, leaves 2–5; stems of fertile branches developing from lateral branches, 0.5–10(25) cm high with 1–7 spathes. Leaves dark green, 2–25 cm long, (0.5)1.5(2.5) mm wide, 3-nerved, with many parallel translucent veins, midrib widening at apex, apex asymmetrically rounded or notched, eventually emarginate; sheath membranous, 1–5(7) cm long, open, with membranous flaps, more enveloping proximally, biauriculate, 3-nerved. Spathes pedunculate, peduncle joined with axis for up to 3.5 cm, free for 1.4–3.3 mm, amplexicaul; spathe-blade to 6.5(10) cm long, deciduous after flowering; spadix 1–1.5 cm long, 1–3 mm wide, with 6–10 flowers, bract-like lobes obovate to triangular, 0.5–1.5 mm long, 0.5–1 mm wide. Stamens shed after liberation of pollen. Pistillate flowers with ovary 1–2 mm long, style 1–1.5 mm long, stigmatic arms 1.5 mm long. Fruit indehiscent, yellowish-brown, narrowly ovoid, 2–2.5 mm long, 1 mm wide, faintly 20–24-striate; seed ellipsoid, flattened at one end, 2–2.5 mm long with reddish brown testa.

Mozambique. N: Ilha de Moçambique, 28.xi.2007, *Massingue-Manjate* 204 (LMU); Mecúfi village, 11.xii.2007, *Bandeira* 2682 (LMU). M: Maputo Bay, Inhaca Is., x–xi.1962, *Mogg* 30099 (K); Inhaca Is., Saco da Inhaca, 26°03.60'S, 32°55.28'E, 3.vii.1995, *Bandeira* 1142 (K, LMU).

Also found in Kenya, Tanzania, Madagascar and South Africa. In shallow calm waters, on mud flats, edges of channels, occasionally on sand; exposed at very low tide; at sea-level.

Conservation notes: Widespread species but distribution limited to a threatened habitat. Globally it has been reported as being Vulnerable, or even Endangered.

After extensive searching, no fertile material has been found on the Mozambique coast. Apparently, variability in salinity and temperature trigger flowering. The description and illustration of inflorescences are from South African material, but there is no reason to believe this to be a different taxon. As Setchell (1933) states, species are distinct in both vegetative and sexual characters, and the vegetative features of material from Mozambique certainly correspond with that of *Zostera capensis* in South Africa. Fertile material has also not been found in Tanzania and Kenya despite extensive searches.

Zostera capensis can be confused with *Halodule wrightii*, which has very similar leaves and a usually sympatric occurrence. However, *Z. capensis* leaves have rounded emarginated tips and translucent transverse veins, while *Halodule wrightii* has a tridentate apex and is without translucent transverse veins.

Excluded species

Zostera stipulacea Forssk., Fl. Aegyt.-Arab.: 158 (1775) is a synonym of *Halophila stipulacea* (Forssk.) Asch. (see Simpson in F.T.E.A., Hydrocharitaceae: 27–28, 1989).

ZANNICHELLIACEAE

by M.C. Duarte

Annual or perennial submerged aquatic herbs found in fresh or brackish water, never marine. Plants monoecious or (sometimes in *Lepilaena*) dioecious. Roots fibrous, non-septate, unbranched, arising singly or in groups from rhizomes or at lower nodes. Stems glabrous, much-branched, thread-like, arising from a creeping, sympodially (rarely monopodially) branched rhizome. Leaves entire, sessile, alternate, opposite, in 2 rows or pseudo-whorled, glabrous, linear, 1(3)-veined, those on the rhizome scale-like, membranous and without vascular tissue; foliage leaves with a basal membranous sheath (sometimes interpreted as stipules) extended beyond insertion of blades as a ligule-like often auricular structure, or (in *Zannichellia*) free from all or most of the blade, or obsolete. Intravaginal scales mostly filiform, usually 2, inserted laterally at each node. Inflorescence sympodial, usually with 2–several small, axillary flowers. Flowers inconspicuous, short-pedicellated, unisexual. Male flower without perianth (in *Zannichellia* and *Pseudalthenia*), or with a minute 3-lobed perianth (in *Althenia* and *Lepilaena*). Stamen 1 (often interpreted as fused stamens) with 2-, 4-, 8- or 12-sporangiate anthers, opening by one or more longitudinal slits; connective sometimes prolonged into a short appendage; pollen grains globose, trinucleate, without an aperture, monosulcate, often in a gelatinous matrix. Female flowers with (1)4–5(8) separate and short-stalked carpels, surrounded by a membranous tubular perianth (in *Zannichellia* and *Pseudalthenia*) or by 3 separate scale-like segments (in *Althenia* and *Lepilaena*), the segments sometimes bifid. Style simple, short to long; stigma enlarged, peltate, funnel-shaped, cup-shaped, or feathery. Ovule solitary, bitegmic, pendulous (placentation apical), anatropous, crassinucellar. Fruit a drupetum; fruitlets indehiscent (in *Lepilaena*, *Zannichellia*) or dehiscent (in *Althenia*, *Pseudalthenia*); outer layers of the pericarp soft and becoming eroded, the inner persisting. Seed solitary; embryo curved, with a simple coiled cotyledon, endosperm development helobial, absent in mature seed. Germination epigeal.

The morphology of the flower is controversial. Some authors interpret the group of carpels as a flower, hence the cupular sheath can be considered as the perianth and the fruit is multiple (a drupetum), while others interpret the group of carpels as an inflorescence (each carpel corresponding to a flower) with the cupular sheath being considered an involucre or tubular bract and with the fruit simple (a drupe). The first interpretation is adopted here.

The classification of the 'fruits' is doubtful. Several authors consider the 'fruit' to be an achene, but the pericarp being composed of soft outer layers (becoming eroded) and a hardened (persistent) inner wall points to its classification as a drupe. However, the fruit, usually referred to as indehiscent, is dehiscent in *Althenia* and *Pseudalthenia*.

The taxonomic position and affinities of Zannichelliaceae are uncertain. Tomlinson & Posluszny (in Taxon **25**: 273–279, 1976), Tomlinson (Anatomy Monocot., VII Helobiae: 340, 1982) and more recently Kubitzki (Fam. Gen. Vasc. Pl., Monocot.: 472, 1998), restrict the family to genera from fresh or brackish water with spheroidal or isobilateral pollen. Authors such as Hutchinson (Fam. Flow. Pl., ed.2, **2**: 559, 1959) also include the wholly marine genera with thread-like pollen, *Amphibolis*, *Cymodocea*, *Halodule*, *Syringodium* and *Thalassodendron*, now generally placed in the Cymodoceaceae (as in F.S.A.). However, the phylogenetic relationship between Zannichelliaceae and Potamogetonaceae has been recently reinforced (Les *et al.* in Syst. Bot. **22**: 443–463, 1997; Angiosperm Phylogeny Group in Ann. Missouri Bot. Gard. **85**: 531–553, 1998; Chen *et al.* in Bot. Bull. Acad. Sin. **45**: 33–40, 2004). As there is still not full agreement, the more conservative concept is retained here and Zannichelliaceae is recognized as a separate family, as in F.T.E.A.

The family comprises 4 genera: *Zannichellia* (cosmopolitan with an uncertain number of species; some consider there to be 1 or 2 highly polymorphic species while others accept as much as 6), *Althenia* (1 species in W Mediterranean and N and S Africa), *Lepilaena* (4 species in Australia and New Zealand), and *Pseudalthenia* (= *Vleisia*) (1 species in South Africa).

ZANNICHELLIA L.

Zannichellia L., Sp. Pl. **2**: 969 (1753).

Annual, rarely perennial, herbs of fresh or brackish water. Stems not clearly differentiated into rhizomes and erect shoots. Roots single or in pairs at the nodes. Leaves opposite, alternate or the furthest in pseudowhorls of 3(4), mostly less than 1 mm wide, 1(3)-veined, the furthest younger leaves with a membranous sheath free from the blade. Monoecious. Inflorescence usually of 1 male and 1 female flower. Male (staminate) flowers naked; anther (2)4(8)-sporangiate; connective prolonged into a blunt appendage. Female (carpellate)flowers with (1)4–5(8) carpels surrounded basally by a membranous tubular perianth. Style short, less than 1 mm long; stigma asymmetric, funnel or cup-shaped, oblique-peltate with wavy margins. Fruit an elongated drupelet, crowned by the persistent style. Endocarp often with a irregularly toothed or warty ridge on the dorsal and occasionally on the ventral side. Chromosome numbers: 2n = 12, 24, 28, 32, 34 and 36.

Two species, 1 with a worldwide distribution (sometimes regarded as up to 6 polymorphic species) and 1 endemic to the W Cape of South Africa.

Zannichellia palustris L., Sp. Pl.: 969 (1753). —Bennett in F.T.A. **8**: 225 (1901). — Obermeyer in F.S.A. **1**: 77 (1966). —Beentje in F.T.E.A., Zannichelliaceae: 1, fig.1 (2000). —Clarke & Klaassen, Water Pl. Namibia: 66 (2001). —Cook, Aq. Wetl. Pl. Sthn. Africa: 265 (2004). Type: Sweden, "Habitat in Europae, Virginiae fossis, fluviis" (LINN 1085.1 lectotype). FIGURE 12.2.**26**.

Submerged annual brittle herb, forming dense lawn-like mats. Leaves 2–5 cm long, with pointed apices. Axillary scales filiform, sometimes lanceolate. Anther 4 or 8-sporangiate, filament elongating at anthesis. Carpels 1–6. Style accrescent in fruit to 1 mm long. Drupelet slightly incurved, c.2 mm long (excluding beak), stalked, laterally compressed. Endocarp with a irregularly toothed or warty ridge in the dorsal side, ventrally inconspicuous.

Botswana. N: NW Ngamiland, Qangwa village, fl.& fr. 29.iv.1980, *P.A. Smith* 3537 (K, LISC). SE: Boteti Delta area, NE of Mopipi, 850 m, fr. 21.iv.1973, *Tyers* 20 (K, SRGH).

Widespread in temperate and tropical latitudes worldwide, including east tropical Africa (Tanzania), N and C Namibia and over much of South Africa. Can be locally dominant, forming large lawn-mats in still or slow-moving, shallow fresh or brackish water, sometimes associated with stoneworts; c.1000 m.

Water-pollinated, flowering Oct–May. A useful food source and refuge for aquatic wildlife.

Conservation notes: A widespread species, although in the Flora area apparently localized to areas of inland brackish water; not threatened.

The species is highly variable and some authors have recognized infraspecific taxa from some regions. The scarce material available from the Flora area prevents a more detailed treatment.

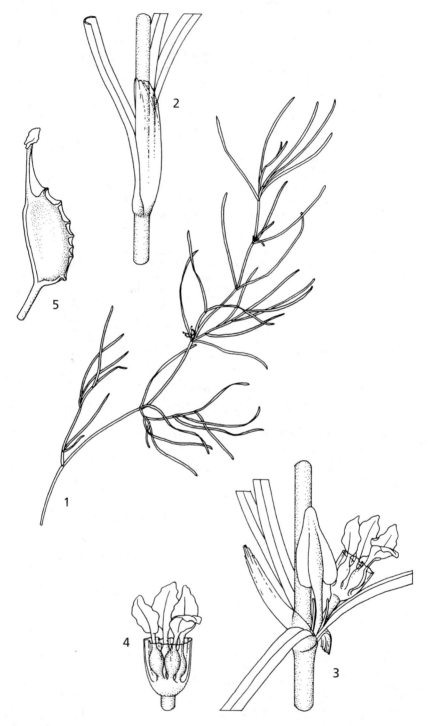

Fig. 12.2.**26**. ZANNICHELLIA PALUSTRIS. 1, habit (× ³/₄); 2, node showing membranous stipular sheath (× 8); 3, inflorescence showing staminate flower, carpellate flowers and scales (× 18); 4, carpellate flowers within membranous tubular perianth (× 20); 5, fruit, with persistent style (× 14). All from *P.A. Smith* 3537. Drawn by Deborah Lambkin.

CYMODOCEACEAE

by M.C. Duarte

Perennial, dioecious, submerged marine herbs ('sea-grasses'). Rhizomes creeping, leafy or scale-bearing, herbaceous and monopodial (*Cymodocea, Syringodium, Halodule*) or woody and sympodial (*Amphibolis, Thalassodendron*). Scales scarious, marked with ± small, dark, longitudinal stripes and dots (tannin cells). Roots branched or unbranched, with few to many root hairs. Leaves alternate, in 2 rows, with distinct blade and sheathing base; leaf sheath broad, open, embracing the stem, leaving open or closed circular scars when shed, auriculate; scarious flaps with numerous dark, longitudinal stripes and dots (tannin cells); ligule (at junction of sheath and blade) present. Leaf blade linear, flat or needle-like (*Syringodium*), with 3 to several parallel or pseudoparallel (*Amphibolis*) nerves, with ± short, dark, parallel longitudinal stripes and dots (tannin cells); apex very variable. Squamules (axillary scales) usually present at nodes. Flowers small, hydrophilous, without perianth, enclosed by leaflike bracts, usually solitary and terminal on erect shoots or branches of erect shoots, or in cymose inflorescences (*Syringodium*). Male flowers subsessile or stalked (the stalk elongating at anthesis), consisting of two tetrasporangiate anthers (8 microsporangia per flower) paired on a common filament; anthers dorsally joined over at least a part of their length and attached either at the same height or at a slightly different level (*Halodule*), extrorsely dehiscent by longitudinal slits, each anther with or without an apical appendage; pollen grains filiform, threadlike, tightly coiled within the anther, trinucleate, and without an exine. Female flowers sessile or shortly stalked (the stalk never elongating at anthesis), each consisting of 2 free ovaries with either a long unbranched style (*Halodule*) or a short style divided into 2–3 filiform stigmas; carpels 1-ovulate, usually only 1 developing into a fruit; ovule bitegmic, ± orthotropous or anatropous (in *Thalassodendron*), pendulous. Fruit indehiscent, either unspecialized with a stony pericarp (*Cymodocea, Halodule, Syringodium*) or viviparous with a stony endocarp and a fleshy exocarp out of which 4 spreading pectinate lobes grow (*Amphibolis*), or consisting of a fleshy bract enclosing the fertilised ovaries (*Thalassodendron*). Seed 1, without endosperm. Embryo often very specialized.

Family consisting of 5 genera with 16 species, mainly occurring in tropical or subtropical seas (*Cymodocea, Halodule, Syringodium* and *Thalassodendron*) or in temperate waters (*Amphibolis*, restricted to S Australia and Tasmania).

Sea-grasses from this family are an important food source for fish and dugongs.

The family delimitation here follows Kuo & McComb in Kubitzki (Fam. Gen. Vasc. Pl., Monocot.: 133–140, 1998), a delimitation also proposed by Tomlinson (in Metcalfe, Anatomy of Monocots., 1982) and Dahlgren, Clifford & Yeo (Families of Monocots., 1985) in contrast to authors such as Hutchinson (Fam. Fl. Pl., Monocot., ed.2: 559, 1959), who suggest the association of Cymodoceaceae with Zannichelliaceae.

1. Rhizome herbaceous, monopodial, with a short erect stem at each node; leaf sheath persisting longer than leaf blade; anthers stalked · · · · · · · · · · · · · · 2
– Rhizome ligneous, sympodial, with elongate, unbranched or ± branched erect stems arising at intervals of 4 internodes; leaf blade shed with its sheath; anthers subsessile · **4. Thalassodendron**
2. Leaves flat; flowers solitary · 3
– Leaves needle-like; flowers in a cymose inflorescence · · · · · · · · **1. Syringodium**
3. Roots unbranched; leaf blade 3-veined; anthers not attached at the same height; style undivided · **2. Halodule**
– Roots branched; leaf blade 7–17-veined; anthers attached at the same height; style divided into 2 stigma · **3. Cymodocea**

1. SYRINGODIUM Kütz.

Syringodium Kütz. in Hohenacker, Meervalgen (Alg. Mar. Sicc.) **9** n° 426 (1860). — Hartog in Sea-Grasses of World (Verh. Kon. Ned. Akad. Wetensch., Afd. Natuurk., Tweede Sect. **59**): 176 (1970).

Marine herbs, rhizome monopodial, with 1–4 branched or unbranched roots and a short erect stem bearing (1)2–3 leaves at each node; internodes 1–5 cm. Scales ovate or elliptic, acute. Leaf sheath 1–7 cm long, open, persisting longer than blade; leaf scars open, circular; auricles obtuse. Leaf blade terete, often narrowed at base. Flowers in distinct terminal inflorescences on erect shoots, inflorescence cymose, the lower branches dichasial, the higher ones monochasial, so forming a cymose inflorescence. Each flower enclosed in a sheathing bract resembling a foliage leaf, sheath up to 9 × 3 mm, elliptic to ovate and inflated, lamina of bract 5–20 mm long, gradually decreasing towards top of inflorescence. Male flower on a stalk up to 10 mm long, with 2 anthers dorsally united on their lower parts, attached at the same height, without apical appendages. Female flower sessile, with 2 free carpels, each with a short style and two slender stigma. Fruit ellipsoidal, quadrangular in cross-section, with a stony pericarp, dorsal inconspicuous ridge and a short beak.

A genus with 2 species, *S. isoetifolium* in the Indo-Pacific Ocean, and *S. filiforme* Kütz. confined to the Caribbean and Gulf of Mexico.

Syringodium isoetifolium (Asch.) Dandy in J. Bot. **77**: 116 (1939). —Macnae in Macnae & Kalk, Nat. Hist. Inhaca Is.: 28 (1969). —Sartoni in Fl. Somalia **4**: 19, fig.12 (1995). —Bandeira in Guide Seashores E Africa: 66 (1997). —Beentje in F.T.E.A., Cymodoceaceae: 5, fig.2 (2002). Type: India, ?near Madras, presumably Tuticorin (Tuticoreen) Is, 2.ix.1871, *Wight* 2413 (K holotype). FIGURE 12.2.**27**.
  Cymodocea isoetifolia Asch. in Sitzungsber. Ges. Naturf. Freunde Berl. **1867**: 3 (1867). —Bennett in F.T.A. **8**: 229 (1901).

Marine herb. Rhizome slender, 1–2.5(3) mm in diameter; roots densely covered with root-hairs. Leafy shoots with internodes 1–4 cm; leaf sheath 1–4 cm long, 1.5–4 mm wide, cylindrical, enlarging towards top; blade fleshy, 4–30 cm × (0.5)1–2.5 mm, narrowed at base, pale green, apex somewhat flattened and conduplicate, minutely toothed. Scales (squamules) absent. Sheath of inflorescence bract up to 7 mm long. Male flower on 7 mm long stalk; anthers 4 mm long, ovate. Female flower sessile; ovary 3–4 mm long, ellipsoid; style 2 mm; stigma 4–8 mm long. Fruit obliquely ellipsoid, 3.5–4 × 1.75–2 × 1.5 mm; beak 2 mm long, bifid. [Flower and fruit descriptions from Hartog 1970].

Mozambique. N: Pemba, Wimbe beach, 23.x.1993, *Bandeira & Boane* 452 (LMU). GI: Inhambane, Inhanombe estuary, 20.i.1954, *Schelpe* 4482 (BM). M: Inhaca Is., Saco, 31.viii.1959, *Watmough* 337 (BM, K, LISC, SRGH).

Widely distributed in the Indian Ocean from the Red Sea south to Mozambique and Madagascar, Mauritius and Seychelles, Persian Gulf and southern coast of India; also in Malaysia far into the W Pacific from the Ryukyu Islands southwards to New Caledonia and E and W coasts of Australia. In sandy and muddy habitats, on coral beds and sandstone conglomerate rocks, below the low-water mark (2–3 m deep, occasionally up to 7 m). Forming marine meadows.

Conservation notes: Widespread species; not threatened.

Although flowering occurs in Mozambique (S. Bandeira, pers. comm.), no flowering specimens have been seen.

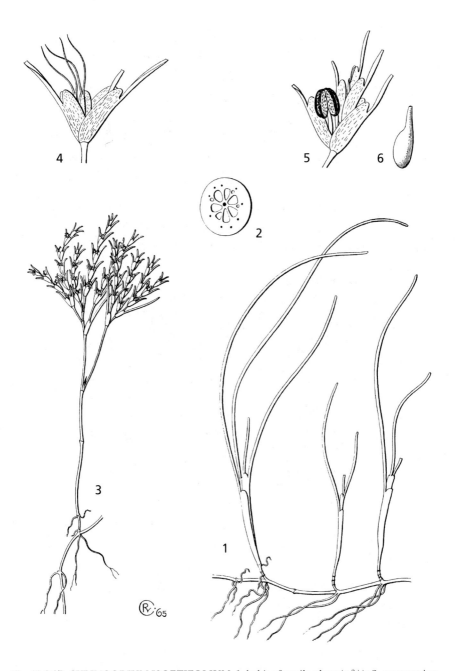

Fig. 12.2.**27**. SYRINGODIUM ISOETIFOLIUM. 1, habit of sterile plant ($\times$ ^{2}/$_3$); 2, cross-section of leaf ($\times$ 15); 3, habit of flowering plant ($\times$ 1); 4, female flower ($\times$ 3); 5, male flower ($\times$ 3); 6, fruit ($\times$ 3). 1–2 from *Schodde* 3015 (Port Moresby), 3 from *Gardiner* CP 2380 (Galle), 4, 6 from *Doty & Velasquez* 16941 (Albany Gulf), 5 from *Børgesen* s.n. (Tuticorin). Drawn by Ruth van Creval. From den Hartog, Sea-grasses of the World, figs. 50 & 51. Reproduced with permission of the Royal Netherlands Academy of Arts and Sciences.

2. HALODULE Endl.

Halodule Endl., Gen. Pl., Suppl. **1**: 1368 (1841). —Hartog in Blumea **12**: 289 (1964). —Hartog in Sea-Grasses of World: 146 (1970).

Diplanthera Thouars, Gen. Nov. Madagasc. **2**: 3 (1806), nom. illegit.

Marine herbs, rhizome monopodial, the nodes with 1-several unbranched roots and a short erect stem bearing 1–4 leaves; internodes 0.5–6 cm. Scales scarious. Leaf sheath 1–6 cm long, scarious, shed after leaf blade leaving a circular scar; auriculate; ligule a narrow ridge; leaf blade linear, entire, often narrowed at base, 3-veined, with conspicuous midrib, widened or forked at apex; leaf apex very variable in outline, rounded or with teeth. Scales (squamules) 2. Flowers solitary, terminal on short erect shoots, subtended by a terminal leaf similar to others; bud in axil of penultimate leaf develops into a prolongation of main axis giving rise to a sympodium. Male flower subsessile or stalked, the 2 anthers attached at different levels. Female flower subsessile, consisting of 2 free carpels each with a long simple style; stigma filiform. Fruit with a stony pericarp, globose, subglobose or ovoid, somewhat compressed, with a short beak.

Genus with about 7 species, widely distributed along the coasts of all tropical seas.

Halodule uninervis (Forssk.) Boiss., Fl. Orient. **5**: 24 (1882). —Obermeyer in F.S.A. **1**: 76, fig.22 (1966). —Macnae in Macnae & Kalk, Nat. Hist. Inhaca Is.: 28 (1969). —Sartoni in Fl. Somalia **4**: 16, fig.9 (1995). —Bandeira in Guide Seashores E. Africa: 66 (1997). —Beentje in F.T.E.A., Cymodoceaceae: 6, fig.3 (2002). —Cook, Aq. Wetl. Pl. Sthn. Afr.: 81 (2004). Type: Yemen, near Al Mukha (Mocha), *Forsskål* s.n. (no material found). FIGURE 12.2.**28**.

Zostera uninervis Forssk., Fl. Aegypt.-Arab.: 157 (1775).

Diplanthera uninervis (Forssk.) F.N. Williams in Bull. Herb. Boiss., sér.2 **4**: 221 (1904).

Halodule wrightii sensu Hartog in Sea-Grasses of World: 154 (1970) in part. —sensu Bandeira, Guide Seashores E. Africa: 66 (1997), non Asch.

Perennial marine herb. Rhizome rather fleshy, 0.5–1.5 mm thick, roots 1–8, densely covered with root-hairs, internodes 0.5–5 cm long. Leaf sheath 1–4 cm long; auricles c.0.5 mm long; ligule becoming thick and dark after leaf-blade shed; leaf blade (2)3.5–10(17) cm × 0.2–1 mm wide in narrow-leaved forms and 1–2.5 mm wide in wide-leaved forms, sometimes falcate in the latter; midrib widening and sometimes forked near apex; leaf apex very variable: in narrow-leaved forms apex usually tridentate with 2 acute lateral teeth (equal or differing in size) and a median tooth (as long as or shorter than lateral teeth, rarely slightly longer) ending the midrib, tooth sometimes absent and apex bicuspidate; in wide-leaved forms apex frequently rounded or almost truncate with ± irregular serrations and 2 small acute lateral teeth and a minute median apiculum in which the midrib ends, lateral tooth usually shorter than apex tip, rarely equal or longer. Male flower on stalk up to 23 mm long; anthers 4–6 mm long, unequal, upper attached 0.25–0.5 mm above lower. Female flower with globose, subglobose or ellipsoid ovary, 1–2 mm wide; style 10–12 mm long, sub-terminal or lateral. Fruit subglobose, 1.5–2 mm wide, with an apical beak up to 1.5 mm long.

Mozambique. N: Cabo Delgado, Mecúfi, 21.x.1993, *Bandeira & Boane* 363 (LMU). GI: Inhambane Estuary, 20.i.1954, *Schelpe* 4481 (BM). M: W coast of Inhaca Is, ♀ fl., 17.i.1952, *Moss* 20793 (K).

Widely distributed along the coasts of the Indian Ocean, from the Red Sea to KwaZulu-Natal in South Africa, Madagascar, Mauritius and Seychelles, southern coast of Asia (Oman, Persian Gulf, India, Sri Lanka) and the western Pacific (from Japan as far south as Tonga and Queensland in Australia). On sandy or muddy substrates on coral reefs in the sublittoral to intertidal zone and can descend to a considerable depth. A locally abundant pioneer species, often associated with several other sea-grasses. An important species in stabilizing sediments.

Conservation notes: Widespread species; not threatened.

As in other geographic areas, two morphological forms based on leaf width are recognized. Data collected from Mozambican specimens suggest the possible relationship of these variants with ecological factors: specimens with narrow leaves

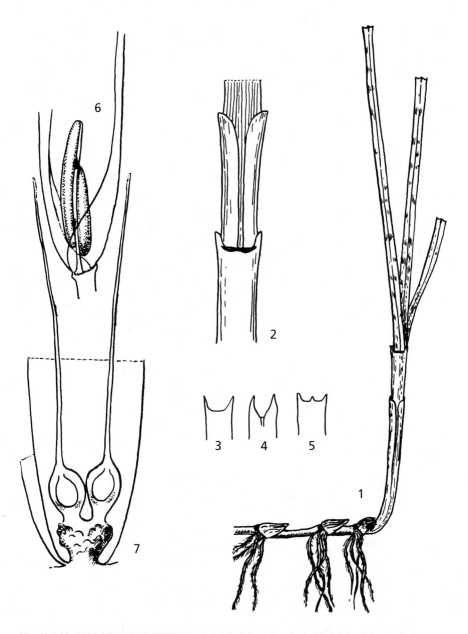

Fig. 12.2.**28**. HALODULE UNINERVIS. 1, habit; 2, leaf sheaths (× 2); 3–5, leaf tips (× 6); 6, young male flower (× 9); 7, female flower (× 10). 3–4 from *Verdcourt* 1066, 5 from *Isaac* A123. Reproduced with permission from Flora of Southern Africa (3–5 drawn by Henk Beentje in F.T.E.A.).

occur particularly in shallow pools and are exposed at low tide, whereas those with broader leaves are related mainly with deeper waters and are always submersed. Further studies, including the analysis of rare reproductive structures, are needed to clarify the taxonomy.

Halodule wrightii Asch. was reported to East African coasts including Mozambique by Hartog (1964, 1970), Macnae (1969) and Bandeira (1997). However, the presence of this species in the Indian Ocean was rejected by authors such as Phillips & Meñez (Smithsonian Contr. Marine Sci. **34**: 1–104, 1988) and Beentje (2002) based on differences in leaf apex. The wide range of variation of leaf tip morphology, and also genetic and chemotaxonomic data support this conclusion. *H. wrightii* only occurs in the Atlantic Ocean, and records of its presence on the East African coast probably refer to *H. uninervis*.

3. CYMODOCEA K.D. Koenig

Cymodocea K.D. Koenig, Ann. Bot. (König & Sims) **2**: 96 (1805).
—Hartog in Sea-Grasses of World: 160 (1970).

Marine herbs, rhizome monopodial, with 1–5 ± branched roots and a short erect stem with 2–7 leaves at each node; internodes 1–6 cm long. Scales ovate. Leaf sheath compressed, 1.5–10 cm long, persisting longer than leaf blade, leaving an open or closed circular scar when shed; ligule up to 1 mm long. Leaf blade linear, up to 30 cm long, leaf apex obtuse or sometimes emarginate, ± serrulate or denticulate; veins 7–17, joined at apex by a marginal commissure. Scales variable in number (up to 10) in two groups at each node. Flowers solitary and terminal, subtended by a leaf similar to the others; an axillary bud at the penultimate leaf develops into a prolongation of the main axis giving rise to a sympodium. Male flowers with a stalk elongating considerably at anthesis, the 2 anthers united dorsally and attached at the same height. Female flowers sessile or shortly stalked; each ovary with a short style divided into 2 slender stigmas. Fruit with a stony pericarp, semi-circular to semi-ovate or elliptic, laterally compressed, with dorsal ridges and a beak.

A genus of 4 species with a wide but disjunct distribution in tropical and subtropical seas (*Cymodocea serrulata* and *C. rotundata* in the Indo-Pacific, *C. nodosa* in the Mediterranean extending to S Portugal and the E Atlantic as far south as the Canary Islands and Mauritania, and *C. angustata* in NW and W coasts of Australia).

Leaf scar closed; leaf sheath cylindrical; leaf blade 1.5–4 mm wide, 9–15-nerved; leaf
  apex obtuse, occasionally serrulate; fruit with coarsely dentate dorsal ridges · ·
  · **1.** *rotundata*
Leaf scar open; leaf sheath obconical; leaf blade 3–8 mm wide, 13–17-nerved; leaf
  apex obtuse with conspicuous marginal teeth; fruit with dorsal blunt ridges · · ·
  · **2.** *serrulata*

1. **Cymodocea rotundata** (Ehrenb. & Hempr.) Aschers. in Sitzungsber. Ges. Naturf.
   Freunde Berlin **1870**: 84 (1870). —Bennett in F.T.A. **8**: 230 (1901). —Macnae in
   Macnae & Kalk, Nat. Hist. Inhaca Is.: 28 (1969). —Sartoni in Fl. Somalia **4**: 17,
   fig.10 (1995). —Bandeira in Guide Seashores E. Africa: 66 (1997). —Beentje in
   F.T.E.A., Cymodoceaceae: 7 (2002). Type: Sudan, Bei Suakin, 5.ix.1868,
   Schweinfurth 188 (K holotype, BM).
   Phucagrostis rotundata Kuntze, Rev. Gen. Pl. **2**: 744 (1891).

Marine herb. Rhizome fleshy, with 1–3 ± branched roots and a short erect stem, bearing up to 3(4) leaves at each node; internodes 1–5.5 cm long. Scales c.0.5(1) cm long. Leaf sheath cylindrical, 1.5-6.5 cm long; auricles acute, sometimes obtuse, c.0.5 mm long; old sheaths forming a scarious mass, leaving closed circular scars on stem when shed; ligules c.0.5 mm high. Leaf blade sometimes slightly falcate, 3–10(16) cm × 1.5–4 mm, enlarging slightly in apical area; apex obtuse, occasionally slightly emarginate, faintly serrulate; margins entire, rarely remotely serrulated; nerves 9–15, accessory veins between each pair generally inconspicuous; marginal nerves reaching apical area. Scales 2 or more in two groups at each node. Male flower with stalk 2–3 cm long; anthers 11–15 mm long, crowned by a needle-like appendage. Female flower sessile, ovary very small, gradually passing into style; stigma spirally coiled, at least 30 mm long. Fruits 1–2, 9–10 × 5–6 mm, 1.5 mm thick, semicircular and laterally compressed, with 3 dorsal, parallel ridges, middle ridge with 6–8 conspicuous teeth and ventral with 3–4 teeth; apical beak somewhat oblique, 2 mm long. [Flower and fruit descriptions mainly from Hartog 1970.]

Mozambique. N: Cabo Delgado, Mecúfi, 21.x.1993, *Bandeira & Boane* 338 (LMU). GI: Inhambane estuary, 22.i.1954, *Schelpe* 4497 (BM). M: Inhaca Is., Saco, E side, 31.viii.1959, *R.Watmough* 328 (BM, K, SRGH).

Widely distributed the Indian Ocean coasts from the Red Sea to Mozambique and Madagascar, the southern coast of India, Andaman and Nicobar Is, and extending through Malaysia into the W Pacific from the Ryukyu Is. to Queensland and New Caledonia. Occurring on sandy or muddy substrata and in coral areas at low tide zone (2–3 m deep). Can form dense mats and occurs with other sea-grasses. Flowering and fruiting specimens have not been recorded in Mozambique.

Conservation notes: Widespread species; not threatened.

Vegetative parts of *Cymodocea rotundata* are very similar to those of *Thalassia hemprichii* (Ehrenb.) Asch. in the Hydrocharitaceae; *Cymodocea* can be distinguished by the presence of leaf ligules and by the branched roots without fine hairs.

2. **Cymodocea serrulata** (R. Br.) Asch. & Magnus in Sitzungsber. Ges. Naturf. Freunde Berlin **1870**: 84 (1870). —Macnae in Macnae & Kalk, Nat. Hist. Inhaca Is.: 28 (1969). —Sartoni in Fl. Somalia **4**: 18 (1995). —Bandeira in Guide Seashores E. Africa: 66 (1997). —Beentje in F.T.E.A., Cymodoceaceae: 9, fig.4 (2002). Type: Australia, S coast, *Brown* 5813 (BM holotype). FIGURE 12.2.**29**.

   Caulinia serrulata R. Br., Prodr. Fl. Nov. Holl.: 339 (1810).

   Cymodocea acaulis Peter in Abl. Ges. Wiss. Gottingen **13**: 39 (1928). Type: Tanzania, Tanga Dist., Kigombe, *Peter* 39812b (B holotype, GOET).

Perennial marine herb. Rhizome fleshy, with 2(3) sparsely branched roots and short erect stem with up to 5 leaves at each node; internodes 1–4 cm long. Leaf sheath obconical, 1.5–4 cm long; auriculae acute, 1–2.5 mm long; old sheaths leaving open circular scars on stem when shed; ligules 0.5–1 mm high. Leaf blade often falcate, 4–19 cm × 3–8 mm wide; apex obtuse, serrulate to denticulate; margins finely serrulate towards apex; nerves 13–17, with very fine accessory veins between each pair of nerves, marginal nerves not reaching the apical area; midrib more conspicuous than other nerves. Scales c.2 in two groups at each node. Male flowers with a stalk c.2 cm long; anthers 6–9 mm long, not crowned by a needle-like appendage. Female flowers sessile; ovary 1.5 mm long, with curved flattened style 2–4 mm long, divided in 2 slender stigmas 23–27 mm long. Fruit sessile, elliptic, laterally compressed, 7–9 × 3–4.5 mm, 2 mm thick, with 3 dorsal, parallel, very blunt ridges; beak apical, straight, c.1 mm long. [Flower and fruit descriptions mainly from Hartog 1970].

Mozambique. N: Angoche Dist., Mafamede Is., 28.x.1965, *Mogg* 32539 (LISC). GI: Bazaruto archipelago, between Santa Carolina and mainland, xi.1969, *Tinley & Costa* 1870 (K, SRGH). M: Inhaca Is., Ponta Torres, 2.ix.1985, *Groenendijk & Maite* 1904 (LMU).

Common along the Indian Ocean coast from the Red Sea along Eastern Africa as far south as Mozambique and Madagascar. Also in the Seychelles, southern coast of India, Sri Lanka and Malaysia into the W Pacific from the Ryukyu Is south to Queensland. Common on sandy or muddy substrata and on coral areas, below the low-tide level (2–7 m deep); can form marine meadows.

Conservation notes: Widespread species; not threatened.

Flowering specimens have not been recorded in Mozambique.

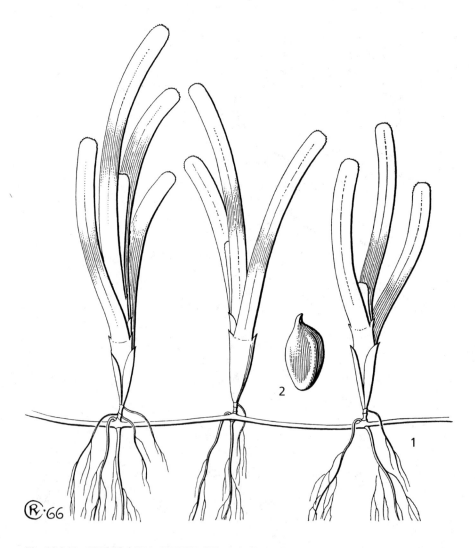

Fig. 12.2.**29**. CYMODOCEA SERRULATA. 1, habit (× 1); 2, fruit (× 3). 1 from *Beccari* 11823 (Wokam), 2 from *den Hartog* 1001 (Thursday Is.). Drawing by Ruth van Creval. From den Hartog, Sea-grasses of the World, fig. 48. Reproduced with permission of the Royal Netherlands Academy of Arts and Sciences.

4. THALASSODENDRON Hartog

Thalassodendron Hartog in Sea-Grasses of World: 186 (1970).

Marine herbs, rhizomes robust, ligneous, sympodial, roots 1–10, ± branched, at the internodes preceding the stem-bearing (3rd) internode and at the stem-bearing (4th) internode. Erect stems 1–2 at every 4th internode, unbranched or little branched, with (5)6–8(9) leaves; internodes up to 3 cm long. Scales deciduous. Leaf sheath compressed, 1–4.5 cm long, open, ligulate, narrowed at base, clasping stem, leaving a closed circular scar when shed. Leaf blade linear, shed together with the sheath, margin ± serrate; nerves 7–29, parallel, connected by fine accessory cross-veins; apex obtuse, sometimes slightly emarginate, denticulate. Scales present in leaf axils. Flowers solitary, terminal on short lateral shoots, enclosed by 3–4 leafy bracts, differing in shape and nervation depending on sex. Male flowers consisting of 2 anthers dorsally united over their whole length, attached at the same height, each with a terminal appendage. Female flowers with 2 free ovaries, each with a short style divided into 2 slender stigmas. Fruit of 1(2) fertilized ovaries and a fleshy innermost bract; germination tending to be viviparous producing a free-floating seedling.

Genus with two species, *T. ciliatum*, widespread in the tropical Indo-West Pacific, and *T. pachyrhizum*, restricted to warm-temperate Western Australia.

Thalassodendron ciliatum (Forssk.) Hartog in Sea-Grasses of World: 188 (1970) for most part. —Sartoni in Fl. Somalia **4**: 18, fig.11 (1995). —Bandeira in Guide Seashores E. Africa: 66 (1997). —Beentje in F.T.E.A., Cymodoceaceae: 2, fig.1 (2002). Type: Yemen, Red Sea, Al Mukha (Moccha), *Forsskål* 26 (C holotype, BM). FIGURE 12.2.**30**.

 Zoostera ciliata Forssk., Fl. Aegypt.-Arab.: 157 (1775).

 Cymodocea ciliata (Forssk.) Asch., Sitzungsber. Ges. Naturf. Freunde Berlin **1867**: 3 (1867). —Obermeyer in F.S.A. **1**: 74 (1966). —Macnae in Macnae & Kalk, Nat. Hist. Inhaca Is.: 28, fig.21 (1969).

Perennial marine herb. Rhizome reddish-brown, angled, up to 0.8 cm wide, vertical rhizomes sometimes present; roots up to 10, little or much-branched. Internodes 0.5–3 cm long. Scales 5–10 mm long, broadly ovate, clasping stem, early deciduous, dark brown; apex obtuse and apiculate. Leafy stems up to 35 cm × 1.5–3.5(4) mm; leaf scars to 20 mm apart at base, the upper successively closer. Leaf sheath 10–17 mm wide, (10)15–30(35) mm long, tapering at base, cream to pink, with obtuse auricles, tannin cells conspicuous, margins scarious; ligule obtuse, 1–1.5 mm high. Leaf blade falcate, 5–15 cm × (5)10–14 mm, often dark brown when dry; nerves 15–29, united below leaf apex, median vein slightly prominent; distal margins irregularly serrate; apical teeth 0.5–1 mm long, acute or truncate, rarely bi- or trifurcate. Scales in two opposite groups of (2)3–5 each, unequal in size, the longer about 2 mm. Flowers on abbreviated side branches near base of leaf-clusters; bracts greenish, often tinged with red at maturity. Outer bracts: 1st bract usually with sheath only, sometimes with a small blade; 2nd bract with a sheath and blade smaller than the sheath, ligulate. Inner bracts: in male plants, 3rd bract shorter than 2nd, with sheathing flaps and without ligule; 4th bract 2 mm long, scale-like, membranous, with 1 median vein; in female plants, 3rd bract same shape as 2nd, ligulate with sheathing flaps, becoming somewhat longer than 2nd; 4th bract fleshy, differentiated into a sheath and slightly longer blade, as long as 3rd bract in full-grown condition. Male flowers: anthers 6–10 mm long, each with a short terminal appendage, sometimes bilobed. Female flowers: ovary ellipsoid, 1–2 × 0.5–0.7 mm wide, stalk 0.3–0.5 mm long; style divided into 2 stigmatic arms 20–30 mm long, strap-shaped. Pollinated under the water surface. Fruit 3.5–6 cm long, oblong; free-floating when ripe; sometimes developing when still attached to the mother plant.

Mozambique. N: Moma Dist., Moma Is., 16°51'S, 39°31'E, ♂ fl. 27.x.1965, *Mogg* 32579 (K, LISC). MS: Mouth of Rio Luabo, 16.v.1858, *Kirk* 10 (K). GI: Bazaruto

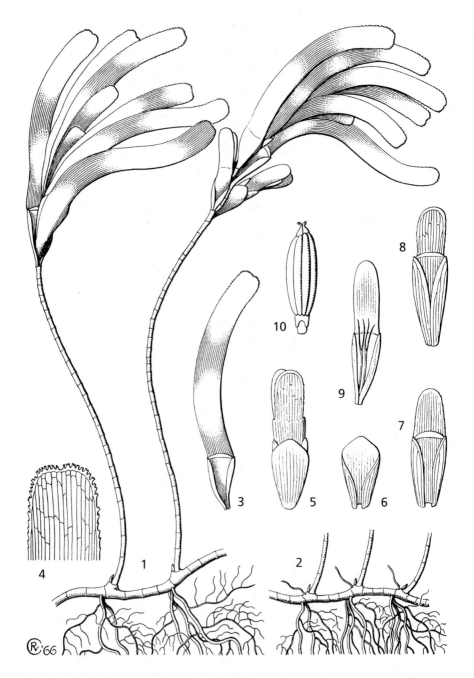

Fig. 12.2.**30**. THALASSODENDRON CILIATUM. 1, habit (× ²/₃); 2, rhizome (× ²/₃); 3, vegetative leaf (× ²/₃); 4, leaf-tip (× 2); 5, cluster of bracts around female flower (× 1); 6–9, bracts around female flower (× 1); 10, male flower (× 4). Drawn by Ruth van Creval. From den Hartog, Sea-grasses of the World, fig. 52. Reproduced with permission of the Royal Netherlands Academy of Arts and Sciences.

archipelago, between Santa Carolina Is. and Inhassoro, x.1969, *Tinley & Costa* 1871 (K, LISC). M: Inhaca Is., ♂ fl.& fr. 15.i.1952, *Moss* 20794 (K).

The species occurs in two disjunct areas: on the coasts of the Red Sea and western Indian Ocean as far as South Africa (KwaZulu-Natal), Madagascar and the Indian Ocean islands (Seychelles, Comoros, Mascarenes, Aldabra, Maldives, Chagos), and on the tropical coasts of the Pacific Ocean in E Malaysia, Solomon Islands and E Australia (Queensland). On sandy bottoms, in the lower intertidal zone, always submerged (3–7 m deep); also in rock pools. Gregarious, forming extensive meadows. Avoids areas where freshwater discharges into the sea.

Conservation notes: Widespread species; not threatened.

Study of *Thalassodendrum ciliatum* specimens has shown that some material previously ascribed to this taxon should be assigned to a new one. This will be published soon.

DIOSCOREACEAE

by P. Wilkin

Rhizomatous perennials, rhizome often reduced to a small crown on top of 1–several annual to perennial starch-rich tubers. Stems herbaceous, twining, lacking tendrils, rarely erect or absent in perennials, with a basal rosette of leaves. Leaves opposite to alternate, entire to compound, base often cordate, veins arising at insertion point of blade on the petiole, spreading then converging towards apex; petiole usually with a basal and apical pulvinus, except in acaulescent species. Inflorescence usually axillary, sometimes on a leafless, herbaceous scape; flowers solitary or in panicles, cymes, spikes, or racemes. Flowers trimerous, almost all actinomorphic, bisexual or more commonly unisexual, usually dioecious. Tepals 6. Stamens usually 6. Ovary inferior, 3-locular, rarely 1-locular. Fruit a capsule, rarely berry-like or samaroid; dehiscence loculicidal, irregular or indehiscent. Seeds flattened and winged, or wingless and either globose and smooth, irregular and ruminate, or prismatic-reniform and longitudinally ridged.

A medium-sized family with 4 genera. *Dioscorea*, with at least 400 species, is by far the largest. *Trichopus* Gaertn. and *Stenomeris* Planch. have 2 species each, while *Tacca* has 13.

Generic diversity in the family is highest in South East Asia – *Trichopus* and *Stenomeris* are found in humid forests, and the former also occurs in E Madagascar. Species diversity is highest in the seasonally dry tropics, especially in C South America, Mexico, the Caribbean, South Africa, Madagascar and Indochina, where *Dioscorea* species numbers are highest, but the range of *Dioscorea* extends into temperate regions.

More than 50 species of *Dioscorea* have edible tubers, and 3 are major cultigens – *D. alata* L., *D. cayenensis* Lam. (including *D. rotundata* Poir.) and *D. esculenta* (Lour.) Burkill.

The family belongs in Dioscoreales with Nartheciaceae and Burmanniaceae. It is most easily recognised by broad leaves where the main veins originate at the leaf base, curve outwards towards the margins, then converge towards the apex, petiolar pulvinii, the lack of tendrils, and trimerous flowers with inferior ovaries. *Tacca* is atypical in some characters due to its stemless habit, but is easily identified by its pseudoumbellate inflorescence bearing both broad inflorescence bracts and pendent, filiform floral bracts. Authors prior to Caddick et al. (Taxon **51**: 103–114, 2002) almost universally placed *Tacca* in its own family, Taccaceae.

Stems elongate, twining, plants dioecious; inflorescence racemose or spicate, simple or compound, never umbelliform, floral bracts not long, conspicuous and filiform; fruit a dry, dehiscent, 3-winged capsule · · · · · · · · · · · · 1. **Dioscorea**
Acaulescent herb, leaves in basal rosette; infloresence a pseudo-umbel of bisexual flowers with both broad inflorescence bracts and long, conspicuous filiform floral bracts; fruit berry-like, indehiscent · · · · · · · · · · · · · · · · · · · 2. **Tacca**

1. DIOSCOREA L.

Dioscorea L., Sp. Pl.: 1032 (1753). —Knuth in Engler, Pflanzenr. IV **43**: 45 (1924). —N'Kounkou, Lejoly & Geerinck in Fragm. Florist. Geobot., suppl. **2**: 139–182 (1993). —Wilkin in Kew Bull. **56**: 361–404 (2001).
Testudinaria Burch., Trav. S. Africa **2**: 148 (1824). —Salisbury, Gen. Pl.: 12 (1866). —Baker in F.C. **6**: 248 (1897).

Plants rhizomatous, in the Flora area rhizome reduced to a crown at the tuber apex bearing thicker roots than the tuber. Indumentum present or not, hairs simple to stellate or T-shaped, most prevalent on young vegetative and reproductive shoots. Stems annual, twining to left or right (i.e. sinistrorse/dextrorse), rarely creeping or erect, terete or winged, often armed with prickles or similar structures, especially towards base; bulbils present in axils of some species. Leaves opposite to alternate; petiole usually thickened at base and apex (hard to see in dried material as pulvinii are fleshy, but there is often a difference in colour), lateral nodal flanges, spines or other outgrowths present at petiole bases in some species ("stipules"); blade simple, entire, lobed or compound, base often cordate, narrowly oblong cystoliths often visible; cataphylls sometimes present towards stem base, especially in right-twining species. Inflorescences axillary, pendent, spicate or racemose, flowers sometimes in cymules, lax to dense and sometimes partially concealing inflorescence axis. Dioecious, flowers always appearing unisexual in the Flora area but vestigial male/female parts usually present. Male infloresence simple or compound with up to 3 ranks of branching; female inflorescence simple, rarely with a single basal branch. Flowers usually with 2 floral bracts, inner bracteole often narrower; tepals 6, in 2 whorls, scarcely to strongly differentiated, free to basally fused, inserted on a variable ring-shaped cylinder (torus), erect to recurved. Male flowers with 6 stamens (except in a few populations of *D. praehensilis*) or 3 stamens and 3 staminodes. Female flowers pendent at first, later spreading, capsules usually ascending when mature; ovary 3-angled, immediately inferior to torus. Fruit a 3-winged capsule, dry and dehiscent at maturity, can be fleshy when immature, wings oblong to rounded. Seeds 6, winged all round margin or at base only, apically winged in 1 species in the Flora area, rarely wingless.

Dioscorea is found in all tropical and subtropical regions, with a number of species extending into temperate areas. Knuth (in Engler, Pflanzenr. IV, **43**: 1–387, 1924) and Burkill (in J. Linn. Soc., Bot. **56**: 319–412, 1960) have produced infrageneric treatments to deal with the large number of species; the limited number encountered in the Flora Zambesiaca region makes this unnecessary here.

Two non-native species, *Dioscorea alata* L. and *D. cayenensis* Lam., are encountered infrequently as cultigens, but are excluded from this treatment except in the keys. *D. alata* has highly distinctive right-twining square stems with a wing at each angle. Female flowering collections have been made in in Mozambique (vernacular names *jiba* and *lipeta*) and it is grown in Zambia, where it is known as *tulungwa, masempe* or *chilungu*. Cultivated for its starchy tuber it has been become pantropical following its domestication, which is thought to have occurred in Indochina. *D. cayenensis* (*njambwa* in Zambia) is thought to be West African in origin and is very close in morphology to *D. praehensilis* but differs in its thicker male inflorescence axes, with fewer per axil and bearing more lax male flowers. It is a marginal crop in the Flora area as it requires high and reliable rainfall. More research is needed on the systematics of this complex, in which human selection and possible multiple origins obscure the patterns of morphological variation.

Three keys are given – one based on male flowers, one on female flowering/ fruiting material, and one solely on vegetative characters and distribution. A key to material with only female flowers is not given as they have few useful taxonomic characters. Immature capsules are often present on such plants which will allow them to be identified.

Key to species based on male floral and vegetative characters

1. Leaves compound · 2
 – Leaves entire · 5
2. Leaflets 3 or more, with 1 main vein per leaflet; inflorescence compound or simple; stamens and staminodes 3 · **12.** *quartiniana*
 – Leaflets 3, with 3 or more main veins per leaflet; inflorescence compound; stamens 6, staminodes 0 · 3
3. Ultimate inflorescence branches lax, flowers 2–10 mm apart at anthesis, axis wholly visible; flowers in cymules of 2–3; inner tepals considerably larger than outer · **7.** *dregeana*
 – Ultimate inflorescence branches dense, flowers separated only by floral bracts at anthesis, axis concealed; flowers in dense spikes; inner tepals scarcely larger than outer · 4
4. Ultimate inflorescence branches on slender peduncles more than 5 mm long · **5.** *cochleari-apiculata*
 – Ultimate inflorescence branches usually sessile, rarely on stout peduncles less than 5 mm long · **8.** *dumetorum*
5. Vegetative parts pubescent, at least on young growth · · · · · · · · · · · · · · · · 6
 – Vegetative parts glabrous · 8
6. Hairs T-shaped, flowers in cymules · · · · · · · · · · · · · · · · · · · **11.** *preussii*
 – Hairs stellate, flowers solitary · 7
7. Leaves ovate to orbicular; stellate hairs dense on leaf undersurface; tepals 1.3–2.5 mm long · **14.** *schimperiana*
 – Leaves lanceolate to ovate or elliptic; stellate hairs lax on leaf undersurface; tepals 2.4–4 mm long · **9.** *hirtiflora*
8. Stem left-twining; lacking cataphylls or wings; leaves usually alternate · · · · · · 9
 – Stem right-twining; possessing cataphylls towards base or strongly winged; leaves usually opposite · 14
9. Leaf apex 1.1–6.3 cm long, thickened, acuminate, with inrolled margins; axillary bulbils present, globose, glossy black · · · · · · · · · · · · · · · · · · **13.** *sansibarensis*
 – Leaf apex to 3 cm long, if acuminate not thickened with inrolled margins; axillary bulbils (if present) not globose and black · · · · · · · · · · · · · · · · · · 10
10. Axillary bulbils brown to purple, verrucate, subglobose to angular; leaf blade entire, thinly chartaceous in texture, apex acuminate · · · · · · · · · · · · · · · · 11
 – Axillary bulbils absent; leaf blade entire to lobed, thickly chartaceous to coriaceous, apex apiculate to short-acuminate · 12
11. Flowers patent to inflorescence axis, pedicellate; tepals spreading to reflexed · **1.** *asteriscus*
 – Flowers turned toward inflorescence apex, sessile; tepals erect (male plants not yet encountered in Flora area) · **4.** *bulbifera*
12. Pedicels 2–5 mm long; tepal apex acute to short-acuminate, not cucullate; torus saucer-shaped, to 4 mm diameter · **3.** *buchananii*

– Pedicels 0.8–2.1 mm long; tepal apex rounded to acute, cucullate; torus less than 1 mm diameter · 13

13. Tepals erect at anthesis, flower urn-shaped; outer tepals ovate or elliptic; pedicels 0.8–1.3 mm long, obconic, fleshy · · · · · · · · · · · · · · · · · · **6.** *cotinifolia*

– Tepals spreading at anthesis; outer tepals narrowly oblong to narrowly oblanceolate or spathulate; pedicels 1.1–2.1 mm long, terete to clavate, not fleshy · **15.** *sylvatica*

14. Stems 4-angled and usually winged; inflorescence axis flexuous · *alata* (cultigen)

– Stem terete, with cataphylls towards base; inflorescence axis straight · · · · · · 15

15. Flowers dense, concealing inflorescence axis, falling after anthesis; tepals wholly dry and horny · **2.** *baya*

– Flowers lax, inflorescence axis visible, persisting after anthesis; tepals dry and horny below, petaloid above · 16

16. Inflorescences 2–8 per axil; flowers dense, less than own diameter apart on a slender axis · **10.** *praehensilis*

– Inflorescences 1–2(3) per axil; flowers relatively lax, at least their own diameter apart on a thick axis · *cayenensis* (cultigen)

Key to species based on female floral/fruit and vegetative characters
(see also Figs. 12.2.**31** & **32**)

1. Capsule at least 1.5 times as long as broad, oblong to pyriform or obovate in outline; base and apex usually acute to truncate; stems left-twining; tepals 2.9–4.5 mm long · 2

– Capsule broader than long, usually transversely elliptic in outline with an emarginate to truncate base and apex; stems right-twining; tepals not exceeding 2.1 mm long · 12

2. Seed towards centre of wing, winged all round margin (e.g. Fig. 12.2.**32.7**) · · 3

– Seed winged at base or apex only, wing absent from at least half of seed margin or very narrow on basal or apical side (e.g. Fig. 12.2.**32.1**) · · · · · · · · · · · · · 5

3. Leaf apex acuminate, thickened or not; seed wing more than 2.7 cm long, lengthened towards capsule base and apex · 4

– Leaf apex usually acute to short-acuminate; seed wing 1–2 cm long, broadly oblong to rotund · **3.** *buchananii*

4. Leaf apex acuminate and thickened with inrolled margins; plant glabrous; capsule 1-layered, lacking a marginal wing along each zone of dehiscence; seed wing 3.3–3.5 cm long · **13.** *sansibarensis*

– Leaf apex acuminate but not thickened; plant with T-shaped indumentum but capsule glabrescent; capsule 2-layered, the outer layer forming a 1–2 mm wide marginal wing along each zone of dehiscence; seed wing 2.7–3.2 cm long · **11.** *preussii*

5. Leaves compound; simple indumentum present on vegetative parts and capsule · 6

– Leaves entire; plant with vegetative parts and capsule wholly glabrous · · · · · · 9

6. Leaflets 3–9, each leaflet with a single main vein; capsule 1.5–3.4 cm long, seed wing 3.5–7 mm long · **12.** *quartiniana*

– Leaflets 3, each leaflet with 3 or more main veins; capsule 3.1–5.8 cm long, seed wing 9–22 mm long · 7

7. Capsule narrowly obovate to elliptic (Fig. 12.2.**32.1**) · · · · · **5.** *cochleari-apiculata*

– Capsule oblong to weakly oblong-obovate or oblong-elliptic · · · · · · · · · · · · 8

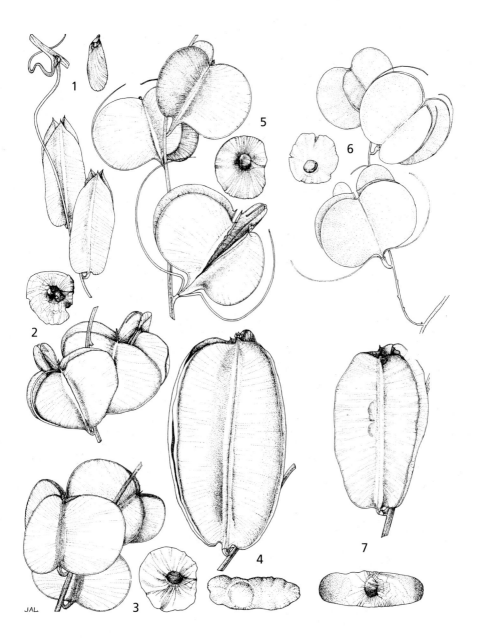

Fig. 12.2.**31**. DIOSCOREA CAPSULES AND SEEDS (1). 1, *Dioscorea asteriscus* ($\times$ $^3/_4$); 2, *Dioscorea schimperiana* ($\times$ $^3/_4$); 3, *Dioscorea hirtiflora* ($\times$ $^3/_4$); 4, *Dioscorea preussii* subsp. *hylophila* ($\times$ $^3/_4$); 5, *Dioscorea praehensilis* ($\times$ $^3/_4$); 6, *Dioscorea baya* ($\times$ $^2/_3$); 7, *Dioscorea sansibarensis* ($\times$ $^3/_4$). 1 from *Batty* 1090, 2 from *Mutimushi* 2258 and *Verdcourt* 2517, 3 from *Faulkner* K670, 4 from *Robertson* 1060 and *Gossweiler* 8637, 5 from *Tweedie* 3625, *Bullock* 3149 and *Newbould & Harley* 4309, 6 from *Louis* 12173, 7 from *Faulkner* K550 and *Vaughan* 917. Drawn by J.A. Lowe (1–5, 7) and Liz Caddick (6). From Kew Bulletin (2001).

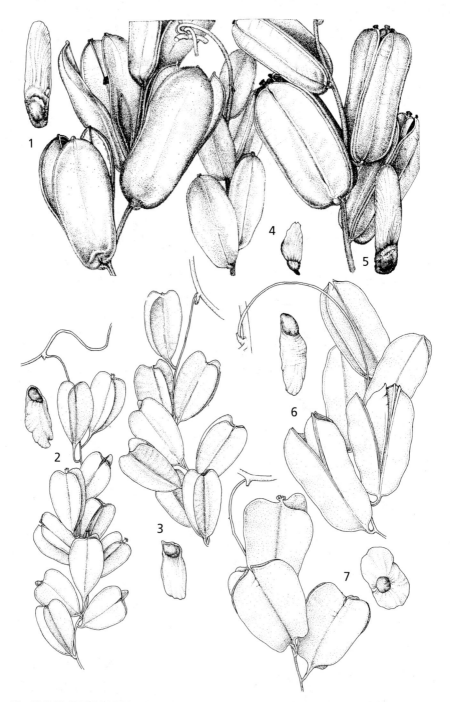

Fig. 12.2.**32**. DIOSCOREA CAPSULES AND SEEDS (2). 1, *Dioscorea cochleari-apiculata* (× 1); 2, *Dioscorea cotinifolia* (× ³/₄); 3, *Dioscorea sylvatica* (× ³/₄); 4, *Dioscorea quartiniana* (× 1); 5, *Dioscorea dumetorum* (× 1); 6, *Dioscorea dregeana* (× ³/₄); 7, *Dioscorea buchananii* (× ³/₄). 1 from *Lewalle* 3747, 2 from *Büchner & Wilson* 142, 3 from *Wilson* 257 and *Loveridge* 1671, 4 from *Peal* 5, 5 from *Archbold* 1133, 6 from *Gerstner* 6716, 7 from *Faulkner* K295. Drawn by J.A. Lowe (1, 4–5) and Liz Caddick (2–3, 6–7). From Kew Bulletin (2001).

8. Capsule as in Fig. 12.2.**32.6**; seed 9–11 mm long · · · · · · · · · · · · · · **7.** *dregeana*
– Capsule as in Fig. 12.2.**32.5**; seed 6–10 mm long · · · · · · · · · · · · **8.** *dumetorum*
9. Leaf apex acuminate, to 3 cm long, lacking a 1–2 mm long triangular apiculus; capsule apex subtruncate to acute; seed wing 4–6 mm wide, basal · · · · · · · · 10
– Leaf apex variable, with a 1–2 mm long triangular apiculus; capsule apex truncate to emarginate or retuse; seed wing 6–9.5 mm wide, apical · · · · · · 11
10. Tepals spreading when stigma is receptive; capsule apex acute, often somewhat triangular (Fig. 12.2.**31**.1) · **1.** *asteriscus*
– Tepals suberect when stigma is receptive; capsule apex rounded to subtruncate · **4.** *bulbifera*
11. Tepals 1.6–2.9 mm long; capsule as in Fig. 12.2.**32.3** · · · · · · · · · · · **15.** *sylvatica*
– Tepals 1.2–1.6 mm long, inner tepals 1–1.2 mm long; capsule as in Fig. 12.2.**32.2** · **6.** *cotinifolia*
12. Vegetative and reproductive parts with stellate to dendroid pubescence, fruits sometimes glabrescent; cataphylls absent; capsule chartaceous to thickly so · · 13
– Vegetative and reproductive parts glabrous; cataphylls present towards stem base; capsule coriaceous · 14
13. Capsule usually transversely elliptic in outline (e.g. Fig. 12.2.**31**.3), chartaceous · **9.** *hirtiflora*
– Capsule usually broadly obovate in outline (e.g. Fig. 12.2.**31**.2), thickly chartaceous · **14.** *schimperiana*
14. Leaf blade base usually truncate to rounded, rarely cuneate or very shallowly cordate; capsule not glaucous · **2.** *baya*
– Leaf blade base truncate to cordate; capsule glaucous even when mature · **10.** *praehensilis* / *cayenensis*

Key to species based on vegetative characters and distribution

1. Stem left-twining, lacking cataphylls or wings; leaves usually alternate · · · · · · 2
– Stem right-twining, possessing cataphylls towards base or strongly winged or indumentum stellate to dendroid; leaves usually opposite · · · · · · · · · · · · · 11
2. Leaves compound · 3
– Leaves entire · 6
3. Leaflets 3–9, each leaflet with a single main vein · · · · · · · · · · · **12.** *quartiniana*
– Leaflets 3, each leaflet with 3 or more main veins · · · · · · · · · · · · · · · · · · · 4
4. Anterior leaflet usually obtuse to truncate (Flora region only) and apiculate · **5.** *cochleari-apiculata*
– Anterior leaflet acute to obtuse and apiculate · 5
5. South Africa, with 1 collection from S Mozambique · · · · · · · · · · · **7.** *dregeana*
– Throughout region except extreme S Mozambique, S Zimbabwe and Botswana · **8.** *dumetorum*
6. Indumentum of T-shaped hairs; stems 6-winged or 6-angled · · · · · · **11.** *preussii*
– Indumentum absent · 7
7. Leaf apex long-acuminate, thickened with inrolled margins; petiole with pair of fleshy oblong lateral nodal organs to 1 cm long at base · · · · · · · **13.** *sansibarensis*
– Leaf apex and petiole not as above · 8
8. Leaf blade thinly chartaceous, apex long-acuminate, lateral nodal organs often present as pair of membranous semicircular projections clasping stem at petiole base; axillary bulbils present · · · · · · · · · · · · · · · · · · **1.** *asteriscus* or **4.** *bulbifera*
– Leaf blade chartaceous to coriaceous, apex short-acuminate to apiculate; bulbils absent · 9

9. Leaf blade entire or with 3, 5 or 7 shallow to deep lobes, apex acute to triangularly short-acuminate, strongly differentiated apiculus not present, chartaceous · **3.** *buchananii*
– Leaf blade entire or shallowly 3-lobed, apex with 1–2 mm long triangular apiculus, usually thickly chartaceous to coriaceous · · · · · · · · · · · · · · · · · ·10
10. Leaves usually opposite, blade entire, ovate to orbicular or transversely elliptic, thickly chartaceous, oblong cystoliths present in blade · · · · · · · · · **6.** *cotinifolia*
– Leaves alternate, blade entire to 3–lobed, cordiform to deltoid, chartaceous to coriaceous, cystoliths not visible, sometimes glaucous or greyish, especially below · **15.** *sylvatica*
11. Indumentum of stellate to dendroid pubescence · 12
– Indumentum absent · 13
12. Leaves ovate to orbicular, stellate hairs dense on lower surface · **14.** *schimperiana*
– Leaves usually lanceolate to ovate or elliptic in Flora area, stellate hairs lax on lower surface · **9.** *hirtiflora*
13. Stems 4-angled and usually winged · *alata*
– Stems terete, never winged · 14
14. Leaf apex acuminate to 1 cm long, thickened and tough, especially at apex, usually not folded in herbarium specimens, cataphylls to 1 cm long; only in NW Zambia · **2.** *baya*
– Leaf apex acuminate to long-acuminate, somewhat thickened (especially distal 3–4 mm), often longitudinally folded along midrib in herbarium specimens, cataphylls to 2 cm long; widespread · · · · · · · · · · · · **10.** *praehensilis / cayenensis*

1. **Dioscorea asteriscus** Burkill in Bull. Jard. Bot. État **15**: 356 (1939). —Sölch in Merxmüller, Prodr. Fl. SW Afrika, fam.153: 1 (1969). —Milne-Redhead in F.T.E.A., Dioscoreaceae: 9 (1975). Type: Malawi, no locality, ♂ fl., no date, *Buchanan* 11 (K holotype, BM). FIGURE 12.2.**31**.1.

    Dioscorea sativa L. sensu Rendle in J. Linn. Soc., Bot. **40**: 212 (1911), in part. —Eyles in Trans. Roy. Soc. S. Africa **5**: 329 (1916).

Herbaceous twining vine to at least 5 m, stems annual from a fleshy tuber. Tuber solitary, short-lived, ± 10 cm in diameter, spherical, periderm glossy black. Indumentum absent. Stems left-twining, to 6 mm in diameter, unarmed, terete; bulbils usually present in axils, 0.4–1 cm in diameter, ovoid and lacking angles, tuberculate, dark purple, rarely red, parenchyma yellow. Leaves alternate, blade entire, apex acuminate, to 3 cm long but never thickened; lateral nodal flanges to c.5 mm in diameter, often present at petiole base as a pair of membranous semicircular projections clasping stem, margins sometimes with small, irregular teeth, usually pale green, sometimes translucent, cataphylls absent. Male flowers often in compound inflorescences, pedicellate, patent to inflorescence axis at anthesis. Tepals recurving (especially outer whorl) to expose stamens, outer tepals 1.7–3 × 0.7–1.2 mm, narrowly ovate to elliptic or lanceolate, inner tepals 1.6–3.1 × 0.5–1 mm, lanceolate to narrowly so or narrowly elliptic. Female flowers also spreading to patent on inflorescence axis, not turned towards apex, tepals. Capsule 1.7–3 × 1.1–1.5 cm; seeds basally winged.

Botswana. N: Samocima, 18°25.6'S, 21°54.3'E, ♂ fl. 20.ii.1979, *P.A.Smith* 2700 (K, MN, PRE, SRGH). **Zambia**. B: Kalomo Dist., Machile, ♂ fl. 14.ii.1965, *Fanshawe* 9110 (K, NDO); Kataba, ♀ fr. 11.vi.1963, *Fanshawe* 7823 (K, SRGH). N: Mansa Dist., Lake Bangweulu, just above Samfya beach, ♀ fl. 8.ii.1959, *Watmough* 223 (K, SRGH). W: Kitwe, ♂ fl. 1.iii.1958, *Fanshawe* F4334 (K); Ndola, ♀ fr. 12.iii.1972, *Fanshawe* 11394 (K, NDO). C: Lusaka Dist., Nachitete R., 31 km SE of Lusaka, 15°36'S, 28°31'E, ♂ fl. 19.iii.1995, *Bingham, Harder & Lwiika* 10446 (K); Kabwe (Broken Hill), Mpima Forest Reserve, ♀ fl. & imm.fr. 13.iii.1962, *Morze* 251 (K). S: Namwala Dist., Ngoma, Kafue Nat. Park, ♂ fl.

20.ii.1963, *Mitchell* 16/67 (K, SRGH); Mapanza E, 1070 m, ♀ fl. 21.ii.1954, *Robinson* 545 (K). **Zimbabwe**. N: Hurungwe Dist., Zambezi valley, Rifu R., 490 m, ♂ fl. 25.ii.1953, *Wild* 4105 (K, LISC, SRGH). W: Hwange Dist., Deka R., 610 m, ♂ fl. 1.iii.1963, *Wild* 6069 (K, LISC, PRE, SRGH). E: Mutare, hilltop on Harare road, 1220 m, ♀ fr. 6.iii.1955, *Chase* 5513 (COI, K, LISC, P, SRGH); Chipinge, Chirinda Forest, ♂ fl. 26.v.1906 & 13.vi.1906, *Swynnerton* 531 (BM, K). **Malawi**. S: Limbe, 1200 m, ♂ fl. 22.ii.1948, *Goodwin* 64 (BM). **Mozambique**. Z: Mopeia Dist., Shamo, near mouth of Shire R., ♂ fl. iii.1859, *Kirk* s.n. (K). T: Boroma, ♀ fl.& fr., ♂ fl. 1890–92, *Menyharth* 677b (B, K, UPS, W). MS: Cheringoma Dist., Inhansato, 50 m, ♂ fl. 13.iii.1956, *Gomes e Sousa* 4287 (COI, K, PRE).

Also in Congo (Katanga), Tanzania, Kenya and possibly Uganda and the Central African Republic. In forests and on forest margins on sandy and lateritic soils, termite mounds, riverine forest and in disturbed habitats; 500–1650 m, rarely lower in Mozambique. Flowering from January to March; fruiting from March onwards.

Conservation notes: Widespread species; not threatened.

This species and *D. bulbifera* are hard to separate. The major difference lies in the male flowers, which are patent to the inflorescence axis and pedicellate in *D. asteriscus*, and turned towards the inflorescence apex and sessile in *D. bulbifera*. The male flowers of *D. asteriscus* are also yellow at all times, while in *D. bulbifera* they are pale green at first, pale pink on opening, and finally maroon-purple. Burkill (1939) alludes to leaf character differences but these are not clear. Variation in these two species needs further study, especially as *D. bulbifera* has been domesticated for its edible bulbils; those of *D. asteriscus* do not appear to be eaten. Specific rank has been retained for both taxa as they appear to be sympatric, although *D. bulbifera* is doubtfully native to the Flora region.

2. **Dioscorea baya** De Wild. in Ann. Mus. Congo Belge, sér.5, **3**: 357, t.51.1 (1912); in Bull. Jard. Bot. État **4**: 328 (1914). —Burkill in Bull. Jard. Bot. État **15**: 391 (1939). —Milne-Redhead in F.T.E.A., Dioscoreaceae: 13 (1975). Type: Congo, Bomaneh, by Congo R., ♂ fl. vii.1910, *Claessens* 695 (BR holotype).

Robust twining vine to 30 m, stems perennial from a large tuber covered by a woody plate; roots from crown woody, to ± 5 mm wide, with tough prickles to 1 cm long. Indumentum absent. Stems right-twining, to 1.5 cm diameter, woody and quite rigid with age, with stout brown or black curved prickles to 1 cm long at base; bulbils absent. Leaves to 1 cm long, usually opposite, especially on fertile shoots, sometimes subopposite or alternate, blade entire, apex acuminate, appearing thickened and tough, especially at extreme apex; cataphylls present on basal stems, usually to c.1 cm long, ovate, acuminate, coriaceous with a thinner translucent margin, to 2.5 cm long at extreme stem base, thickened and highly coriaceous, base of cataphyll often clasping node and axillary branch; lateral nodal flanges absent. Inflorescence spicate, flowers sessile; male inflorescence simple, with flowers falling after anthesis; outer tepals 0.8–1.3 × 0.8–1.3 mm, broadly to very broadly ovate or very broadly oblong or sometimes transversely elliptic, dry and horny, inner tepals 0.6–1.1 × 0.4–0.8 mm, ovate, apex acute to acuminate, sometimes hooded, fleshy. Capsule 1.5–2.3 × 2.5–3.8 cm, not ascending at dehiscence but pendent to patent to infructescence axis, each lobe with a c.1 mm wide semicircular opercular strip along margin which reflexes at dehiscence; wing around seed margin.

var. **baya**. —N'Kounkou in Belg. J. Bot. **126**: 57 (1993). —N'Kounkou et al. in Fragm. Florist. Geobot., suppl. **2**: 170 (1993). FIGURE 12.2.**31**.6.

Dioscorea baya var. *subcordata* De Wild. in Bull. Jard. Bot. État **4**: 328 (1914). Types: Congo, Dundusana, ♂ fl. vi.1913, *De Giorgi* 972 (BR syntype) & Dundusana, 1913, *Mortehan* 198 (BR syntype).

Male inflorescence densely flowered, 2–5 cm long, axis concealed by flowers except where the basal ones fall.

Zambia. W: Mwinilunga, ♂ fl. 10.x.1955, *Holmes* 1250 (K).

Also in N Angola, Congo, Congo (Brazzaville), Gabon, Cameroon, Burundi and Uganda. Mushitu forest; normally a forest species but like all yams is found in open areas and persists in areas of disturbance; c.1500 m, but from c.500 m in Central Africa.

Conservation notes: Widespread species; not threatened.

D. baya var. *kimpundi* De Wild. from Congo, Cameroon, Ivory Coast and Liberia differs from the type variety in having lax-flowered male inflorescences 5–8 cm long, with an axis visible for its entire length. As in var. *baya*, the flowers fall after anthesis. Both varieties have edible tubers.

3. **Dioscorea buchananii** Benth. in Hooker's Icon. Pl. **14**: 76, t.1397, 1398 (1882). — Durand & Schinz, Consp. Fl. Afr. **5**: 273 (1893). —Harms in Engler, Pflanzenw. Ost-Afrikas **C**: 146 (1895). —Baker in F.T.A. **7**: 415 (1898). —Milne-Redhead in F.T.E.A., Dioscoreaceae: 19 (1975). —Wilkin in Kew Bull. **56**: 377 (2001). Types: Malawi, Shire Highlands, ♂ fl. 1881, *Buchanan* 173 (K syntype), and ♀ fr. 1881, *Buchanan* 358 (K syntype). FIGURE 12.2.**32**.7.

   Dioscorea buchananii var. *ukamensis* R. Knuth in Engler, Pflanzenr. IV **43**: 185 (1924). Type: Tanzania, Morogoro Dist., Ukami, no date, *Stuhlmann* 8283 (B† holotype).

Twining vine to 10 m, stems annual from a globose to ovoid, perennial, woody tuber. Stems left-twining, herbaceous, unarmed, to 5 mm diameter, bulbils absent. Indumentum absent. Leaves alternate, blade entire or with 3, 5 or 7 shallow to deep lobes, apex acute to triangularly short-acuminate, rarely obtuse or truncate and apiculate; cataphylls and lateral nodal flanges absent. Inflorescence simple; male racemose, male flower tepals on a scarcely thickened, saucer-shaped torus to 4 mm in diameter, outer tepals 2.5–4.7 × 1.1–2.3 mm, inner tepals 2.7–4.5 × 1.3–2.3 mm, conical in bud. Capsule 2.2–3 × (1.8)2–3.2 cm; seeds 2.5–4 × 3–4 mm with a wing 1–2 × 0.7–1.3 cm extending all around seed margin.

Zambia. C: Muchinga escarpment, near Mutamba's Village, 12°45'S, 31°27'E, 1200 m, ♂ fl. 18.iii.1988, *Phiri* 2192 (K, UZL). S: Mazabuka, 1600 m, st. ii.1933, *Trapnell* 1117 (K). E: Chipata (Fort Jameson), Jumbe, 520–600 m, ♂ fl. 17.iii.1951, *Gilges* 60 (PRE, SRGH). **Zimbabwe**. N: Makonde Dist., Mangula, Whindale Farm, ♂ fl. 24.iii.1969, *Biegel* 2893 (K, LISC, PRE, SRGH). C: Harare (Salisbury), ♀ fl. 6.iii.1924, *Eyles* 3847 (K, SRGH); Harare, 1460 m, ♂ fl. 14.iii.1926, *Eyles* 4479 (K, SRGH). E: Mutare, Imbeza Valley, La Rochelle, 1220 m, ♂ fl. 17.i.1955, *Chase* 5430 (BM, COI, K, LISC, SRGH); Odzi (Otzi) R., c.29 km W of Mutare on main Harare road, ♀ fl.& fr. 26.iii.1958, *Wilson* 255 (K, SRGH). S: Gwanda Dist., Marangudzi, 730 m, ♀ fr. 10.v.1958, *Drummond* 5755 (K, LISC, SRGH); Gwanda Dist., granite hill c.8 km NE of Doddieburn Dam on Tsibizini R., ♂ fl. 8.v.1972, *Smit* 8 (SRGH). **Malawi**. C: Dowa Dist., Kongwe Forest Reserve, E side, UTM WV 998987, 1500 m, ♀ fl. 7.iii.1982, *Brummitt, Polhill & Banda* 16392 (K, LISC); Dedza Dist, Dedza–Mphunzi road, ♂ fl. 18.iii.1968, *Salubeni* 1014 (K, LISC, MAL, PRE, SRGH). S: Zomba Dist., Naisi Hill, 1 km E of Zomba, UTM YT519994, ♂ & ♀ fl. 10.iii.1982, *Brummitt & Polhill* 16405 (K). **Mozambique**. N: Ribáuè, serra de Ribáuè, Mepáluè, c.1000 m, ♀ fl. 23.i.1964, *Torre & Paiva* 10194 (LISC). Z: Lugela Dist., Namagoa, ♂ & ♀ fr. 22.iv.1949, *Faulkner* K295 (K, SRGH); Lugela Dist., Namagoa, ♀ fl. 22.iv.1949, *Faulkner* K297 (K). T: Cahora Bassa Dist., Mágoé, Serra de Songo, 900 m, ♂ fl. 16.iii.1970, *Torre & Correia* 18277 (LISC, LMU, SRGH); Songo, near cemetery, old track to Marueira, ♀ fl. 24.iii.1972, *Macedo* 5087 (LISC, LMU). MS: Cheringoma, Durunde sawmill, ♀ fr. 26.v.1942, *Torre* 4226 (LISC); Manica Dist., Serra Garuso, ♂ fl. 5.iii.1948, *Barbosa* 1136 (LISC). GI: Magaruge Is., near Vilanculos, ♀ fr. ix.1963, *Guy* in *SRGH* 155443 (LISC); Pomene, some km past airstrip, near mangroves, ♀ fr. 24.ix.1980, *Jansen, de Koning & Zunguze* 7532, (K, LMU, WAG).

Also in Tanzania, Angola, S and E Congo. Frequently associated with rocky habitats, often in *Brachystegia* woodland, but also on termitaria, in riverine forest and near mangrove swamps, on limestone and granite substrates; 0–1600 m. Flowering between January and April; fruiting from April onwards.

Conservation notes: Widespread species; not threatened.

Specimens from Zimbabwe and Mozambique have occasionally been misidentified as *D. rupicola* Kunth., which differs in its floral morphology – the male flowers have 3 rather than 6 stamens, inserted on the inner whorl of tepals only, and the female flowers have 3 staminodes; its tepals are smaller (to 2.7 mm in male flowers), as is the style (1–1.4 mm long), and the torus is thickened and discoid or doughnut-shaped with the staminodes inserted at its base in female flowers. *D. rupicola* is only known from South Africa in montane areas of KwaZulu-Natal, Swaziland and Western Cape Province, and rarely below 600 m. *D. buchananii*, by contrast, has been collected at very low elevations in coastal Mozambique.

4. **Dioscorea bulbifera** L., Sp. Pl.: 1033 (1753). —Harms in Engler, Pflanzenw. Ost-Afrikas **C**: 146 (1895). —Burkill in Bull. Jard. Bot. État **15**: 357 (1939). —Miège in F.W.T.A., ed.2 **3**(1): 152 (1968). —Milne-Redhead in F.T.E.A., Dioscoreaceae: 9 (1975). —N'Kounkou et al. in Fragm. Florist. Geobot., suppl. **2**: 159, fig.7 (1993) —Wilkin in Kew Bull. **56**: 379 (2001). Type: Plate facing p.217 in Hermann, Paradisus Batavus (1698).

Dioscorea sativa sensu Baker in F.T.A. **7**: 415 (1898), in part. —De Wildeman in Bull. Jard. Bot. État **4**: 347 (1914), in part. —Eyles in Trans. Roy. Soc. S. Africa **5**: 329 (1916). — Gomes e Sousa in Bol. Soc. Estud. Colón. Moçamb. **32**: 298 (1936).

Dioscorea anthropophagorum A. Chev., Fl. Afr. Centr., Énum. Pl. Récolt.: 309 (1913). Type: Central African Republic, Haute Ombella, Diouma, ♂ fl. no date, *Chevalier* 5931 (P lectotype, selected by Chevalier 1936).

Dioscorea bulbifera var. anthropophagorum (A. Chev.) Summerh. in Trans. Linn. Soc., Zool. **19**: 293 (1931). —Milne-Redhead in F.T.E.A., Dioscoreaceae: 10 (1975).

Climber, stems annual to 5 m. Tuber perennial, irregularly ovoid to subglobose, sometimes absent. Bulbils 0.6–3 cm wide, subglobose to ovoid in non-cultivated forms, flattened, angular to 10 cm long in cultivated forms, grey-brown to purplish-brown, warty. Male inflorescence simple or compound, flowers sessile, turned toward inflorescence apex in both genders; male flower tepals (sub)erect at anthesis, pale green in bud, pale pink on opening and at anthesis, maroon-purple post-anthesis, outer tepals 1.7–2.3 × 0.4–0.8 mm, lanceolate to narrowly elliptic, narrowly oblong or rarely lorate, inner tepals 1.4–2.2 × 0.3–0.6 mm, narrowly oblong to narrowly oblong-elliptic or more narrow; female flowers with erect tepals. Capsule as *D. asteriscus* but 1.5–2.2(3) × 0.95–1.5 cm, oblong to oblong-elliptic in outline, apex not triangularly acute but rounded to subtruncate; seeds basally winged.

Caprivi Strip. 3.2 km SW of Katima Mulilo, on Kongola road, c.1000 m, st. 15.ii.1969, *de Winter* 9187 (K, PRE), probably referable to *D. bulbifera* as has edible bulbils; may be cultivated. **Zambia**. B: Mongu Dist., Simbundu, 1040 m, cult., ♀ fl. vii.1933, *Trapnell* 1246 (K). **Zimbabwe**. C: Harare Expt. Station, cult., 1500 m, ♀ fl. 8.iii.1926, *Eyles* 1156 (K). E: Mutare, SE forested portion of Murahwa's Hill Commonage, 1130 m, ♀ fl. 5.iii.1970, *Chase* 8590 (K, P, SRGH). **Malawi**. C: Lilongwe, Kawai Hill, Dzalanyamas, ♀ fl. 29.iv.1958, *Jackson* 2227 (K, PRE, SRGH). S: Mulanje Dist., Mulanje Mt, foot of Great Ruo Gorge, between hydroelectric station and dam, 870–1060 m, ♀ fl. 18.iii.1970, *Brummitt & Banda* 9206 (K, LISC, PRE, SRGH); Kasupe, Liwonde Nat. Park, Namalembo thicket, ♀ fr. 19.vi.1984, *Dudley* 2055 (MAL). **Mozambique**. T: Marávia Dist., N of Zambezi R., c.400 m from Cabora Bassa dam towards mountain, 230–330 m, ♀ fr. 19.iv.1972, *Pereira & Correia* 2150 (LMU). MS: Manica, Macequece, ♀ fl. 24.ii.1924, *Eyles* 3393 (K, SRGH as *Eyles* 5441 but same date).

Widespread in tropical Africa and Asia, the only species found on both continents. In various forest and woodland types, including mopane. It may require higher rainfall and nutrient levels than *D. asteriscus*; 200–1300 m. Flowering in February and March; fruiting from March onwards.

Conservation notes: Widespread species; not threatened. Possibly introduced.

Dioscorea bulbifera has been domesticated in Africa and a number of edible races exist that have larger and more palatable bulbils and are eaten after boiling (see Burkill in Bull. Jard. Bot. État **15**: 357, 1939 and Chevalier in Bull. Mus. Natl. Hist. Nat., sér.2, **8**: 524, 1936). It is known as air or aerial yam, air potato or bulbil yam in English, but has many vernacular names. As well as being used as a starchy food source, certain races are also used medicinally, socially or as poisons. The frequency of cultivation of *D. bulbifera* gives rise to doubts as to whether it is native to both continents. No male material of *D. bulbifera* has been collected in the Flora area, perhaps because female races are preferred in cultivation. This raises the intriguing question of how female plants in southern Africa set fruit and supports the assertion that it may not be native.

The lectotypification of *D. bulbifera* by Milne-Redhead in F.T.E.A. (1975) is discussed in Wilkin (2001).

5. **Dioscorea cochleari-apiculata** De Wild. in Bull. Jard. Bot. État **4**: 350 (1914). — Burkill in Bull. Jard. Bot. État **15**: 370 (1939). —Milne-Redhead in F.T.E.A., Dioscoreaceae: 22 (1975). —Wilkin in Kew Bull. **54**: 32 (1999); in Kew Bull. **56**: 381 (2001). Type: Congo, Katanga, Lukafu, ♂ fl. xi.1899, *Verdick* 267b (BR holotype). FIGURES 12.2.**32**.1 & 12.2.**33**.

    Dioscorea stolzii R. Knuth in Engler, Pflanzenr. IV **43**: 136 (1924). Types: Tanzania, Mbeya Dist., Utengule, ♂ fl. 1913, *Stolz* 2366 (B syntype, K) and Rungwe Dist., Bulambia, ♂ fl. 1912, *Stolz* 1716 (B syntype, K).

Herbaceous twining vine to at least 12 m, usually annual shoots from a fleshy tuber. Tuber to 10 cm in diameter, solitary and 4-lobed, or 4–5 in a cluster and each depressed, globose, near soil surface. Indumentum of simple straight hairs 0.2–0.8 mm long, fruit glabrescent at maturity. Stems left-twining, prickles concentrated towards base, sparse to dense, stout, brown, usually present in leaf axils. Leaves alternate, trifoliolate, anterior leaflet apex obtuse to truncate, with apex to 10 mm long; cataphylls and lateral nodal flanges absent. Male inflorescence often present when plant is leafless or leaves only beginning to expand, compound, peduncle of partial inflorescence 2–16 mm long, always more than 5 mm long in basal partial inflorescences of compound inflorescence, often slender; partial inflorescences 6–26 mm long, flowers (sub)sessile, dense towards branch apices and covering axis completely, rarely subdense towards base. Outer tepals of male flowers 0.7–1.0 × 0.5–0.9 mm, ovate to broadly ovate or oblong; inner tepals 0.9–1.4 × 0.9–1.6 mm, very broadly ovate to suborbicular. Capsule 4.4–5.8 × 1.9–2.9 cm; seeds winged at base.

Caprivi Strip. E Caprivi, Mpilila Is., 900 m, ♂ fl. 15.i.1959, *Killick & Leistner* 3398 (K, PRE). **Zambia**. N: Mpulungu Dist., Mpulungu to Mbala (Abercorn) road, close to Mpulungu, 885 m, ♂ & ♀ fl.& imm.fr. 20.xii.1951, *Richards* 109 (K). W: Mufulira, ♂ fl. 15.xi.1968, *Mutimushi* 2813 (K, NDO); Kitwe, ♀ fl. 16.xii.1958, *Fanshawe* F5047 (K, LISC). S: Namwala, sandy bush country, 1160 m, ♂ & ♀ fl.& imm.fr. xii.1932, *Trapnell* 1149 (K). **Zimbabwe**. N: Hurungwe Dist., near Mensa Pan, 18 km ESE of Chirundu Bridge, c.460 m, ♂ fl. 4.ii.1958, *Drummond* 5473 (K, LISC, PRE, SRGH); ♀ fl.& imm.fr., *Drummond* 5473a (K, LISC, SRGH). W: Hwange Dist., Hwange Nat. Park (Wankie Game Reserve), ♂ fl.& imm.fr. i.1961, *Davison* in SRGH 115370 (K, LISC, SRGH). C: Harare (Salisbury), 1460 m, ♀ fl. 17.iii.1925, *Eyles* 1193 (K, SRGH). E: Mutare, Burma Valley, Manyera Farm, 850 m, ♂ fl. 28.xii.1960, *Chase* 7242 (BM, K, PRE, SRGH). **Malawi**. N: Ngonga, Rumphi, 16 km on road from Livingstonia, 1070

Fig. 12.2.**33**. DIOSCOREA COCHLEARI-APICULATA. 1, habit of male plant (× ²/₃); 2, male inflorescence (× 3); 3, three male flowers showing dense arrangement, position of floral bracts and limited floral opening (× 8); 4, open male flower showing arrangement of tepals, stamens and torus (× 12); 5, male floral bract (× 12); 6, male flower outer tepal, dorsal view (× 12); 7, male flower outer tepal, ventral view (× 12); 8, male flower inner tepal, dorsal view (× 12); 9, male flower inner tepal, ventral view, with stamen (× 12); 10, female flower with bract and bracteole (× 6); 11, female flower inner tepal, dorsal view (× 12); 12, female flower inner tepal and staminode, ventral view (× 12); 13, female flower outer tepal and staminode, ventral view (× 12); 14, styles and stigmas (× 12); 15, infructescence showing dehisced capsules (× ¹/₄); 16, seed (× ²/₃); 17, mature leaf with necrotic spots (× ¹/₄); 18, section of lower stem showing prickles (× ²/₃). 1 from *Richards* 19293, 2–9 from *Van Rensburg* 1003 KBS, 10–14 from *Herb. Dept. Agric. S. Rhodesia* 1193, 15–16 from *Lewalle* 3747, 17 from *Wilkin & Tawakali* 785, 18 from *Milne-Redhead* 3751. Drawn by Linda Gurr. From Kew Bulletin (2001).

m, ♂ fl. 12.i.1976, *Phillips* 934 (K, MO). C: Mangochi, Cape Maclear, above hotel, ♂ & ♀ fl., 16.xii.1978, *Blackmore* 44 (K). S: Machinga, Machinga Forest Reserve, near Zomba-Liwonde road on Liwonde side, 15°08.54'S, 35°14.34'E, ♀ imm.fr. 24.i.1995, *Wilkin & Tawakali* 772 (K, MAL); ♂ fl. 24.i.1995, *Wilkin & Tawakali* 773 (K, MAL). **Mozambique**. N: Ribáuè Dist., Posto Agrônomico de Ribáuè, Mt Namatupurro, 700 m, ♀ fr. 30.i.1964, *Torre & Paiva* 10335 (LISC); Monapo, 11 km from Namialo to Meserepane, 150 m, ♂ fl. 25.xi.1963, *Torre & Paiva* 9280 (LISC). Z: Milange, slopes of Sabelua Mt Ilulassa, ♀ fr. 19.ii.1972, *Correia & Marques* 2727 (BR, LMU). T: Cabora Bassa, R. Mucangádzi, 5 km from dam, near Posto Policial no.3, new road to Meroeira, 470 m, ♂ fl.& ♀ imm.fr. i.1973, *Torre, Carvalho & Ladeira* 18945 (LISC). MS: Gorongosa Dist., Mt Gorongosa, ♀ fr. 27.vii.1941, *Torre* 3144 (LISC); Between Púnguè and Chimoio (Vila Pery), ♂ fl. 4.xi.1941, *Torre* 3765 (LISC, WAG).

Also in Tanzania, Burundi and S Congo. On forest margins, termite mounds, rocky areas, and edges of watercourses, often on sandy soils, rarely in heavily disturbed places; 400–1600 m.

Conservation notes: Widespread species; not threatened.

Mature male flowers are usually seen when the leaves are only just starting to develop, usually late October to early December in the Flora area. This has also been noted infrequently in the closely-related *D. dumetorum* (e.g. *Salubeni & Patel* 3613 (MAL)). Circular necrotic spots are frequently seen on the leaves of these two closely-related species.

Chase 7242 is a mixed collection; the young shoot on the sheet at Kew is *D. dumetorum*. Young female material is infrequently collected as it is probably not very conspicuous; the female flowers are described from a single collection (*Fanshawe* 5047 (K)).

The tubers are eaten as a famine food after washing and drying to remove alkaloidal toxins, which are used to make arrow poison in Mozambique.

6. **Dioscorea cotinifolia** Kunth, Enum. Pl. **5**: 386 (1850). —Durand & Schinz, Consp. Fl. Afr. **5**: 273 (1893). —Wilkin in Kew Bull. **56**: 383 (2001). Type: South Africa, Cape Province, Morley, 2.ii.1832, *Drège* 4500 (KIEL† holotype, K). FIGURE 12.2.**32**.2.

> *Dioscorea malifolia* Baker in J. Bot. **27**: 1 (1889). —Durand & Schinz, Consp. Fl. Afr. **5**: 274 (1893). —Baker in F.C. **6**: 248 (1896). —Schinz in Mém. Herb. Boiss. **10**: 30 (1900). Type as for *D. cotinifolia*.

Twining vine to 10 m, stems annual. Tubers several per plant, ovoid to pyriform, to 5 cm long, usually on short to long stout roots. Indumentum absent. Stems left-twining, wiry, unarmed, axillary bulbils not observed. Leaves usually opposite, blade entire, ovate to orbicular or transversely elliptic, thickly papery, apex obtuse to acuminate, rarely emarginate, with a 1–2 mm long triangular brown apiculus, oblong cystoliths present in blade; cataphylls and lateral nodal flanges absent. Inflorescence simple, male simple or compound, flowers 1–4 mm apart, solitary or in pairs, axis often flexuous. Male flower urn-shaped; outer tepals 1.3–2.1 × 0.5–1.4 mm, ovate to broadly so or elliptic, thickly papery; inner tepals 1.1–1.5 × 0.5–0.8 mm, narrowly elliptic or narrowly oblong to lanceolate or spathulate. Capsule 1.6–3.3 × 1.4–2.4 cm; seeds winged at apex.

Mozambique. M: Moamba Dist., 10 km from Moamba towards Chinhanguanine, ♀ fl.& fr. 28.xi.1979, *de Koning* 7662 (K, LMU, MAL); Boane Dist., Umbeluzi, near Boane, ♂ fl. 17.xi.1944 *Torre* 6843 (LISC).

Also in South Africa and Swaziland. In open dry forest, forest margins, scrubby vegetation and rocky places; sea level to 400 m (1800 m in South Africa). Flowering in November and December; fruiting from late November onwards.

Conservation notes: Widespread species; not threatened.

The relatively small, opposite leaves with a triangular brown apiculus, urn-shaped male flowers, and apically winged seeds in oblong to pyriform capsules are distinctive. In S Mozambique the pulp of the tuber is regarded as edible.

7. **Dioscorea dregeana** (Kunth) T. Durand & Schinz, Consp. Fl. Afr. **5**: 274 (1893). —Baker in F.C. **6**: 250 (1896). —Burkill in Bull. Jard. Bot. État **15**: 371 (1939). —von Teichman in Aloe **18**: 69, figs.1–4 (1980). —Wilkin in Kew Bull. **54**: 34 (1999); in Kew Bull. **56**: 384 (2001). Type: South Africa, Eastern Cape, Pondoland, between St. John's R. and Untsikaba R., ♀ fl.& fr., *Drège* 4502b (KIEL† holotype, K, P). FIGURE 12.2.**32**.6.

   Helmia dregeana Kunth, Enum. Pl. **5**: 437 (1850).
   Dioscorea dregeana var. *hutchinsonii* Burkill in Bull. Jard. Bot. État **15**: 371 (1939), invalid name.

Herbaceous twining vine to 12 m, stems annual from a fleshy tuber. Tuber solitary and lobed, or main tuber with several smaller subsidiary ones in a dense cluster near soil surface, persisting for several years, 2.5–7 cm in diameter, larger with age, ovoid to spheroidal. Indumentum of simple straight hairs 0.1–1.6 mm long, sometimes dense and velutinous but always sparse on leaf upper surface, stems and fruit sometimes glabrescent with age. Stems left-twining, often with ridged ascending prickles, concentrated towards the base; prickles to 15 mm long, fleshy to hard, similar in colour to stem. Leaves alternate, trifoliolate; anterior leaflet rounded to acute or rarely truncate at apex and bearing an apex to 36 mm long; cataphylls and lateral nodal flanges absent. Male inflorescence compound, partial inflorescences with axis visible for entire length, flowers usually in 2(3)-flowered cymules, sometimes solitary towards inflorescence apex, occasionally reduced to a single cymule. Male flower sessile or on a very short pedicel; outer tepals 1.1–1.9 × 0.5–1.4 mm, lanceolate to narrowly ovate; inner tepals 1.2–2.3 × 1–2.3 mm, broadly obovate to orbicular, rarely ± spathulate. Capsule 3.8–5 × 1.8–2.7 cm; seeds winged at base.

Mozambique. M: Namaacha, ♂ fl. 13.i.1948, *Torre* 7137 (LISC).

Also in South Africa. In forest or at scrub margins and openings, usually in humus-rich or stony soils, often among rocks in ravines; c.400–1400 m. Both male and female flowers open towards start of the rainy season from October to February with a peak in December; fruiting from December onwards.

Conservation notes: Widespread species; not threatened globally, but may be Vulnerable in Flora area.

The collection from Mozambique, the only one from the Flora area, was from a watercourse margin. The closely allied *D. dumetorum* is also found near water.

8. **Dioscorea dumetorum** (Kunth) Pax in Engler & Prantl, Nat. Pflanzenfam. **5**: 134 (1887). —Durand & Schinz, Consp. Fl. Afr. **5**: 273 (1893). —Harms in Engler, Pflanzenw. Ost-Afrikas **C**: 147 (1895). —Baker in F.T.A. **7**: 419 (1898). —De Wildeman in Bull. Jard. Bot. État **4**: 349 (1914). —Chevalier in Bull. Mus. Natl. Hist. Nat., sér.2, **8**: 529 (1936). —Burkill in Bull. Jard. Bot. État **15**: 371 (1939). —Miège in F.W.T.A., ed.2, **3**(1): 151 (1968). —Milne-Redhead in F.T.E.A., Dioscoreaceae: 21 (1975). —N'Kounkou et al. in Fragm. Florist. Geobot., suppl. **2**: 150 (1993). —Wilkin in Kew Bull. **54**: 35 (1999); in Kew Bull. **56**: 385 (2001). Types: Ethiopia, Tigray, near Djeladjeranne, ♂ fl. 15.viii.1840, *Schimper* 1449 (P syntype) & R. Taccaze, ♀ fr. 12.ix.1838, *Schimper* 786 (BM, P syntypes). FIGURE 12.2.**32**.5.

   Helmia dumetorum Kunth, Enum. Pl. **5**: 436 (1850).
   Dioscorea buchholziana Engl. in Bot. Jahrb. Syst. **7**: 333 (1886). Type: Cameroon, Mungo, ♂ fl. ix.1874, *Buchholz* 9 (B† holotype).

Dioscorea triphylla L. var. *tomentosa* Rendle, Cat. Afr. Pl. Welw. **2**: 40 (1899). Types: Angola, Golungo Alto, near Bango, ♂ fl. & ♀ fl.& fr. ii & vi.1855, *Welwitsch* 4052 (BM syntypes) and Sierra Leone, near Freetown, ♂ fl.& ♀ fl. ix.1853, *Welwitsch* 4051 (BM syntype).
Dioscorea triphylla L. var. *dumetorum* (Kunth) R. Knuth in Engler, Pflanzenr. IV **43**: 132 (1924). —Weimer in Bot. Not. **1937**: 161 (1937).

Twining vine to 5 m or more, stems usually annual from a fleshy tuber, possibly persisting for more than one season. Tuber solitary and deeply lobed, or main tuber with several smaller subsidiary tubers in a dense cluster near soil surface, 2.5–7 cm in diameter, larger with age, ovoid to spheroidal, roots from rhizomatous crown thicker than those from tubers. Indumentum of simple, straight or curved hairs, 0.1–1 mm long, stems and leaf upper surfaces with coarser hairs, those on leaf lower surface finer, usually forming a soft tomentum, capsule wings and stems glabrescent. Stems left-twining, with fleshy to hard prickles, especially towards base, similar in colour to stem. Leaves alternate, trifoliolate, anterior leaflet apex obtuse to rounded, rarely truncate, apiculate to long-acuminate; cataphylls and lateral nodal flanges absent. Male inflorescence usually only present when the leaves are fully expanded, compound, partial inflorescences usually sessile or rarely on relatively stout peduncles to 5 mm long, flowers dense and covering axis completely. Male flower sessile, outer tepals 0.7–1.2 × 0.5–0.9 mm, oblong to very broadly obovate, inner tepals 0.7–1.3 × 0.6–1.1 mm, oblong to orbicular. Capsule 3.1–5.3 × 1.5–2.5 cm; seeds winged at base.

Zambia. N: Mporokoso Dist., Kalungwishsi R., Lumangwe Falls, 900 m, ♂ fl. 8.i.1960, *Richards* 12306 (K); Mpulungu Dist., Lunzua (Lauzua) R. Hydro by Electric Station, 1080 m, ♀ fl. 10.i.1964, *Richards* 18758 (K). W: Chiengi Dist., Lake Mweru, Kafulwe Mission, ♀ fr. 4.xi.1952, *White* 3599 (FHO, K); Ndola, Ndola West Forest Reserve, ♂ fl. 29.xii.1951, *White* 1817 (FHO, K). C: Kafue, Mumbwa, ♀ fr. 25.iii.1964, *van Rensburg* 2879 (K). S: Machile (Machili), ♀ fl. 6.i.1961, *Fanshawe* 6089 (K, NDO); ♂ fl. 6.i.1961, *Fanshawe* 6090 (K, NDO). **Zimbabwe**. N: Chinhoyi, N of Lion's Den, E bank of Angwa R., 1230 m, ♂ fl. 14.xii.1994, *Wilkin* 704 (K, SRGH). C: Wedza Dist., 30 km S of Marondera (Marandellas), Monte Christo Farm, 1250 m, ♂ fl. 5.xii.1971, *Biegel* 3672 (K, LISC, SRGH); Harare (Salisbury), ♀ fl. 1.ii.1933, *Eyles* 7283 (K, SRGH). E: Chipinge Dist., Gungunyana Forest Research St., ♂ fl. ii.1976, *Goldsmith* 2/76 (K, PRE, SRGH); Gungunyana Forest Reserve, 950 m, ♀ fr. iv.1964, *Goldsmith* 12/64 (K, LISC, PRE, SRGH). S: Masvingo, Great Zimbabwe, ♂ fl. 9.ii.1963, *West* 6293 (SRGH). **Malawi**. N: Rumphi Dist., Rumphi, Chelinda R., 1070 m, ♀ fr. 4.v.1974, *Pawek* 8597 (K, MAL, MO, SRGH, UC); Rumphi Dist., Njakwa Gorge, 3.2 km E of Rumphi, 1070 m, ♂ fl. 30.xii.1973, *Pawek* 7668 (K, MAL, MO). C: Lilongwe Dist., Lilongwe Nature Sanctuary, ♂ fl. 30.i.1985, *Patel & Banda* 2011 (K, MAL); Lilongwe Dist., new Capital site, ♀ fr. 23.iii.1970, *Brummitt & Little* 9331 (K, MAL). S: Mulanje Dist., Machemba North, top of Machemba Hill, ♀ fl.& imm.fr. 9.ii.1980, *Patel & Morris* 476 (K, MAL); Liwonde Dist., Liwonde Nat. Park, banks of Likwenu R. near Park entrance, 560 m, ♂ fl. 24.i.1995, *Wilkin & Tawakali* 776 (K, MAL). **Mozambique**. N: Monapo Dist., Forest Reserve of Sr. Wolf, ♂ fl. 10.ii.1984, *Groenendijk, de Koning & Dungo* 988 (K, LMU); Montepuez Dist., Chipembe, along track towards river, ♀ fr. 29.iii.1983, *Nuvunga* 1289 (LMU). Z: Namacurra (Inhamacurra), road to Maganja da Costa, ♂ fl. 23.iii.1943, *Torre* 4991 (LISC). MS: Chimoio, between Braunstein sawmill and Amatongas, ♀ imm.fr. 14.ii.1948, *Garcia* 223 (LISC); Mossurize Dist., lower Zona R. (Jihu Dist.), near Espungabera (Spungabera), 660–1230 m, ♂ fl.& ♀ imm.fr. ii–7.iii.1906, *Swynnerton* 215 (BM, K, SRGH).

Found across most of Subsaharan Africa from Ethiopia to Senegal, extending south into the Flora area but not South Africa. On riverbanks, forest margins, termite mounds and rocky areas, often persisting in quite extensively disturbed areas; 500–1400 m. Both male and female plants usually flower from late December to February; fruiting from January onwards.

Conservation notes: Widespread species; not threatened.
This species has a number of vernacular names. In English it is often referred to as the cluster, bitter or 3-leaved yam. The tubers are used as a famine food after washing and drying to remove toxins. A survey of semi-domesticated and less toxic varieties and their conservation/sustainable use is desirable.

9. **Dioscorea hirtiflora** Benth. in Hooker, Niger Fl.: 537 (1849). —Baker in F.T.A. **7**: 416 (1898).—De Wildeman in Bull. Jard. Bot. État. **4**: 323 (1914). —Chevalier in Bull. Mus. Natl. Hist. Nat., sér.2, **8**: 531 (1936). —Burkill in Bull. Jard. Bot. État **15**: 374 (1939). —Miege in F.W.T.A. **3**(1): 152 (1968). —Sölch in Merxmüller, Prodr. Fl. SW Afrika, fam.153: 2 (1969). —N'Kounkou et al. in Fragm. Florist. Geobot., suppl. **2**: 155 (1993). Type: Nigeria, R. Niger (Quorra) opposite Stirling, ♂ fl. no date, *Vogel* s.n. (K holotype, BM). FIGURE 12.2.**31**.3.

Dioscorea rubiginosa Benth. in Hooker, Niger Fl.: 538 (1849). —Durand & Schinz, Consp. Fl. Afr. **5**: 274 (1893). —Chevalier in Bull. Mus. Natl. Hist. Nat., sér.2, 8: 532 (1936). Type: Sierra Leone, ♀ fl. no date, *Don* s.n. (BM holotype).

Dioscorea polyantha Rendle, Cat. Afr. Pl. Welw. **2**: 27 (1899). Type: Angola, Golungo Alto, Mt Queta above N-delle, ♂ fl. iv.1856, *Welwitsch* 4039 (BM holotype, K, P).

Dioscorea hirtiflora var. *trapnellii* Burkill in Bull. Jard. Bot. État **15**: 375 (1939), nom. illegit. in key, described from *Trapnell* 1150 (Busala, Zambia).

Twining vine to 3(5) m, stems annual from a fleshy tuber, often slender in the Flora area. Tubers 1–6, to 5 cm diameter, usually narrower, cylindric, descending vertically into substrate, replaced annually. Indumentum of vegetative parts and inflorescences stellate, rarely dendroid, dense only on immature vegetative parts and inflorescences, pubescence of leaf ventral surface subdense to lax. Stems right-twining, unarmed; bulbils not seen. Leaves usually opposite, blade entire, cataphylls and lateral nodal flanges absent. Inflorescence simple, spicate or racemose, flowers lax. Male flowers subsessile to pedicellate; outer tepals 2.4–3.4(4) × 0.7–1.5 mm, lanceolate to narrowly oblong; inner tepals 2.1–2.8(3.4) × 0.4–1.1 mm, narrowly elliptic or oblong to lanceolate or oblanceolate; outer staminal whorl fertile, inner whorl staminodial. Capsule 1–1.9 × 2.3–3.7 cm, thickly chartaceous; seed wing extending all around seed margin.

The species is divided into three subspecies, two of which are found in the Flora area; subsp. *hirtiflora* is found from Guinea to Cameroon and south to Angola. The stems are usually clothed with a short, persistent stellate indumentum, and the leaf blade is usually of equal length and width. In the other two subspecies the leaf blade is about twice as long as wide, though this can vary on a single plant. However, vegetative characters are not sufficiently consistent to permit the identification of the 3 subspecies in sterile or female material; the male flower pedicel length and outer tepal pubescence characters are necessary.

subsp. **orientalis** Milne-Redh. in Kew Bull. **26**: 574 (1972); in F.T.E.A., Dioscoreaceae: 16 (1975). —Wilkin in Kew Bull. **56**: 388 (2001). Type: Tanzania, Korogwe Dist., 16 km from Korogwe on Handeni road, ♂ fl. 12.vi.1969, *Faulkner* 4238a (K holotype).

Dioscorea lindiensis R. Knuth in Notizbl. Bot. Gart. Berlin-Dahlem **12**: 703 (1935). Type: Tanzania, Lindi Dist., c.60 km W of Lindi, Rondo plateau, ♂ fl. iii.1935, *Schlieben* 5971 (B holotype, K, LISC).

Stems usually glabrous or rapidly glabrescent. Pedicels of male flowers 0.05–0.4 mm long. Outer tepals glabrous or sparsely pubescent at base only on dorsal surface.

Malawi. C: Nkhotakota (Kotakota), Lani Hill, ♂ fl. 22.ii.1953, *Jackson* 1088 (B, K). **Mozambique**. N: Monapo Dist., forest of Sr. Wolf, ♂ fl. 13.ii.1984, *de Koning, Groenendijk & Dungo* 9633 (K, LMU). MS: Dondo, 28 km NW of Beira, ♂ fl.

23.iii.1960, *Wild & Leach* 5228 (COI, K, SRGH).
Also in E Tanzania and Kenya. In forests, open woodland and rocky areas; sea level to c.600 m. Flowering from January to March.
Conservation notes: Coastal distribution; possibly Lower Risk near threatened.

subsp. **pedicellata** Milne-Redh. in Kew Bull. **26**: 573 (1972); in F.T.E.A., Dioscoreaceae: 16 (1975). —Wilkin in Kew Bull. **56**: 388 (2001). Type: Zambia, Chingola, ♂ fl. 19.xii.1955, *Fanshawe* 2661 (K holotype, SRGH).

 Dioscorea hirtiflora var. *nyassica* Burkill in Bull. Jard. Bot. Belg. **15**: 375 (1939), nom. illegit. in key, described from *Buchanan* 1506 (Malawi, no locality).

Stems often pubescent, glabrescent only when mature. Pedicels of male flowers 0.45–1.05 mm long. Outer tepals pubescent on dorsal surface.

Caprivi Strip. Singalamwe area, c.16 km from Singalamwe on road to Katima Mulilo, 1000 m, ♂ fl. 3.i.1959, *Killick* 3272 (BM, K, PRE, SRGH). **Zambia**. B: Sesheke Dist., Lonze Forest, N of Machile, ♂ fl. 19.xii.1952, *Angus* 950 (BM, FHO, K, PRE). N: Mpulungu Dist., Kambole, extreme edge of Kambole escarpment, 1525 m, ♂ fl. 2.i.1955, *Richards* 3861 (K). W: Mwinilunga Dist. between Matonchi R. and Kewumbo (Kaoomba) R., ♂ fl. 15.xii.1937, *Milne-Redhead* 3671 (B, BM, K, PRE). C: Kabwe Dist., 14 km from Lusaka along main road to Kabwe, 1200 m, ♂ fl. 9.ii.1973, *Strid* 2819 (K). S: Mazabuka, Siamambo Forest Reserve, ♂ fl. 17.i.1960, *White* 6302 (FHO, K, SRGH). **Zimbabwe**. N: Miami Dist., K.34 Experimental Farm, ♂ fl. 7.iii.1947, *Wild* 1857 (K, SRGH). W: Hwange Dist., Victoria Falls Rainforest, ♂ fl. 18.ii.1977, *Moyo* 21 (K, PRE, SRGH). C: Harare, Norton, granite kopje by Hunyani R., 1370 m, ♂ fl. 4.i.1948, *Wild* 2270 (K, SRGH). E: Chipinge Dist., 3 km N of confluence of Musiriswi–Bwazi rivers, QDS 2032D1, ♂ fl. 28.i.1975, *Biegel, Pope & Russell* 4820 (K, SRGH). **Malawi**. S: Liwonde Nat. Park, ♂ fl. 24.i.1995, *Wilkin & Tawakali* 778 (K, MAL). **Mozambique**. Z: Mocuba, rocky hill, 32 km on Mocuba road towards Mugeba, 200 m, ♂ fl. 5.iii.1966, *Torre & Correia* 15044 (LISC). T: Cabora Bassa, L bank of R. Mucangádzi, 9 km from Posto Policial no.3 on Bero road, 570 m, ♂ fl. 29.i.1973, *Torre, Carvalho & Ladeira* 18909 (LISC).

Also in S Congo, W Tanzania and Uganda. In riverine vegetation, on termite mounds, and in open woodland and grassland with scattered trees; 200–1700 m. Flowering from January to March.
Conservation notes: Widespread taxon; not threatened.

The tubers are said to be a famine food by the Lozi in Zambia. Bulbils of the West/ Central African subspecies are edible.

Some intermediate populations between the two subspecies are found in Mozambique and Malawi. Female specimens and those males that cannot be ascribed to subspecies are given below.

Intermediate females: **Zambia**. B: Mongu Dist., 77 km from Mongu on road to Kaoma, ♀ fr. 30.i.1975, *Brummitt, Chisumpa & Polhill* 14177 (K, SRGH); Kabompo, 85 km from Zambezi (Balovale) on Kabompo road, ♀ fl.& imm.fr. 26.iii.1961 *Drummond & Rutherford-Smith* 7382 (K, SRGH). N: Mpulungu Dist., Kalambo Falls, ♀ fr. 15.ii.1964, *Richards* 19029 (K, SRGH). W: Mwinilunga Dist., between Matonchi R. and Kaoomba R., ♀ fl. 15.xii.1937, *Milne-Redhead* 3672 (K); Chingola Dist., Luano Forest Reserve, ♀ fl. 3.ii.1958, *Fanshawe* 4247 (K, NDO). C: Mumbwa, near Lutale R., ♀ fr. 25.iii.1964, *van Rensburg* in *KBS* 2868a (K); Chilanga, Mt. Makulu Research Station, Makulu stream, ♀ fl. 25.i.1964, *Angus* 3835 (K, FHO, LISC). S: Kalomo, Twin Fountains Farm, ♀ fr. 3.iv.1989, *Klöckner* 4 (K). **Zimbabwe**. N: Mazowe Dam, ♀ fr. 24.v.1934, *Gilliland* 234 (K). W: Victoria Falls Rainforest, ♀ fr. 31.v.1930, *Milne-Redhead* 410 (K, SRGH). S: Mberengwa Dist., Bukwa Mt, lower NW foothills, ♀ fr.

30.iv.1973, *Pope* 1028 (K, LISC, SRGH). **Malawi**. N: Nyika Plateau, 2440 m, ♀ fl. ii–iii.1903, *McClounie* 144 (K). **Mozambique**. Z: Chamo (Shamo), ♀ fr. iii.1859, Kirk *s.n.* (K). MS: Chimoio (Vila Pery), 2 km towards Garuso, ♀ fl.& imm.fr. 21.v.1949, *Pedro & Pedrógão* 5407 (LMU).

Intermediate males: **Malawi**. No locality, ♂ fl. 1891, *Buchanan* 1506 (K). N: Nkhata Bay, Chikale Beach, ♂ fl. 4.ii.1973, *Pawek* 6420 (CAH, K, MAL, MO, SRGH, UC). **Mozambique**. N: Mutuáli, near R. Nalume, ♂ fl. 14.ii.1954, *Gomes e Sousa* 4192 (K). Z: Gúruè, Lioma, Mt Come, 3 km from Lioma on road to Mutuali, ♂ fl. 25.ii.1966, *Torre & Correia* 14894 (LISC). MS: Mossurize, Maconi, Madanda, ♂ fl. 5.ii.1907, *Johnson* 101 (K).

10. **Dioscorea praehensilis** Benth. in Hooker, Niger Fl.: 536 (1849). —Durand & Schinz, Consp. Fl. Afr. **5**: 275 (1893). —Baker in F.T.A. **7**: 418 (1898) as *prehensilis*. —De Wildeman in Bull. Jard. Bot. État **4**: 335 (1914). —Miège in F.W.T.A., ed.2 **3**(1): 153 (1968). —N'Kounkou et al. in Fragm. Florist. Geobot., suppl. **2**: 165 (1993). —N'Kounkou in Belg. J. Bot. **126**: 52 (1993). —Wilkin in Kew Bull. **56**: 390 (2001). Type: Sierra Leone, ♂ fl. vi.1841, *Vogel* 21 (K lectotype, chosen by Wilkin 2001). FIGURES 12.2.**31**.5 & 12.2.**34**.

    Dioscorea cayenensis Lam. var. *praehensilis* (Benth.) A. Chev. in Bull. Mus. Natl. Hist. Nat., sér.2, **8**: 537 (1936).

    Dioscorea odoratissima Pax in Bot. Jahrb. Syst. **15**: 146 (1892). —Harms in Engler, Pflanzenw. Ost-Afrikas **C**: 146 (1895). —De Wildeman in Bull. Jard. Bot. État **4**: 335 (1914). —Burkill in Bull. Jard. Bot. État **15**: 381 (1939); in J. Linn. Soc., Bot. **56**: fig.53G (1960). —Milne-Redhead in F.T.E.A., Dioscoreaceae: 11 (1975). Type: Togo, Bismarckburg, ♂ fl. viii.1890, *Büttner* 104 (B†, K lectotype chosen here).

    Dioscorea angustiflora Rendle, Cat. Afr. Pl. Welw. **2**: 39 (1899). Type: Angola, Golungo Alto, Sange, ♀ fl. no date, *Welwitsch* 4036 (BM holotype).

    Dioscorea liebrechtsiana De Wild. in Bull. Herb. Boiss., sér.2, **1**: 53 (1900); in Bull. Jard. Bot. État **4**: 333 (1914). —Burkill in Bull. Jard. Bot. État **15**: 381 (1939). Type: Congo, Kisantu, ♀ fl. 1900, *Gillet* 384 (BR holotype).

Herbaceous twining vine to at least 10 m, less when climbing on low vegetation, usually dying back annually. Tuber 1, often present alongside withered tuber of previous season, up to 5 cm diameter, at least 60 cm long, cylindric, descending vertically into substrate, white when young but periderm later turning brown. Indumentum absent. Stems right-twining, concolorous to pale brown prickles concentrated on larger, cataphyll-bearing stems, most dense and largest near stem base. Leaves opposite, rarely alternate; blade entire, apex acuminate to long-acuminate, somewhat thickened (especially distal 3–4 mm), usually longitudinally folded along midrib in herbarium specimens, cataphylls present on basal stems, usually to c.2 cm long, ovate, acuminate, thick and coriaceous with a thinner translucent margin, base often clasping node and axillary branch; lateral nodal flanges absent. Inflorescence spicate, simple. Male flowers patent to axis and 1–2(4) mm apart; outer tepals 1.4–2.1 × 0.8–1.6 mm, elliptic to elliptic-oblong or ovate, sometimes broadly so, thinly papery, rigid and somewhat translucent when dry, basal ⅓ often thicker but more translucent, especially in older flowers where base may appear grey in herbarium specimens; inner tepals 1.1–1.9 × 0.6–1.4 mm, elliptic or ovate to broadly so, occasionally obovate, erect and concave, thickly papery and opaque, stamens (5)6(9). Capsule not ascending at dehiscence but pendent to patent to infructescence axis, 1.5–2.4 × 2.8–4.3 cm, each lobe with a c.1 mm wide, semicircular opercular strip along margin which reflexes at dehiscence, pedicel swollen and obconic in mature capsule; seed wing extending all around seed margin.

    Zambia. B: Zambezi Dist., ♂ fl. 25.ii.1964, *Fanshawe* 8338 (K, NDO); Kabompo R., 1000 m, ♀ fr. 22.vi.1954, *Gilges* 392 (K). N: Mansa (Fort Roseberry), ♀ fr. 5.v.1964, *Fanshawe* 8572 (K, NDO); Mporokoso Dist., W of Mporokoso, Kalungwishi R., Kundabwika Falls, 1130 m, ♂ fl. 14.iv.1961, *Phipps & Vesey-Fitzgerald* 3154 (K, PRE, SRGH). C: Lusaka, ♀ fl.& imm.fr. 26.ii.1965, *Fanshawe* 9220 (K); Kabwe (Broken

Hill), ♂ fl. 9.ii.1964, *Fanshawe* 8269 (K, NDO). E: Chipata (Fort Jameson), Kapatamoyo, 1150 m, ♂ fl. 5.i.1959, *Robson* 1033 (BM, BR, K, LISC, PRE, SRGH); ♀ fl. 5.i.1959, *Robson* 1033a (BM, K, LISC, PRE, SRGH). **Zimbabwe**. E: Chimanimani Dist., Haroni R., ♀ imm.fr. 21.iv.1973, *Mavi* 1433 (K, SRGH). **Malawi**. N: Nkhata Bay Dist., Nkwazi Forest, 14.5 km S of Nkata Bay junction, ♂ fl. 22.ii.1976, *Pawek* 10865 (BR, K, MAL, MO, SRGH, UC, WAG); Nkhata Bay, Nkuwadzi Forest Reserve, c.31 km N of Chintheche, ♀ fr. 11.vi.1993, *Chifumbi & Nangoma* 67 (MAL). C: Dowa, Chimombo Hill summit, near Post Office transmitter station, 13°39.96'S, 33°56.08'E, ♂ fl. 30.i.1995, *Wilkin & Tawakali* 793 (K, MAL); Dedza, Mua Livulezi Forest, fire protected plot EP2, ♀ imm.fr. 22.iii.1962, *Adlard* 462 (MAL, SRGH). S: Mt. Mulanje, Likabula Valley, where track up first crosses stream, 15°56.09'S, 35°30.93'E, ♂ fl. 25.i.1995, *Wilkin & Tawakali* 782 (K, MAL); ♀ fl. 25.i.1995, *Wilkin & Tawakali* 781 (K, MAL). **Mozambique**. N: Ribáuè, Serra do Ribáuè, Mepalué, ♂ fl. 23.i.1964, *Torre & Paiva* 10143 (LISC); Ribáuè, Serra do Ribáuè, Mepalué, ♀ fl.& imm.fr. 23.iii.1964, *Torre & Paiva* 11367 (LISC). Z: Lugela Dist., M'guluni Mission, ♂ fl. no date, *Faulkner* K126 (K, P); Mocuba Dist., road Milange–Quelimane, between 1st and 2nd crossings for Mocuba, 5.3 km from 1st, ♀ fr. 20.v.1949, *Barbosa & Carvalho* 2762 (K). T: Cahora Bassa Dist., Songo Mt, 10 km from camp to Chitima (Estima), 700 m, ♂ fl. 10.ii.1970, *Torre & Correia* 17851 (K, LISC); 2 km from Songo to Maroeira, ♀ fr. 31.iii.1972, *Macedo* 5131 (LISC, LMU). MS: Gondola Dist., road from Chimoio (Vila Pery) to Buzi, 600 m, ♀ fr. 4.vi.1941, *Torre* 2791 (LISC); Sussendenga Dist., Mavita, between Chicuizo and school, ♂ fl. 21.i.1948, *Barbosa* 859 (LISC).

Widespread in W, C and E Africa, in a broad range of forest and grassland habitats, including cultivated areas; 400–1700 m. Flowering in January and February; fruiting from March onwards.

Conservation notes: Widespread species; not threatened, although local varieties or races may be at risk.

D. praehensilis is the most important edible wild yam of southern Africa and is harvested quite extensively (e.g. in Malawi) from forest and forest-margin habitats. A piece of tuber and the crown is often replanted to produce a yam for subsequent years. The leaves may also be eaten. Sometimes termed bush yam, forest yam or white yam in English.

Following study of the type specimens of *D. odoratissima* Pax and *D. praehensilis*, they appear to differ only in the presence of a spiny root cluster around the tuber apex in W African populations of *D. praehensilis*. Since much human selection has taken place, and possibly introgression with other *Dioscorea* species, it is believed that the spiny and non-spiny forms are conspecific, as also suggested by Burkill (1939) and N'Kounkou (1993).

The broader relations of *D. praehensilis* with the cultivated yam species and other wild allies, including *D. cayenensis* Lam., *D. abyssinica* Kunth and *D. liebrechtsiana* De Wild., are not clear. The observed variation in stamen number in some specimens examined from the Flora area may be a result of hybridisation.

Dioscorea cayenensis is not native to the Flora Zambesiaca area, but is encountered infrequently as a cultigen. It is thought to be of West African origin and is morphologically very close to *D. praehensilis*, although it differs in the thicker male inflorescence axes with more lax male flowers. It is a marginal crop in the Flora area as it requires high and constant rainfall levels. The name Njambwa in Zambia may be attributable to *D. cayenensis*. Macaulay 1113 from Zambia and *Torre & Correia* 17460 from Mozambique may be attributable to *D. sagittifolia* Pax. Pan-African research is needed on the systematics of this complex of *Dioscorea* species, in which human selection and possible multiple origins obscure the patterns of morphological variation and species limits.

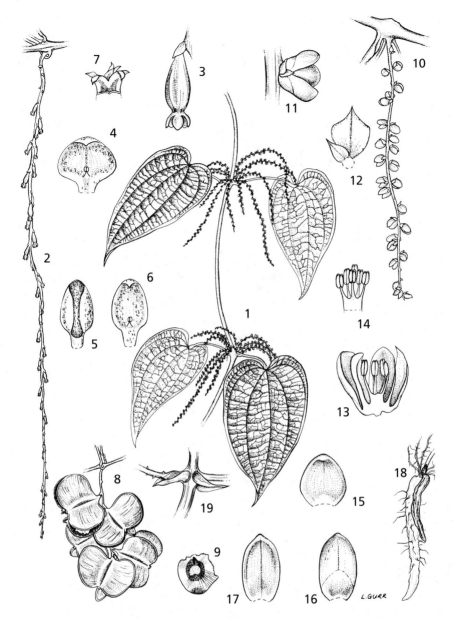

Fig. 12.2.**34**. DIOSCOREA PRAEHENSILIS. 1, habit of male plant (× ¹/₂); 2, female inflorescence (× ²/₃); 3, female flower and floral bract (× 3); 4, female flower outer tepal and staminode, ventral view (× 10); 5, female flower inner tepal, dorsal view (× 10); 6, female flower inner tepal and staminode, ventral view (× 10); 7, stigmas and styles (× 10); 8, infructescence (× ¹/₂); 9, seed (× ²/₃); 10, male inflorescence (× 2); 11, male flower and floral bracts (× 6); 12, male floral bract and bracteole (× 12); 13, male half flower, with inner and outer tepals, torus and stamens (× 12); 14, part of torus and stamens of male flower (× 10); 15, male flower outer tepal, ventral view (× 10); 16, male flower inner tepal, dorsal view (× 10); 17, male flower inner tepal, ventral view (× 10); 18, current and previous seasons tubers (× ¹/₆); 19, cataphylls at stem basal node (× ¹/₂). 1 from *Bingham & Harder* 10327 and *Wilkin & Tawakali* 783, 2–7 from *Wilkin & Tawakali* 781, 8 from *Bingham & Harder* 10445 and *Trapnell* 1430, 9 from *Trapnell* 1430, 10 from *Wilkin & Tawakali* 793, 11–17 from *Bingham & Harder* 10327, 18 from slide, 19 from *Wilkin & Tawakali* 783. Drawn by Linda Gurr. From Kew Bulletin (2001).

11. **Dioscorea preussii** Pax in Bot. Jahrb. Syst. **15**: 147 (1892). —Durand & Schinz, Consp. Fl. Afr. **5**: 275 (1893). —Baker in F.T.A. **7**: 417 (1898). —De Wildeman in Bull. Jard. Bot. État **4**: 318 (1914). —Chevalier in Bull. Mus. Natl. Hist. Nat., sér.2, **8**: 533 (1936). —Burkill in Bull. Jard. Bot. État **15**: 351 (1939). —Miège in F.W.T.A., ed.2, **3**(1): 152 (1968). —Milne-Redhead in F.T.E.A., Dioscoreaceae: 20 (1975). —N'Kounkou et al. in Fragm. Florist. Geobot., suppl. **2**: 150 (1993). Type: Cameroon, between Barombi and Kumba, ♂ fl. 11.xi.1890, *Preuss* 504 (K lectotype, chosen by Wilkin 2001).

> *Dioscorea andongensis* Rendle, Cat. Afr. Pl. Welw. **2**: 37 (1899). Type: Angola, Pungo Andongo, between Catete and Luxillo, ♀ fl.& fr. i.1857, *Welwitsch* 4040 (BM holotype, K).
> *Dioscorea preussii* var. *glabra* Burkill in Bull. Jard. Bot. État **15**: 351 (1939), nom. illegit.

Herbaceous twining vine to at least 6 m, stems annual. Tubers 1–2 (previous and current year), at least 50 × 2 cm, narrowly cylindric, tapering gradually towards crown, sometimes lobed, descending vertically into substrate. Indumentum of medifixed T-shaped hairs. Stems left-twining, terete or hexagonal in cross-section, sometimes with 6 membranous, dry undulate wings. Leaves alternate, entire; cataphylls and lateral nodal flanges absent. Male inflorescence simple, racemose, with flowers single and pedicellate or in cymules of 2–3(6) flowers. Male flowers with 3 stamens and 3 staminodes. Capsule 4.7–5 × 2.7–3 cm, wall 2-layered, outer layer exceeding inner and forming a 1–2 mm wide, papery wing along each zone of dehiscence. Seed wings oblong, extending all around margin but lengthened towards truncate base and apex.

Subsp. **preussii**. —Wilkin in Kew Bull. **56**: 392 (2001).

Subglabrous to laxly pubescent, with a few hairs on stems, inflorescence (mainly towards apex) and leaf apex. Male inflorescences with single, pedicellate flowers, rarely in pairs but not forming cymules of more than 2 flowers. Male flowers on 0.9–1.1 mm long pedicels, tepals more spreading than those of subsp. *hylophila* and not ventrally concave, such that the flower is rotate at anthesis; outer whorl glabrous, 1.7–2.1 × 0.8–1.3 mm, ovate to broadly so, inner whorl 1.8–2.4 × 0.9–1.3 mm, broadly ovate to broadly elliptic.

Mozambique. N: Ribáuè, Serra da Ribáuè, Mepalué, c.1000 m, ♂ fl. 23.i.1964 *Torre & Paiva* 10178 (LISC).

Common in West and Central Africa as far east as Uganda and south to Angola. In riverine vegetation and open woodland; c.1000 m. Flowering in January and February; fruiting from February onwards.

Conservation notes: Widespread across Africa, but local in the Flora area; not threatened.

Subglabrous plants, such as the collection cited above, are rare, but are also encountered in Angola and were placed in a distinct variety, var. *glabra*, by Burkill (1939). This taxon has not previously been recorded for the Flora area and more material should be sought.

Subsp. **hylophila** (Harms) Wilkin in Kew Bull. **56**: 392 (2001). Type: Tanzania, W Usambara Mts, Lutindi, ♂ fl., no date, *Holst* 3423 (B holotype, K). FIGURE 12.2.**31**.4.

> *Dioscorea hylophila* Harms in Engler, Pflanzenw. Ost-Afrikas **C**: 146. (1895). —Baker in F.T.A. **7**: 416 (1898). —Engler, Pflanzen. Afrikas **2**: 363: fig.258 (1908). —Burkill in Bull. Jard. Bot. État **15**: 352 (1939). —Milne-Redhead in F.T.E.A., Dioscoreaceae: 20 (1975).

Indumentum usually dense on stem, lower surface of leaf blade and inflorescence. Male inflorescence with flowers in cymules of 2–3(6) flowers, rarely 1 or 2 solitary flowers per inflorescence. Male flowers on pedicels to 2 mm long (single flowers) or peduncle to 3 mm long, pedicels 0.4–1 mm long (cymules); tepals suberect and ventrally concave at anthesis forming a cup-shaped flower, outer whorl dorsally pubescent; 1.3–2.2 × 0.8–1.8 mm, broadly ovate to orbicular, inner whorl 1.1–2 × 0.8–2.1 mm, ovate to broadly so.

Zambia. N: Mpika, ♂ fl. 10.ii.1955, *Fanshawe* F2042 (K). **Malawi**. C: Lilongwe, Bunda Hill, 1250 m, ♂ fl. 7.ii.1959, *Robson* 1503 (BM, K, LISC, PRE, SRGH); Lilongwe, Ngara Hill, 1250 m, ♀ imm.fr. 7.ii.1959, *Robson* 1503a (K). S: Mangochi, hills E of Mangochi on Mangochi–Chowe Road, c.4 km beyond end of tarmac, 14°25.63'S, 35°21.76'E, ♂ fl. 26.i.1995, *Wilkin & Tawakali* 784 (K, MAL). **Mozambique**. N: Marrupa, c.15 km on road to Nungo, slopes of Mt. Kuwanku, 850 m, ♂ fl. 22.ii.1982, *Jansen & Boane* 8034 (K, LMU, PRE).

Also in Tanzania. In riverine vegetation and open woodland; 300–1700 m.

Flowering in January and February; fruiting from February onwards.

Conservation notes: Widespread taxon; not threatened.

Wilkin & Tawakali 788 is atypical in that its female flowers have floral bracts to 15 mm long, narrowly elliptic to lanceolate or linear and leaf-like in texture.

In Mozambique, the tuber of this subspecies is said to be edible. However, there are also reports that this species is poisonous and that it is used as a cure for snakebite.

D. preussii is easily identified by its unique T-shaped hairs and oblong capsule with a wing along the dehiscence zone. As in *D. hirtiflora*, the distribution of infraspecific variation suggests that subspecific rank is the most appropriate. If more populations of *D. preussii* subsp. *preussii* are discovered in Mozambique, however, there would be grounds for reducing the two subspecies to varietal rank. The single record from Mozambique may be a relict population or a relatively recent dispersal. The females of the two subspecies are indistinguishable in fruit or flower.

12. **Dioscorea quartiniana** A. Rich., Tent. Fl. Abyss. **2**: 316, t.96a (1851). —Durand & Schinz, Consp. Fl. Afr. **5**: 275 (1893). —Harms in Engler, Pflanzenw. Ost-Afrikas **C**: 146 (1895). —De Wildeman in Bull. Jard. Bot. État **4**: 351 (1914). —Chevalier in Bull. Mus. Natl. Hist. Nat., sér.2, **8**: 531 (1936). —Burkill in Bull. Jard. Bot. État **15**: 362 (1939). —Miège in F.W.T.A., ed.2, **3**(1): 151 (1968). —Sölch in Merxmüller, Prodr. Fl. SW Afrika, fam.153: 2 (1969). —Wilkin in Kew Bull. **54**: 16 (1999); in Kew Bull. **54**: 27 (1999); in Kew Bull. **56**: 394 (2001). Type: Ethiopia, Tigray, near Aderbati, ♀ fr. 26.ix.1839, *Quartin-Dillon & Petit* s.n. (P holotype). FIGURE 12.2.**32**.4.

 Dioscorea crinita Hook.f. in Bot. Mag. **111**: t.6804 (1885). —Durand & Schinz, Consp. Fl. Afr. **5**: 273 (1893). —Baker in F.C. **6**: 251 (1896). —Burkill in J. Linn. Soc., Bot. **56**: 379, fig.4L (1960). Type: South Africa, KwaZulu-Natal, Umhloti, ♂ fl. 3.ii.1882, *Wood* 1618 (K lectotype, chosen by Wilkin 1999).

 Dioscorea beccariana Martelli, Fl. Bogos.: 83 (1886). —Baker in F.T.A. **7**: 420 (1898). — Rendle et al. in J. Linn. Soc., Bot. **40**: 212 (1911). —Eyles in Trans. Roy. Soc. S. Africa **5**: 329 (1916). —Burkill in Johnston, Brit. Centr. Africa: 272 (1897). Type: Eritrea, Keren, Mt Deban, ♂ fl. viii.1870, *Beccari* 303 (FT holotype, K).

 Dioscorea forbesii Baker in J. Bot. **27**: 2 (1889). —Durand & Schinz, Consp. Fl. Afr. **5**: 274 (1893). —Baker in F.C. **6**: 251 (1896). —Schinz in Mém. Herb. Boiss. **10**: 30 (1900). Type: Mozambique, Maputo (Delagoa) Bay, ♂ fl. 1822, *Forbes* s.n. (K holotype).

 Dioscorea stuhlmannii Harms in Engler, Pflanzenw. Ost-Afrikas **C**: 146 (1895). —Baker in F.T.A. **7**: 421 (1898). —De Wildeman in Bull. Jard. Bot. État **4**: 356 (1914). Type: Tanzania, Usaramo, Bagamoyo, ♂ fl. no date, *Stuhlmann* 6793 (B† holotype, K).

 Dioscorea dinteri Schinz in Mém. Herb. Boiss. **20**: 11 (1900). —Sölch in Merxmüller, Prodr. Fl. SW Afrika, fam.153: 2. Type: Namibia, Grootfontein, Streydfontein, ♀ fr. 21.v.1899, *Dinter* 705 (Z holotype, B†?).

 Dioscorea apiculata De Wild. in Ann. Mus. Congo Belge, Bot., sér.4, **1**: 14, t.5 (1902); in Bull. Jard. Bot. État **4**: 356 (1914). Type: Congo, Katanga, Lukafu, ♂ fl. xi.1899, *Verdick* 269 (BR holotype).

 Dioscorea excisa R. Knuth in Engler, Pflanzenr. IV **43**: 155 (1924). —Suessenguth in Proc. & Trans. Rhod. Sci. Assoc. **43**: 78 (1951). Type: Zimbabwe, Bulawayo, ♂ fl. xii.1902, *Eyles* 1141 (B holotype, BM, SRGH).

Dioscorea schliebenii R. Knuth in Notizbl. Bot. Gart. Berlin-Dahlem **11**: 659 (1932). Type: Tanzania, Ulanga Dist., Mahenge, Sali, c.35 km S of Mahenge, ♂ fl. v.1932, *Schlieben* 2247 (B holotype)

Dioscorea quartiniana var. *apiculata* (De Wild.) Burkill in Bull. Jard. Bot. État **15**: 365 (1939).

Dioscorea quartiniana var. *dinteri* (Schinz) Burkill in Bull. Jard. Bot. État **15**: 365 (1939).

Dioscorea quartiniana var. *excisa* (R. Knuth) Burkill in Bull. Jard. Bot. État **15**: 365 (1939).

Dioscorea quartiniana var. *schliebenii* (R. Knuth) Burkill in Bull. Jard. Bot. État **15**: 365 (1939). —Milne-Redhead in F.T.E.A., Dioscoreaceae: 24 (1975).

Dioscorea quartiniana var. *stuhlmannii* (Harms) Burkill in Bull. Jard. Bot. État **15**: 365 (1939). —Milne-Redhead in F.T.E.A., Dioscoreaceae: 25 (1975).

Dioscorea retusa sensu Mogg in Macnae & Kalk, Nat. Hist. Inhaca Is.: 143 (1958), non Mast.

Dioscorea quartiniana var. *quartiniana* – Milne-Redhead in F.T.E.A., Dioscoreaceae: 24 (1975).

Slender, herbaceous twining vine to 5 m, stems annual. Tubers several, clustered near soil surface, to c.10 cm long, ovoid, fleshy, with dense fine fibrous roots. Indumentum of simple, straight, 0.1–0.6 mm long hairs, quite coarse. Stems left-twining, unarmed; small brown tuberculate bulbils sometimes present in axils. Leaves alternate, digitately 3–7(9) foliolate with one main vein per leaflet, simple leaves encountered very occasionally; cataphylls and lateral nodal flanges absent. Male inflorescence simple or compound; partial inflorescences spicate, subdense to dense, catkin-like, axis covered by flowers and floral bracts. Male flower sessile to pedicellate, outer tepals 0.9–2.3 × 0.4–1.5 mm, ovate to lanceolate, inner tepals 0.9–2.4 × 0.3–1.2 mm, ovate to lanceolate, outer whorl of 3 stamens subtended by outer perianth whorl, inner whorl of 3 staminodes subtended by inner perianth whorl. Capsule 1.5–3.4 × 0.9–1.8 cm. Seeds winged at base.

Caprivi Strip. Singalamwe area, c.96 km from Katima Mulilo on road to Singalamwe, c.1000 m, ♂ fl. 30.xii.1958, *Killick & Leistner* 3200 (K, PRE, SRGH); Lizauli, c.1000 m, ♀ fl. 2.i.1959, *Killick & Leistner* 3259 (PRE). **Botswana**. N: 21 km S of Botswana/Zimbabwe border, ♂ fl. 14.i.1960, *Leach & Noel* 28 (K, SRGH). **Zambia**. B: Sesheke, Kazu Forest near Machile, ♂ fl. 20.xii.1952, *Angus* 977 (FHO, K, PRE); Masese, ♀ fr. 12.iii.1960, *Fanshawe* 5474 (K). N: Mpulungu Dist., Mbala (Abercorn), Katula Gorge, 1460 m, ♂ fl. 8.i.1952, *Richards* 402 (K); Mbala (Abercorn), top of Kalambo Falls, by top rondaval, 1200 m, ♀ fl. 16.ii.1965, *Richards* 19695 (K, LISC, SRGH). W: Mwinilunga, Kalene Hill, ♀ fl. 15.xii.1963, *Robinson* 6071 (B, K, SRGH); ♂ fl. 15.xii.1963, *Robinson* 6072 (B, K, PRE, SRGH). C: Lusaka, 13 km SE of Lusaka, ♂ fl. xii.1957, *Noak* 293 (K, PRE). S: Mazabuka, 21 km N of Mazabuka, ♂ fl. 9.i.1964, *Angus* 3821 (FHO, K, LISC); Kalomo, ♀ fl. 15.ii.1965, *Fanshawe* 9108 (K, NDO). **Zimbabwe**. N: Gokwe, ♂ fl. 6.ii.1964, *Bingham* 987 (BM, K, SRGH); Darwin, top of hill N of Mavuradonha Mission, directly above Mission, 915 m, ♂ fl. 30.i.1970, *Pope* 217 (K, LISC, SRGH). W: Hwange (Wankie), no locality, ♀ fl. ii, *Levy* 143 (K, PRE); Matobo Dist., Quaringa Farm, 1460 m, ♂ fl. xii.1957, *Miller* 4829 (COI, K, LISC, P, PRE, SRGH). C: Rusape (Rusapi), ♂ fl. 3.ii.1921, *Hislop* s.n. (K). E: Mutare, Commonage, 1160 m, ♂ fl. 27.ii.1950, *Chase* 2124 (BM, COI, K, LISC, SRGH); Mutare (Umtali), Darlington suburb, municipal sand pits, 1100 m, ♀ fl. 8.iii.1963, *Chase* 7962 (K, LISC, SRGH). S: Beitbridge Dist., Tshiturapadsi (Chiturapazi), ♂ fl. 25.ii.1961, *Wild* 5426 (COI, K, SRGH). **Malawi**. N: Mzimba Dist., Mzuzu, Marymount, 1370 m, ♂ fl. 10.iv.1974, *Pawek* 8316 (K, MAL, MO, SRGH, UC); Nkhata Bay, Vipya, 48 km SW of Mzuzu, 1680 m, ♀ fl. 15.iii.1969, *Pawek* 1844 (K, MAL). C: Salima, road near Lake Nyasa Hotel, 480 m, ♂ fl. 14.ii.1959, *Robson* 1599 (BM, K, LISC, MAL, PRE); Salima, Lake Nyasa Hotel, edge of beach, ♀ fl. 15.ii.1959, *Robson* 1612 (BM, K, LISC, MAL). S: Mt Mulanje, Linji (Upper Lichenya) Plateau, 1900 m, ♂ fl. 20.ii.1986, *Chapman & Chapman* 7235 (K, MAL, MO); Mt Mulanje, just below Thuchila Hut to S, 1980 m, ♀ fr. 7.iv.1970, *Brummitt* 9730 (K, LISC, MAL, PRE).

Mozambique. N: Marrupa, road from Mecula c.20 km to Matiquite, 760 m, ♂ fl. 10.ii.1981, *Nuvunga* 494 (BM, K, LISC, LMU, SRGH); Marrupa, c.15 km on road to Nungo, slopes of Mt Kuwanku, 850 m, ♀ fl. 22.ii.1982, *Jansen & Boane* 8062 (K, PRE). Z: Maganja da Costa, 12 km along road to Namacurra, 40 m, ♂ fl. 26.i.1966, *Torre & Correia* 14127 (LISC); ♀ fr., 26.i.1966, *Torre & Correia* 14128 (LISC). T: Cabora Bassa, from Songo to R. Mucangádzi, c.500 m, ♂ fl. 27.i.1963, *Torre, Carvalho & Ladeira* 18896 (LISC). MS: Between Gondola & Nhamatanda (Vila Machado), ♀ fl. 22.iii.1960, *Wild & Leach* 5222 (K, MAL, SRGH); Gondola Dist., Garuso Mt., ♂ fl. 4.iii.1948, *Barbosa* 1095 (LISC). GI: Masiyena, 32 km N of Limpopo R. mouth, ♂ fl. 1.iii.1928, *Earthy* 54 (PRE). M: Inhaca Is., lighthouse, ♂ fl. 21.xii.1958, *Mogg* 30367 (K, MAL, PRE, SRGH); Inhaca, fields close to Estação de Biologia Marítima, ♀ fr. 3.i.1975, *Moura* 52 (LISC, LMU).

The species is extremely variable and can be found across all of Subsaharan Africa in both forest and grassland; sea level to 2000 m. Flowering from December to March; fruiting from March onwards.

Conservation notes: Widespread species; not threatened.

The tuber of *D. quartiniana* is occasionally said to be eaten. In Inhambane, S Mozambique, the cooked tubers are eaten as a laxative. It is also said to be a drug for bees.

In the Flora area, a number of forms are encountered:

1. Leaflets elliptic or oblanceolate, sparsely pubescent above to pubescent on veins only. Inflorescence usually very dense. This form is found over most of the region except the Caprivi Strip, Botswana and W Zimbabwe.

2. Leaflets obovate or spathulate, with dense soft greyish pubescence above and below. Concentrated in the Caprivi Strip, Botswana, Zimbabwe and S Zambia.

3. Habit and stems very slender, leaflets small, thin, apices acuminate, subglabrous. Male inflorescences often small or quite lax. Only found at higher elevations on Mt Mulanje, Malawi.

4. Leaflets glabrous except on veins, very narrowly elliptic or oblanceolate to linear, 5–7(9)-foliolate. Only in Mozambique.

However, there are intermediates between all which makes an infraspecific classification impossible – see Wilkin (Kew Bull. **54**: 1–18, 1999) for more discussion and a complete synonymy. The lectotype of *D. crinita* was erroneously cited in that paper as *Wood* s.n., rather than *Wood* 1618. Forms with very long floral bracts are sometimes collected in coastal Mozambique; they are sometimes placed under the South African synonym *D. crinita* (e.g. in Reddy & Balkwill's unpublished Inhaca Is. Checklist). The South African species *D. retusa* Mast. is found in KwaZulu-Natal and the former Transvaal; it has yet to be collected in the Flora area but may be found in Mozambique. It differs from *D. quartiniana* in its strongly undulate leaflet margins (rarely found in the latter species, and less markedly undulate when present) and in male plants by the more lax inflorescence, which is not catkin-like due to the 1.5–2.7 mm long floral bracts, usually shorter than the tepals. The pedicels are usually longer, (0.5)0.8–1.8(2) mm, and the outer tepals are 2.4–3(3.3) mm long, narrowly ovate, usually held erect, giving the flower a conical appearance in bud. Female plants, either in flower or fruit, are much harder to distinguish though they do have more erect tepals.

13. **Dioscorea sansibarensis** Pax in Bot. Jahrb. Syst. **15**: 146 (1892). —Durand & Schinz, Consp. Fl. Afr. **5**: 275 (1893). —Harms in Engler, Pflanzenw. Ost-Afrikas **C**: 146 (1895). —Baker in F.T.A. **7**: 418 (1898).— Chevalier in Bull. Mus. Natl. Hist. Nat., sér.2, **8**: 529 (1936). —Burkill in Bull. Jard. Bot. État **15**: 348 (1939). —Miège in F.W.T.A., ed.2, **3**(1): 152 (1968). —Milne-Redhead in F.T.E.A., Dioscoreaceae: 7

(1975). —N'Kounkou et al. in Fragm. Florist. Geobot., suppl. **2**: 157, fig.6 (1993).
—Wilkin in Kew Bull. **56**: 397 (2001). Type: Tanzania, Bagamoyo (Bagamojo), ♂
fl. v.1874, *Hildebrandt* 1284 (B† holotype). FIGURE 12.2.**31**.7.

   Dioscorea macroura Harms in Notizbl. Königl. Bot. Gart. Berlin **1**: 266 (1897). —Baker in
F.T.A. **7**: 416 (1898). —De Wildeman in Ann. Mus. Congo Belge, Bot., sér.5, **2**: 22, t.28
(1907). —Chevalier in Bull. Mus. Natl. Hist. Nat., sér.2, **8**: 523 (1936). Types: Cameroon,
Yaoundé, ♂ fl. ix.1891, *Zenker* 620 (B† syntype) & Yaounde, ♂ fl. ix.1894, *Zenker & Staudt*
414 (B† syntype, BM, K).

   Dioscorea welwitschii Rendle, Cat. Afr. Pl. Welw. **2**: 37 (1899). Type: Angola, Golungo Alto,
Sobado de Quilombo-Quiacatubia, ♂ fl. ii.1855, *Welwitsch* 4041 (BM holotype, K).

Robust herbaceous twining vine to 30 m, stems annual from a large perennial tuber. Tuber
solitary, to 50 cm diameter or more, globose with flattened base and depressed apex; roots
many, close to substrate surface. Indumentum absent. Stems left-twining, usually unarmed;
bulbils often present in axils, 0.6–5.5 cm in diameter, globose to ovoid, sometimes impressed at
apex like tuber, often smooth, glossy and black in Flora area (dark purple-brown when dry), but
may also be glaucous green (immature) or pale brown (mature) to somewhat pink and warty.
Leaves usually opposite, blade entire, apex 1.1–6.3 cm long, acuminate, thickened and with
inrolled margins, darker in colour and containing nitrifying bacteria, immature blade smaller,
membranous, with margins shallowly to quite deeply lobed, especially towards base, lobes
usually 3–9; petiole with a pair of fleshy oblong lateral nodal flanges to c.1 cm long at extreme
base in larger leaves, most frequently seen near stem bases where they appear like cataphylls.
Inflorescence spicate, simple, lax. Male flowers solitary or in pairs, (sub)sessile, campanulate;
outer tepals 2–3.6 × (0.5)0.7–1.6 mm, narrowly elliptic to lanceolate or narrowly oblong, rarely
elliptic or narrowly ovate, inner tepals 2.1–3.5 × (0.5)0.9–1.7 mm. Capsule 3.6–5 × 2–2.8 cm,
glaucous, especially when immature. Seed wings extending all around seed margin, but
elongate towards base and apex.

Zambia. E: Mpika Dist., Luangwa Game Reserve, st. 7.v.1965, *Mitchell* 2879 (K).
Zimbabwe. N: Hurungwe Dist., near Rukomeche Research Station, 550 m, ♀ fl.
ii.1967 *Müller* 579 (K, PRE, SRGH). **Malawi**. N: Nkhata Bay, Makaluwe, Likoma Is.,
♂ fl. 20.iii.1989, *Balaka & Patel* 2011 (K, MAL). C: Ntcheu Dist., Ntonda Mission, st.
16.ii.1972, *Salubeni* 1771 (MAL, PRE, SRGH). S: Blantyre, Mpatamanga gorge, along
Shire R. banks, ♂ fl. 5.ii.1979, *Masiya & Tawakali* 155 (MAL, SRGH); Mangochi, SW
of Maldeco path to Senjele village, ♀ fl. 13.ii.1989, *Tawakali & Nachamba* 1407
(MAL). **Mozambique**. N: Mandimba, Ngami R., st. 9.iii.1942, *Hornby* 4512 (PRE);
Mogovolas, Nametil, ♀ fl. 11.iii.1937, *Torre* 1314 (COI, LISC). Z: Lugela Dist.,
Namagoa, st. 19.iii.1949, *Faulkner* K 419 (BR, K); Milange, 4 km from Sabelua towards
Liciro, crossing for Mongoé after 8.8 km, Mt Sonelafuti, 600 m, ♂ fl. 14.ii.1972,
Correia & Marques 2639 (LMU). T: Lupata, st. 30.i.1859, *Kirk* s.n. (K); Cabora Bassa,
after dam on N bank of R. Zambezi, 230–330 m, ♂ fl. 12.iv.1972, *Pereira & Correia*
1989 (LMU). MS: Gondola, L bank of R. Nhamissanguere, st. 3.ii.1948, *Garcia* 15
(LISC). GI: Between Lumane and Missão de Chucumbane, st. 7.ii.1948, *Torre* 7291
(LISC). M: Maputo, st. 26.xi.1947, *Mendonça* 3537 (LISC).

Found across most of Subsaharan Africa and in Madagascar, but in Zimbabwe
found only in the Zambezi valley. Often growing close to water or in riverine forest;
200–650 m. Rarely seen in flower, which is between February and April; no fruiting
collections from the region.

Conservation notes: Widespread species; not threatened.

14. **Dioscorea schimperiana** Kunth, Enum. Pl. **5**: 339 (1850). —Durand & Schinz,
   Consp. Fl. Afr. **5**: 275 (1893). —Baker in F.T.A. **7**: 419 (1898). —Burkill in Bull.
   Jard. Bot. État **15**: 371 (1939). —Miège in F.W.T.A., ed.2, **3**(1): 152 (1968). —
   Milne-Redhead in F.T.E.A., Dioscoreaceae: 14 (1975). —N'Kounkou et al. in

Fragm. Florist. Geobot., suppl. **2**: 153 (1993). —Wilkin in Kew Bull. **56**: 398 (2001). Type: Ethiopia, Tigray, Djeladjeranne, ♂ fl. & ♀ fr. 7.viii.1840, *Schimper* 1642 (B holotype, BR, K, P, UPS). FIGURE 12.2.**31**.2.

   Dioscorea schimperiana var. *vestita* Pax in Bot. Jahrb. Syst. **15**: 148 (1892). —Durand & Schinz, Consp. Fl. Afr. **5**: 276 (1893). —Harms in Engler, Pflanzenw. Ost-Afrikas **C**: 146 (1895). —De Wildeman in Bull. Jard. Bot. État **4**: 322 (1914). —Eyles in Trans. Roy. Soc. S. Africa **5**: 330 (1916). —Chevalier in Bull. Mus. Natl. Hist. Nat., sér.2, **8**: 533 (1936). — Weimer in Bot. Not. **1937**: 161 (1937). Types: Malawi, Shire Highlands, ♂ fl. 1881, *Buchanan* 112 (K syntype) and Sudan, Equatoria, Gumango Hill, ♂ fl. 17.vi.1870, *Schweinfurth* 3920 (B† syntype, K).

   Dioscorea hockii De Wild. in Bull. Jard. Bot. Etát **3**: 277. (1911). Type: Congo, Katanga, Luembe R., ♂ fl. 11.ii.1910, *Hock* s.n. (BR holotype).

   Dioscorea stellato-pilosa De Wild. in Ann. Mus. Congo Belge, Bot., sér.5, **3**: 369 (1912). Type: Congo, Katanga, Katola, ♀ fl. iv.1908, *Sapin* s.n. (BR holotype).

Herbaceous twining vine to at least 8 m, stems annual from a fleshy tuber. Tuber solitary, replaced annually, at least 60 × 3 cm diameter, cylindric, descending vertically into substrate, periderm brown or red-brown, parenchyma cream-yellow, often with dense blood-red pigmentation. Indumentum consisting of stellate to dendroid hairs, often dense and smaller in diameter on leaf lower surface, inflorescence axis, ovaries and tepals. Stems right-twining, usually stout and lacking prickles; bulbils to 1 cm in diameter, often present in axils, especially when in flower, dark brown globose or ovoid, tuberculate, soon falling. Leaves usually opposite, entire; cataphylls and lateral nodal flanges absent. Inflorescence simple, spicate. Male flowers lax, (sub)sessile, solitary; outer tepals 1.7–2.5 × 0.8–1.8 mm, ovate to lanceolate or narrowly elliptic, inner tepals 1.3–2.5 × 0.6–1.2 mm, elliptic to narrowly elliptic or narrowly oblong, stamens 6. Capsule 1.8–2.4 × 2.5–3.9 cm, papery. Seed wings extending all around seed margin.

Zambia. N: Mpika, ♂ fl. 30.i.1955, *Fanshawe* 1898 (K, NDO); Mpulungu Dist., Lunzua valley, c.5 km from Mpulungu Road, 880 m, ♀ fl. 6.ii.1952, *Richards* 1163 (K). W: Mwinilunga, just S of Matonchi Farm, ♂ fl. 2.i.1938, *Milne-Redhead* 3919 (B, BM, K); ♀ fl. 2.i.1938, *Milne-Redhead* 3921 (B, BM, K). C: Mumbwa, ♂ & ♀ fl. 1912, *Macaulay* 1148 (K, one sheet mixed collection with *D. hirtiflora*). E: Lundazi, ♀ fr. 18.x.1967, *Mutimushi* 2258 (K, NDO, SRGH). **Zimbabwe**. N: Mazoe (Mazowe), Yatagma Valley, 1310 m, ♂ fl. i.1907, *Eyles* 503 (BM, SRGH); Mazoe Dist., Mazoe Dam, ♀ fr. 23.v.1934, *Gilliland* B210 (BM). C: Marondera (Marandellas), Delta, 1370 m, ♂ fl. 11.ii.1942, *Dehn* 589 (K, SRGH); 43 km N of Harare beyond Domboshawa, near Nyakudya, ♀ fl. 2.ii.1997, *Wilkin* 936 (K, SRGH). E: Chipinge Dist., Chirinda, 1130–1220 m, ♂ fl.& ♀ fl.& imm.fr. ii.1906, *Swynnerton* 214 (BM some fr. are *D. dumetorum*, K, SRGH); Nyanga (Inyanga), near Nyamingura R., 915 m, ♀ fr. 23.iv.1958, *Phipps* 1235 (K, LISC, SRGH). S: Bikita Dist., Bikita communal land, Moodie's Pass, Mahagwane Hill, 915 m, ♂ fl. 14.xii.1978, *Hall* 259 (SRGH). **Malawi**. N: Nkhata Bay, 8 km E of Mzuzu, Roseveare's, 1280 m, ♂ fl. 14.ii.1970, *Pawek* 3359 (K, MAL); Nkhata Bay, 4 km SW of Chikangawa, 1750 m, ♀ fr. 18.viii.1978, *Phillips* 3781 (K, MAL, MO, SRGH, WAG). C: Lilongwe, Ngara Hill, 1250 m, ♂ fl. 7.ii.1959, *Robson* 1508 (K, LISC, SRGH); Dedza, Chongoni Forestry School, ♀ fl. 2.iii.1967, *Jeke* 78 (K, LISC, MAL, SRGH). S: Mt. Mulanje, lower slopes, below Cilemba cliffs, 1080 m, ♀ fr. 6.ii.1986, *Chapman & Chapman* 7175 (K, MAL, MO); Zomba, Zomba Plateau, 1830 m, ♂ fl. 23.i.1995, *Wilkin & Tawakali* 763 (K, MAL). **Mozambique**. N: Lago Dist., Maniamba, ravine to E, ♀ fr. 22.v.1948, *Pedro & Pedrógão* 3803 (LMU); Maniamba, base of Serra Jéci, near Malulo, c.60 km from Lichinga, 1450 m, ♂ fl. 3.iii.1964, *Torre & Paiva* 11009 (LISC). T: Tsangano Dist., c.60 km from Nelongue, ♀ fr. 18.ii.1980, *Macuácua, Stefanesco & Mateus* 1073 (MAL). MS: Mossurize Dist., Serra de Espungabera, 1030 m, ♂ fl. 15.ii.1966, *Pereira, Sarmento & Marques* 1373 (LMU).

Widespread across tropical Africa. Among rocks and on riverbanks, termite mounds, forest margins and in *Brachystegia* woodland, generally in more open or disturbed areas; 800–1900 m. Flowering from December to March, with male flowers usually earlier and female flowers later. Both male and female flowers open in the afternoon. Fruiting from February onwards.

Conservation notes: Widespread species; not threatened.

D. schimperiana is closely related to *D. hirtiflora*, with which it shares stellate pubescence. They can usually be differentiated in the vegetative state by the dense pubescence of the lower leaf surface in *D. schimperiana* together with the ovate to orbicular leaf shape; the leaves of *D. hirtiflora* are not densely pubescent below and are usually lanceolate to ovate or elliptic in the Flora area. The male flowers of *D. schimperiana* have 6 fertile stamens and tepals 1.3–2.5 mm long, while *D. hirtiflora* has 3 fertile stamens and 3 staminodes and tepals 2.4–4 mm long. The capsules are also distinctive, the lobes usually being broadly obovate in outline in *D. schimperiana* and transversely elliptic in *D. hirtiflora*; the fruit is thicker-walled and darker brown when dry in the former.

The bulbils are eaten occasionally as a famine food in Zimbabwe.

15. **Dioscorea sylvatica** Eckl. in S. Afr. Quart. J. **1**: 363 (1830). —Drège in Linnea **20**: 234 (1847). —Britten in J. Bot. **46**: 200 (1908). —Burkill in J. S. Afr. Bot. **18**: 188 (1952). —Wilkin in Kew Bull. **56**: 400 (2001). Type: South Africa, Cape Province, Uitenhague, Krakakamma (Kragakamma, between Port Elizabeth and Van Stadenberg), ♂ fl. & ♀ fr. —i.1830, *Ecklon & Zeyher*, sheet labelled "*Dioscorea sylvatica* Ecklon 3.1" (TCD lectotype, BM, P). FIGURE 12.2.**32**.3.

    Testudinaria sylvatica Kunth, Enum. Pl. **5**: 443 (1850). —Baker in F.C. **6**: 253 (1896). Type: Cultivated plant of South African origin (B†).

    Dioscorea rehmanni Baker in F.C. **6**: 248 (1896). Type: South Africa, Limpopo Prov., Zoutpansberg, Hout Bosch, ♂ fl. 1875–1880, *Rehmann* 5783 (K holotype).

    Testudinaria paniculata Dummer in Bull. Misc. Inform., Kew **1912**: 195 (1912). Type: Cultivated plant from South Africa (KwaZulu-Natal), ♂ fl. x.1911, *Dümmer* s.n. (K holotype).

    Testudinaria multiflora Marloth in Trans. Roy. Soc. S. Africa **3**: 127 (1913). Type: South Africa, Limpopo Prov., Zoutpansberg Range, near Santa, ♀ fr. 1912, *Marloth* 5097 (B holotype B, BM).

    Dioscorea brevipes Burtt-Davy in Bull. Misc. Inform., Kew **1924**: 232 (1924). Type: South Africa, Limpopo Prov., Waterburg Dist., Potgietersrust, ♂ fl. 3.xi.1908, *Leendertz* 1510 (K holotype).

    Dioscorea marlothii R. Knuth in Engler, Pflanzenr. IV **43**: 321 (1924), nom. nov. for *Testudinaria multiflora* Marloth. Type as for *T. multiflora*.

    Dioscorea montana (Burch.) Spreng. var. *glauca* R. Knuth in Engler, Pflanzenr. IV **43**: 323 (1924). —Weimer in Bot. Not. **1937**: 161 (1937). Type: South Africa, KwaZulu-Natal, Inanda, ♀ fl. 11.xii.1888, *Wood* 780 (K holotype).

    Dioscorea montana (Burch.) Spreng. var. *lobata* Weim. in Bot. Not. **1937**: 162 (1937). Type: Zimbabwe, Makoni, between Rusape and Nyanga, c.25 km from Rusape, ♂ fl. 28.xi.1930, *Norlindh & Weimarck* 3270 (LD holotype, BM).

    Dioscorea montana (Burch.) Spreng. var. *sagittata* Suess. in Proc. & Trans. Rhod. Sci. Assoc. **43**: 78 (1951). Type: Zimbabwe, Marondera (Marandellas), xii.1947, *Dehn* 721 (M holotype).

Herbaceous twining vine to at least 4 m, stems annual, 1–many per year from a perennial woody tuber. Tuber solitary, subterranean, to 30 cm diameter, entire to deeply lobed, irregularly ovoid to flattened-ovoid (usually horizontally flattened), surface rough and cracked, sometimes in a regular reticulate pattern producing 'scales'; dense, thick roots emerging from tuber, especially on lower surfaces. Indumentum absent. Stems left-twining, unarmed; bulbils absent. Leaves alternate, entire to 3-lobed, heart-shaped to deltoid, papery to coriaceous, sometimes glaucous or greyish, especially below, apex acuminate to obtuse, with a 1–2 mm long triangular apex, cataphylls and lateral nodal flanges absent. Inflorescence racemose; male simple, flowers solitary on pedicels or sometimes in cymose pairs; female inflorescence to 10 cm long. Male flowers reflexed before tepals open; tepals narrowly oblong to narrowly oblanceolate or spathulate, suberect at first but spreading or sometimes recurved at anthesis; outer tepals 1.9–3.2 × 0.5–1.1 mm; inner tepals 1.9–3.3 × 0.5–1.1 mm. Capsule 1.6–2.8 × (1)1.4–2 cm. Seeds winged at apex, occasionally with a very narrow basal wing.

Zambia. N: Mpika, ♂ fl. 4.ii.1955, *Fanshawe* 1972 (K, LISC); ♀ fr. 4.ii.1955, *Fanshawe* 1983 (K, LISC). C: Chilanga, Mt Makulu, ♀ fr. 1.i.1958, *Angus* 1809 (K, LISC); Kabwe (Broken Hill), ♂ fl. 16.xii.1957, *Fanshawe* 4133 (K). S: Mazabuka, 32 km N of Mazabuka on Great North Road, ♂ fl. 29.xii.1962, *Angus* 3459 (FHO, K, LISC, P); ♀ fr. 29.xii.1962, *Angus* 3459A (FHO, K). **Zimbabwe**. N: Gokwe, Sengwa Research St., ♂ fl. 8.xi.1968, *Jacobsen* 275 (K, SRGH). W: Matobo Dist., Besna Kobila, 1400 m, ♂ fl. xi.1956, *Miller* 3937 (K, PRE, SRGH); Matobo Dist., Besna Kobila Farm, ♀ fr. iii.1960, *Miller* 7229 (SRGH, WAG). E: Mutare, Mozambique border, on granite ridge above customs post, c.1370 m, ♂ fl. 28.xi.1947, *Chase* 564 (BM, K, SRGH); Nyanga Dist., Nyamziwa Falls, 1650 m, ♀ fl. 12.i.1951, *Chase* 3684 (BM, K, LISC, SRGH). **Mozambique**. M: Inhaca Is., Cabo de Inhaca, ♀ fr. 22.ix.1957, *Mogg* 31717 (K, PRE, SRGH).

Also in South Africa. Often found in rocky areas and kopjes, in *Brachystegia* or mopane woodland, but may also occur on sandy soils, termite mounds and more disturbed habitats; sea level–1800 m. Flowering from November to February; fruiting from November onwards.

Conservation notes: Widespread species; not threatened.

This species has a complex nomenclatural history, summarised by Burkill (1952) and Wilkin (2001). The specimen(s) of *D. sylvatica* determined by Ecklon & Zeyher as *Testudinaria montana*, and sent to Berlin where Knuth saw them, were probably destroyed. Further studies, especially of a number of South African taxa, are needed to investigate the specific delimitation of *D. sylvatica*. *D. elephantipes*, for example, appears to differ only in its leaf shape. A thorough examination of tuber morphology would be very useful.

D. sylvatica is tolerant of heavy metal-rich soils such as those on the Great Dyke in Zimbabwe. There are records of the use of its tubers as food after soaking in water in Zambia. It was used in the past as a source of steroidal saponins for the pharmaceutical industry (see Coursey, Yams: Nature, Origins, Cultivation & Utilisation: 217, 1967).

2. TACCA J.R. & G. Forst.

Tacca J.R. & G. Forst., Char. Gen. Pl.: 35 (1775), nom. conserv. —Limpricht, Beitr. Taccac.: 43 (1902). —Drenth in Blumea **20**: 374 (1972). —Caddick et al. in Taxon **51**: 110 (2002).

Stemless perennial herbs. Underground parts rhizomatous, rhizome frequently short, at apex of 1–3 globose or elongate tubers. Indumentum of simple, multicellular hairs usually present only on young parts. Leaves in basal rosette, petiolate; laminae large, entire to pinnately-lobed or palmatisect; primary venation palmate or pinnate, secondary venation reticulate; petiole erect, ribbed, weakly sheathing at base, rarely hollow. Plants hermaphrodite; inflorescences on leafless, unbranched scapes arising directly from rhizome, cymose but appearing umbelliform, surrounded by large, ovate involucral bracts in two whorls; floral bracts long, filiform, pendent, falling off after anthesis. Flowers with 2 whorls of 3 tepals, basally fused, green to dark brown or purple in colour. Stamens 6, in 2 whorls, basally fused to tepals, with short dorso-ventrally flattened filaments, connective broad, lower part growing over the thecae in a hood, forcing them to dehisce extrorsely. Ovary inferior, 6-ribbed. Fruit berry-like, 6-ribbed, indehiscent to disintegrating irregularly, or capsular, dehiscing loculicidally. Seeds prismatic to reniform, with 6–20 longitudinal ridges.

A genus of 13 species from the wet to seasonally dry tropics; 10 species in Asia, 2 species in Africa and Madagascar and 1 in South America.

Recent research has suggested that it may be necessary to recognise more species. For example, Phillipson (http://www.efloras.org/florataxon.aspx/flora_id=128&taxon_id =132180) has suggested there are 7 species in Madagascar alone based on differences in leaf morphology, involucral bracts and ecology. This requires further testing, especially in the light of lack of floral character differences between species. A single African species is recognised here, as in Drenth's treatment.

Tacca leontopetaloides (L.) Kuntze, Rev. Gen. Pl. **2**: 704 (1891). —Carter in F.T.E.A., Taccaceae: 1 (1962). —Drenth in Blumea **20**: 375 (1972). —Lisowski, Malaisse & Symoens in F.A.C., Taccaceae: 3 (1976). Type: India, fig. in Amman, Comment. Acad. Sci. Imp. Petrop. **8**: 211, t.13 (1736) (see Merrill in J. Arnold Arbor. **26**: 92, 1945). FIGURE 12.2.**35**.

   Leontice leontopetaloides L., Sp. Pl.: 313 (1753).

   Tacca pinnatifida J.R. & G. Forst., Char. Gen. Pl.: 35 (1775). —Durand & Schinz, Consp. Fl. Afr. **5**: 272 (1893). —Baker in F.T.A. **7**: 413 (1898). —Rendle, Cat. Afr. Pl. Welw. **2**: 36 (1899). —Limpricht, Beitr. Taccac.: 50 (1902). —Chevalier, Fl. Afr. Centr., Énum. Pl. Récolt.: 308 (1913). —Danguy in Bull. Mus. Natl. Hist. Nat. **19**: 493 (1913). —Limpricht in Engler, Pflanzenr. IV **42**: 27 (1928). Type: Tahiti, no locality or date, *Forster* s.n. (K holotype, BM, P).

   Tacca involucrata Schumach. & Thonn. in Schumacher, Beskr. Guin. Pl.: 177 (1827). —Hooker, Niger Fl.: 535 (1849). Type: Ghana ("Danish Guinea"), no locality or date, *Thonning* 293 (C holotype).

   Tacca quanzensis Welw. in Ann. Cons. Ultram., parte não official, sér.1 [Apontamentos Phytogeographicos]: 591 (1859). —Rendle, Cat. Afr. Pl. Welw. **2**: 36 (1899). Type: Angola, Pungo Andongo, 28.x.1858, *Welwitsch* 6475 (BM).

   Tacca abyssinica Baker in F.T.A. **7**: 413 (1898). Type: Ethiopia, no locality, fr. iii.1846, *Schimper* 1946 (K holotype, BM, P, S).

   Tacca pinnatifida var. *acutifolia* H. Limpr., Beitr. Taccac.: 55 (1902). Types: Ethiopia, NW, Wochui, no date, *Steudner* 436 (B syntype); Tanzania, no locality or date, *Stuhlmann* 6081 (B syntype) & 6484 (B syntype); Malawi, Shire Highlands, 1885, *Buchanan* 4 (K syntype).

Erect herb to 2 m, annual from a fleshy tuber, leaves and flowers produced together during the rainy season. Rhizome cylindric, tough, produced at apex of previous season's, with a narrower rhizomatous branch bearing the new tuber; tuber ovoid to globose, sometimes flattened or irregular, up to 4(10) cm in diameter. Indumentum restricted to small, flattened translucent projections on veins of leaf lower surface and capitate glandular hairs on the style base. Leaves usually 1–3 per plant; petiole to 1.3(1.8) m long, erect, terete, longitudinally ridged, usually solid, narrowing towards apex; blade trifid, compound, to c.90 × 60 cm, thin and stiff, with three leaflets to 40 cm long, pinnately divided into lobes, sometimes also palmately branched, lobes variable, often dimorphic in appearance and/or larger towards rachis apex, margins entire or with shallow, rounded lobes; cataphylls 2 or 3 inserted on rhizome, erect, clasping petiole and scape bases, 4–21 × 0.8–3 cm, ovate-acuminate to very narrowly oblong, membranous when dry. Inflorescences 1(2) per plant, scape erect, 30–170 × 0.3–1.5(3) cm, hollow, longitudinally ridged, umbelliform; inflorescence bracts 4–8, forming an involucre with floral buds, erect to spreading at anthesis, shape and size variable but not dimorphic, 8–30(44) × 7–29 mm, oblanceolate or obovate to ovate, elliptic, narrowly elliptic or lanceolate, green with purple margins and apices, sometimes more extensively pigmented, persistent; floral bracts numerous, to 250 × c.1 mm, filiform, pendent, usually dark purple, paler to white towards apex, partially persistent. Flowers 10–40, pale green to white, yellowish

Fig 12.2.**35**. TACCA LEONTOPETALOIDES. 1, habit (× ¹/₆); 2, part of leaf (× ¹/₃); 3, tuber, scape and cataphyll (× ¹/₂); 4, inflorescence at fruiting stage (× ²/₃); 5, flower, side view (× 3); 6, opened flower, ovary and style with 1 inner and 1 outer tepal (× 3); 7, 3 tepals with anthers in hoods (× 3); 8, ovary and style (× 6); 9, capsule, side view (× ²/₃); 10, seed (× 4). 1, 3, 5 from *Pawek* 4171, 2, 4 from *Goyder & Paton* 3511, 6–8 from *Bingham* 10224, 9–10 from *Pawek* 2310. Drawn by Juliette Williamson.

or brown, usually with variably distributed purple pigmentation, pedicels 9–25 × c. 1 mm (larger in fruit); Tepals differentiated into 2 whorls of 3 (outer narrower), thick and fleshy, fused at base and forming a 3–4 mm deep bowl-shaped torus. Filaments 2–2.8 × 2–3 mm, anthers dorsifixed, partially surrounded by a 1–2.6 × 0.8–2.7 mm broadly oblong to broadly oval hood. Ovary obconic to obovoid with 6 longitudinal ridges; style erect, 1.7–2.3 mm long, base expanded and conical. Fruit berry-like, 20–30 × 17–29 mm, pendent, only 2–7 flowers per umbel developing, globose to ovoid, green but becoming orange at maturity, dark brown when dry, base usually rounded. Seeds many, 4–6.2 × (2.5)2.9–3.9 mm, ovoid to ellipsoid, sometimes longitudinally flattened, longitudinally striate, red-brown when dry, surrounded by a thin fleshy aril not visible when dry.

Zambia. N: Kasama Dist., 13 km N of Kasama, fl. 22.xii.1961, *Astle* 1143 (K). W: Ndola, fl. 9.i.1954, *Fanshawe* 655 (K, LISC). C: Lusaka, Protea Hill Farm, 13 km SE of Lusaka, fl. 25.xii.1994, *Bingham* 10224 (K). E: Chipata Dist., Nzamane, fr. 6.i.1959, *Robson* 1044 (BM, K, LISC). S: Victoria Falls, fl. i.1906, *Allen* 251 (K). **Zimbabwe**. N: Bindura Dist., Chinamora communal land, Pote R., fl.& fr. 28.i.1964, *Müller & Smith* 17 (K, SRGH). E: Mutare Dist., Burma Valley, Manyera Farm, fl. 22.xi.1959, *Chase* 7483 (K, SRGH). **Malawi**. N: Chitipa Dist., 6.5 km E of Lufita, 6.5 km W of Kapoka crossroads, fl. 29.xii.1977, *Pawek* 13443 (K, MAL, MO). C: Kasungu Dist., Chimaliro Forest, 1.6 km E on Phaso road, fr. 27.iii.1978, *Pawek* 14123 (K, MO). S: Kasupe Dist., Malosa Forest Reserve, c.3 km E of Kasupe 15°11'S, 35°20'E, 10.i.1992, *Goyder & Paton* 3511 (BR, K, MAL, PRE). **Mozambique**. N: Cabo Delgado, 15 km from Pemba, fl. 17.i.1984, *Groenendijk, Maíte & Dungo* 823 (K, LMU). Z: Lugela Dist., Namagoa, fl. ii–iii.1943, *Faulkner* 286 (K, PRE). T: Moatize Dist., on road Zóbuè–Tete, crossing to Kambuatsitsi (Entroncamento), 400 m, fl.& fr. 14.i.1966, *Torre & Correia* 14041 (LISC). MS: Garuso, Jagersberg Farm, fr. 29.i.1949, *Chase* 1362 (K, SRGH).

Widespread in Subsaharan Africa west to Senegal and south to Zimbabwe; also in Madagascar, Comoros, Mauritius, Seychelles and Tropical Asia to the Pacific. In a wide range of habitats from forest to grasslands, rocky places, termite mounds, dunes and beaches, often found on sandy soils; sea level–1500 m. Flowering in December and January; fruiting from January onwards.

Conservation notes: Widespread species; not threatened, although local forms or cultivars that merit conservation may exist.

Tacca is easily identified when fertile by the umbelliform inflorescence with an involucre of inflorescence bracts and filiform floral bracts. When in a vegetative state it may be confused with *Amorphophallus abyssinicus* (A. Rich.) N.E. Br., but *T. leontopetaloides* has a longitudinally ridged petiole while that of *A. abyssinicus* is smooth.

The tubers are edible although bitter, and contain steroidal taccalonolides. They are grated, washed and cooked for a long time or turned into flour. Records from N Malawi suggest that the tuber is cooked and pounded to obtain a milk substitute. Although now eaten mainly in times of famine, it may have been more widely cultivated in the past. Elsewhere in Africa, its uses include being a fibre source and a cure for oedema. See Burkill (Useful Plants W Tropical Africa, ed.2, **5**: 183, 2000) for a detailed account of its uses as a food, in medicine and rituals in Africa and beyond.

As a cultivated plant, it is questionable whether *T. leontopetaloides* is native to Africa. It may have been brought from Asia in prehistoric times with *Dioscorea alata* and bananas. A phylogeographic study using material from across the species' range could be used to discover its area of origin.

BURMANNIACEAE

by M. Cheek

Herbaceous, terrestrial, annual or perennial herbs, usually with underground rootstocks; partly or completely saprophytic (mycotrophic) with reduced green leaves, or lacking green tissue entirely; rarely epiphytic. Leaves rosette-forming or spirally-arranged on stems, sometimes sparse and reduced to scales and inconspicuous, entire. Inflorescence terminal, peduncled, cymose, bracteate; bracteoles absent. Flowers hermaphrodite, zygomorphic or actinomorphic, trimerous. Perianth tube well-developed, lobes 3–6, in 1 or 2 whorls. Anthers 3–6, included within and attached to perianth tube, bithecal, connective well-developed. Ovary inferior, 3-locular with axile placenta, or 1-locular with parietal or pendulous placenta, sometimes with globose glands embedded around the style base; ovules numerous, minute; style included, stigma 3-lobed or flattened. Fruit a dehiscent capsule. Seeds numerous, minute; endosperm sparse or nil.

A family of 18 genera and c.140 species, mainly in the tropics. Four genera in Africa.

Some authors have segregated Thismiaceae from Burmanniaceae to accommodate those genera (such as *Afrothismia*) that have a style much shorter than the perianth tube and towards which the staminal filaments curve downwards, below an annulus. The fruits lose the perianth tube entirely.

The key, family and generic descriptions here are mainly based on material from the Flora area.

1. Flowers zygomorphic, perianth tube curved, largely horizontal; externally visible perianth lobes 6, filamentous · **1. Afrothismia**
 – Flowers actinomorphic, perianth tube straight, erect; externally visible perianth lobes 3, subovate or elliptic · 2
2. Plant of dambos in full sun; leaves green, in minute basal rosette; flowers blue, persistent, with 3 large conspicuous lateral wings · · · · · · · · · · · · **2. Burmannia**
 – Plant of forest shade; leaves colourless, scattered on stem; flowers white, short-lived, without wings · **3. Gymnosiphon**

1. AFROTHISMIA Schltr.

Afrothismia Schltr. in Bot. Jahrb. Syst. **38**: 138 (1906); in Notizbl. Bot. Gart. Berlin-Dahlem **8**: 43 (1921). —Jonker in Meded. Bot. Mus. Herb. Rijks. Univ. Utrecht **51**: 522 (1938). —Cowley in F.T.E.A., Burmanniaceae: 6 (1988).

Perennial saprophytic herbs entirely lacking green tissue; underground stems with clusters of root bulbils. Leaves alternate, reduced to sparse scales. Inflorescence 1–few-flowered, terminal. Flowers held at ground level, zygomorphic, perianth tube usually ± horizontal, lobes 6, in one whorl, usually filiform; perianth tube separated into two parts, upper and lower, divided by an annulus; lower tube with stamens and style. Stamens 6, lower part of filaments attached to perianth tube, upper parts arching down and attached to stigma. Ovary 1-locular, placentas 3, on a central column. Fruits with circumscissile apical lid.

About 14 species, all in tropical Africa.

Afrothismia zambesiaca Cheek sp. nov.* Type: Malawi, Mua-Livillezi Forest Reserve, fl. 19.iii.1955, *Exell, Mendonça & Wild* 1066 (SRGH holotype, LISC). FIGURE 12.2.**36**.

Afrothismia zambesiaca Cheek sp. nov. ab *A. insigni* Cowley tubo perianthii leviter s-formi (non geniculato), ore perianthii horizontali (non verticali), lobis perianthii valde inequalibus (non aequalibus) differt.

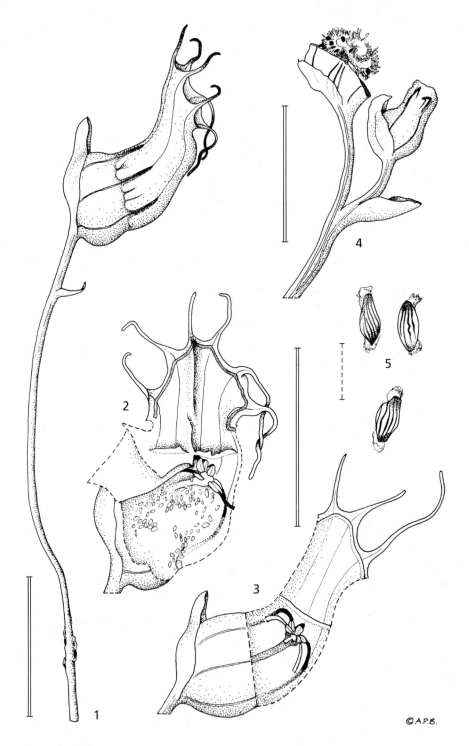

Fig. 12.2.**36**. AFROTHISMIA ZAMBESIACA. 1, habit; 2, flower, dissected; 3, flower, perianth tube in longitudinal section (reconstructed); 4, infructescence with both dehisced and intact fruits; 5, seeds. From *Exell, Mendonça & Wild* 1066. Scale bars: double bar = 1 cm; dashed bar = 500 mm. Drawn by Andrew Brown.

Perennial herb; stem underground, up to 3.6 cm long, with spherical clusters of bulbils (inferred). Stem above ground reduced to the inflorescence, or absent. Leaves colourless, rare or absent, scale-like, spreading, c.2.5 × 1 mm. Inflorescence 1–2-flowered, bracts elliptic, keeled, c.3.5 × 1 mm. Perianth-tube drying dull white with purple lines, glabrous, slightly S-shaped, suberect, c.10 mm long, 3.7 mm wide at midpoint, lower perianth tube c.4.5 mm along main axis; annulus a flap ± 0.02 mm wide, extending ± the whole circumference; upper tube c.5.5 mm along main axis, dilating slightly at mouth; mouth probably elliptic in shape, 6 mm high when dry; width unknown, inner surface with a raised rim c.0.02 mm in diameter, inserted c.0.03 mm from edge. Tepals filamentous, apex acute, base triangular; dorsal and central tepal pairs 3–4 mm long, ventral pair c.9 mm long. Stamens with filaments joined to perianth tube for ± 3 mm, appearing as purple lines on exterior tube surface, distal part free, c.1.5 mm long, arching inward to stigma; anthers ellipsoid c.0.06 × 0.02 mm, appendages absent. Ovary cup-shaped, c.4 × 5 mm, not dissected successfully. Style terete 2.5 × c.0.04 mm; stigma ± 0.9 mm wide, lobation not studied. Fruit capsule c.2 × 6 mm, apex with seeds exposed, the placental mass c.2.2 × 5.5 mm raised from capsule cavity on the axile column. Seeds ellipsoid, ± 0.5 mm long, brown, longitudinally ridged, base and apex with white bodies – possibly elaiosomes. The description is based on sparse, partly-rehydrated herbarium material.

Malawi. C: Dedza Dist., Mua-Livillezi Forest Reserve, 660 m, fl. 19.iii.1955, *Exell, Mendonça & Wild* 1066 (LISC, SRGH).

Known only from the type collection in central Malawi. Growing in the understorey of evergreen forest, it is invisible unless in flower, when it is inconspicuous; 660 m.

Conservation notes: Assessed here as Critically Endangered, CR D1 since it is known from less than 50 individuals at a single site where the forest habitat is threatened by tree cutting; it may already be extinct.

This species was previously determined as both *Afrothismia winkleri* and *A. ?sp. nov.* It differs from *A. winkleri* of West Africa and *A. insignis* of Tanzania in having a gently S-shaped (and not sharply angled) perianth tube and in the perianth lobes, which are strongly unequal in length.

The number immediately prior to this specimen, *Exell, Mendonça & Wild* 1065 (LISC, SRGH) from the same locality, is a *Gymnosiphon*. It is possible that the two species were growing together, as often happens with saprophytes.

2. BURMANNIA L.

Burmannia L., Sp. Pl.: 287 (1753) & Gen. Pl., ed.5: 139 (1754). —Jonker in Meded. Bot. Mus. Herb. Rijks. Univ. Utrecht **51**: 57 (1938).

Erect annual herbs, partly saprophytic (elsewhere in Africa several species are wholly so), lacking underground rootstocks. Stems erect with a basal rosette of small ligulate leaves. Inflorescence terminal, cymose, dichasial. Perianth tube subfleshy, often 3-winged or 3-ridged, outer tepal lobes 3, spreading, inner tepals 3, reduced to flaps in the sinuses. Stamens 3, sessile or subsessile, inserted below inner tepals, thecae lateral, dehiscing transversely, connective with crest above, entire or forked, sometimes with a basal spur. Ovary 3-locular, placentation axile, often 3-winged with perianth; style entire, ± as long as perianth tube; stigmas 1–3, terminal. Capsule with persistent perianth, 3-winged or angled, dehiscing irregularly by slits between the wings. Seeds subellipsoid, minute.

About 50 species in the tropics, either in wetlands in grassland, when typically with blue and yellow-winged flowers and a small basal rosette of green leaves or (not in our area) in forest shade, when typically lacking green tissue entirely, the whole plant dull or bright white, the flowers lacking wings.

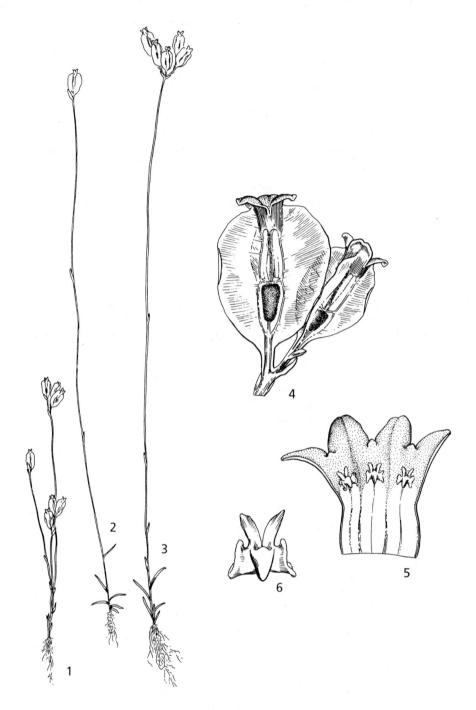

Fig. 12.2.**37**. BURMANNIA MADAGASCARIENSIS. 1–3, habit (× ²/₃); 4, flowers (× 4); 5, perianth tube opened out to show inner tepals and stamens (× 6); 6, anther (× 10). 1 from *Stolz* 2320, 2–3 from *Milne-Redhead & Taylor* 10926, 5–6 from *Richards* 16425. Drawn by Eleanor Catherine. From F.T.E.A.

Burmannia madagascariensis Mart. in Nov. Gen. Sp. Pl. **1**: 12 (1824). —Jonker in Meded. Bot. Mus. Rijks. Univ. Utrecht **51**: 96 (1938). —Obermeyer in Fl. Pl. Afr. **36**: t.1427 (1964). —Geerinck in F.C.B., Burmanniaceae: 4 (1970). —Cowley in F.T.E.A., Burmanniaceae: 2 (1988). Type: Madagascar, *du Petit Thouars* s.n. (BR holotype, P). FIGURE 12.2.**37**.

    Burmannia madagascariensis Baker in J. Linn. Soc., Bot. **20**: 268 (1883), nomen superfl. Types: Madagascar, 21.vi.1866, *Gerrard* 101 (K syntype) & Madagascar, *Baron* 1561 (K syntype).

    Burmannia bicolor Mart. var. *africana* Ridl. in J. Bot. **25**: 85 (1887). Type: Angola, Huíla, Morro de Lopolo, iv.1860, *Welwitsch* 6474 (BM holotype).

    Burmannia bicolor Mart. var. *micrantha* Engl. & Gilg in Warburg, Kunene-Samb.-Exped. Baum: 202 (1903). Type: Angola, Longa, below Chicungo (Chijija), 1200 m, 4.i.1900, *Baum* 619 (B holotype, S).

    Burmannia blanda Gilg in Warburg, Kunene-Samb.-Exped. Baum: 202 (1903). Type: Angola, Cuiriri (Quiriri) swamp, 1300 m, ii.1900, *Baum* 726 (B holotype, K).

    Burmannia inhambanensis Schltr. in Repert Spec. Nov. Regni Veg. **11**: 82 (1912). Type: Mozambique, *Schlechter* 12086 (B holotype).

    Burmannia welwitschii Schltr. in Repert Spec. Nov. Regni Veg. **21**: 84 (1925). Type: Angola, Huíla, Morro de Lopolo, iv.1860, *Welwitsch* 6474 (BM holotype).

Annual herb, 5–35 cm tall, partly saprophytic (leaves highly reduced but still green). Stem erect, minute, with numerous adventitious roots below the leaves. Leaves green, mostly basal in a rosette, a few ascending the stem, linear, 2–12 × 1–2 mm. Inflorescence usually single, rarely 2–3, 1–9-flowered; inflorescence branches each 0–5(28) mm long; bracts minute; pedicels 1–3 mm long. Flowers 1–9, subsessile, 5–8 mm long, 3-winged, wings equidistant, spreading, 1.5–5 mm, wide extending from base of outer tepal lobes to base at apex of pedicel; outer tepal lobes white or reddish white, oblong, 1.5–2.5 × 1.5–2.5 mm, obtuse, slightly fleshy, minutely papillate on upper surface; inner tepals c.0.3 mm, inserted at tube mouth, minute, inconspicuous, variable in length. Stamens sessile, c.1 × 1 mm, connective with 2 variable upper appendages and a basal spur, inserted below inner tepals. Ovary subellipsoid, 2–8 mm long; style cylindric; stigmatic lobes 2–3 mm long, occluding mouth of perianth tube. Seeds brown, numerous, minute.

Zambia. N: Kawambwa Dist., Kawambwa, c.1300 m, fl. 21.vi.1957, *Robinson* 2320 (K, SRGH). W: Zambezi Rapids, Mwinilunga, fl. 18.v.1969, *Mutimushi* 3353 (SRGH). E: Chipata Dist., Luangwa R., fl. 13.vi.1964, *Sayer* 584 (SRGH). **Zimbabwe.** W: Matobo Dist., Matobo Hills, 1500 m, fr. iv.1904, *Eyles* 52 (SRGH). E: Chimanimani Mts, fl. 8.vi.1949, *Wild* 2948 (K, SRGH). **Malawi.** C: Nkhota Khota Dist., Chia area, 480 m, fl. 1.ix.1946, *Brass* 17473 (K, NY). S: Namassi, c.1897, *Cameron* 20A (K). **Mozambique.** Z: Lugela Dist., Namagoa, 100 m, fl. vii.1945, *Faulkner* 253 (K, PRE). MS: Sussundenga Dist., Chimanimani Mts, adjacent to Makurupini Forest, c.300 m, fl. 27.v.1969, *Müller* 1064 (SRGH).

Also in Guinea and countries along the coast east to Cameroon, Congo-Brazzaville, Congo, Uganda, Kenya, Tanzania, Angola, South Africa (KwaZulu-Natal, former Transvaal), also in Madagascar. Dambos and other open areas of shallow water, often with *Xyris, Eriocaulon, Utricularia, Drosera* and Cyperaceae; sea-level–1500 m.

As is indicated by the synonymy, numerous segregate species were previously accepted. In this generic account I have followed Cowley (1988) in recognising only one species, variable in size and in stamen morphology as discussed by Obermeyer (1964).

Conservation notes: Widespread and common within its habitat; Least Concern.

3. GYMNOSIPHON Blume

Gymnosiphon Blume, Enum. Pl. Javae **1**: 29 (1827). —Jonker in Meded. Bot.
Mus. Herb. Rijks. Univ. Utrecht **51**: 168 (1938). —Cowley in F.T.E.A.
Burmanniaceae: 2 (1988).

Perennial herbs, saprophytic (heteromycotrophic), without chlorophyll. Stems erect,
unbranched, with scale-leaves, arising from a cluster of tubers. Inflorescence terminal, cymose.
Flowers actinomorphic, bisexual. Perianth tubular, abscissing below insertion of anthers after
flowering; outer tepals 3, conspicuous; inner tepals minute, alternating with outer tepals,
inserted in perianth tube throat (absent in our species). Stamens 3, alternating with outer
tepals; anthers sessile or subsessile, dithecal thecae lateral, dehiscing transversely; connective
broad. Ovary ovoid or ellipsoid, 1-locular, with 3 parietal placentas, each usually with 2
spherical, apically inserted, (presumed) nectary glands opening around the style base, ovules
numerous; style extending to stamens; stigma 3-lobed, abscissing with distal perianth tube,
filamentous appendages absent in our species. Capsule subglobose, late-dehiscent, seeds
numerous, lacking appendages.

A pantropical genus of around 27 species, all lacking chlorophyll and confined to
moist forest understorey. Five species are known from Africa, two in Madagascar.

Gymnosiphon afro-orientalis Cheek sp. nov.* Type: Malawi, Nchisi Forest, 1500 m, fl.
18.iv.1991, *Cheek* 3175 (K holotype (spirit), K, MAL, MO). FIGURE 12.2.**38**.

Erect perennial herb, 6–10 cm tall, rarely branched, glabrous. Base of stem 2–5 cm
underground, dilated, with an ellipsoid cluster of tubers, 4–9 × 3–4.5 mm, tubers subglobose,
each 1–2 mm wide. Roots produced from between tubers, mostly 6–14 mm long, unbranched,
rarely with 1–2 side branches. Stem purple, terete, c.1 mm wide, entirely vertical, internodes
c.0.6 cm long; scale-leaves subtriangular, c.1.5 × 1.5 mm, clasping stem. Inflorescence single,
terminal, peduncle 3–4 cm long, resembling stem but lacking scale-leaves; cyme initially 1-
flowered, then 2-branched and 3–5(7)-flowered; bracts oblong-triangular, 1.2–3 mm long,
acute, clasping inflorescence axis. Pedicel 2–4 mm long at anthesis, extending to 4 mm in fruit.
Flower buds 7–9 mm long, distal part of perianth tube c.4 mm long, dilated, c.2 mm wide,
proximal part c.3 × 0.7 mm; receptacle-ovary c.2 × 1.2–1.5 mm. Open flowers lacking scent at
noon, white, mouth yellow, with perianth lobes held flat, transversely elliptic-oblong, 3.7–4.3 ×
4.2–6 mm, apex 3-lobed, lobes c.1 mm long, rounded; corolla tube c.4.3 mm long; inner tepals
absent. Anthers inserted c.1.5 below tepal sinuses, attached basally, transversely elliptic, c.0.4 ×
0.6 mm; connective ± 0.4 × 0.4 mm, rounded, lacking appendages. Style c.4 mm long; stigma
flat, c.1 mm across, with 3 descending lobes, lobes concave with dorsal notch, dorsiventrally
flattened. Fruit subellipsoid, ± 2.5 × 2.5 mm, outer surface with 6 low parallel lines, surmounted
by persistent perianth tube, 2–4 mm long, and style 4–5 mm long. Description taken largely
from spirit pressed material.

Zambia. W: Kitwe Dist., Mwekera, fl. 15.ii.1964, *Fanshawe* 8324 (K, NDO, SRGH).
Malawi. C: Lilongwe Dist., Dzalanyama Forest Res. 1375 m, fl. 26.iii.1977, *Brummitt,
Seyani & Patel* 14957 (K, MAL, SRGH). S: Mt. Mulanje, head of Nasoto stream, below
Chilemba cliffs, 1100 m, fl. 6.ii.1986, *J.D. & E.G. Chapman* 7179 (E).
    Also found in S Tanzania and Congo (Katanga). Moist evergreen forest, often near
watercourses; 1100–1400 m.
    Conservation notes: Although known from 7 sites in 4 countries, it is here assessed
as vulnerable VU B2ab(iii) as its forest habitat is threatened by tree cutting.

---

* *Gymnosiphon afro-orientalis* Cheek sp. nov. ab *G. usambaricus* Engl. lobis stigmatis descendentibus
(non ascendentibus), tepalis interioribus carentibus (non praesentibus), bulbis subterraneis
praesentibus (non carentibus) distincta.

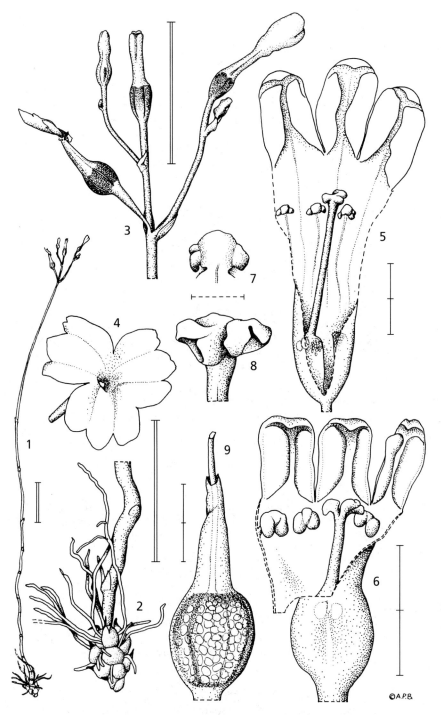

Fig. 12.2.**38**. GYMNOSIPHON AFRO-ORIENTALIS. 1, habit; 2, base of stem showing tubers, roots and scale-leaves; 3, inflorescence; 4, flower at anthesis; 5, flower bud near anthesis, opened longitudinally; 6, flower bud, perianth tube opened; 7, stamen, reverse view; 8, stigma, descending stigmas; 9, fruit. From *Cheek* 3175. Scale bars; double bar = 1 cm; graduated single bar = 2 mm; dashed bar = 500 mm. Drawn by Andrew Brown.

Populations at each site vary from just 2 individuals (*Chapman* 7179) to areas of 10 × 10 m (*Cheek* 3175) with hundreds of individuals.

Specimens of *Gymnosiphon afro-orientalis* from the Flora Zambesiaca area had previously been determined as *G. usambaricus*, a species restricted to the Usambara Mts and Eastern Arc Mts of Tanzania. *G. afro-orientalis* differs from *G. usambaricus* in having more slender and shorter inflorescences, in lacking entirely inner tepal lobes, and in having the stigma lobes facing down, not upwards. It also has an unusual rootstock with a cluster of globose tubers reminiscent of *Afrothismia*, which may be unique in the genus. The absent inner tepals, descending stigma lobes and ridged fruits are also unusual among African *Gymnosiphon*.

Exell, Mendonça & Wild 1065, collected with *Afrothismia zambesiaca*, is anomalous having a very short fruiting perianth, subwinged fruits and occurs at low altitude. It may represent an undescribed species.

PANDANACEAE

by H.J. Beentje

Trees or shrubs, less often lianas or epiphytic shrubs; often with stilt roots or aerial roots from lower trunk, and sometimes from the branches. Leaves spirally arranged in 3 or 2 rows (the latter appearing as 4 rows), crowded towards the apex of the branch, simple, lanceolate or linear, coriaceous, keeled and often 3-plicate, usually with small prickles on the midrib beneath, on the margins, and sometimes also on the distal ventral pleats. Flowers unisexual, crowded in branched and unbranched spadices (panicles in *Sararanga*); inflorescences terminal or terminating lateral shoots, enclosed at first by spathaceous and often coloured bracts. Male flowers without or with vestigial perianth; stamens few to numerous, in umbellate groups, the filaments free or connate, the anthers erect and basifix, 2-thecous, opening lengthwise by slits; ovary vestigial or absent. Female flowers without perianth, with 1 to several ovaries; ovary 1-locular, free or almost joined with adjacent ovaries of the same flower but always with separate stigmas corresponding to the number of locules; style absent or vestigial; stigma appressed or erect, lined with glandular hairs at maturity, often V- to kidney-shaped; staminodes, if present, hypogynous and attached to the ovary wall; ovules anatropous, solitary (in *Pandanus*) or many (*Freycinetia*). Fruiting heads ('cephalia') globose to oblong, solitary or several in a spiciform arrangement, composed of many fruits joined together but with distinct apices: fruits either berries (*Freycinetia*), drupes or clusters of partly fused drupes (*Pandanus*); pericarp thin; mesocarp fibrous with spongy pith, fleshy; endocarp fibrous, enclosing the locules in an integral structure. Seeds free (*Sararanga, Freycinetia*) or inseparable from the endocarp (*Pandanus*); testa membranous; endosperm white, homogenous, with sub-basal embryo.

A tropical and subtropical family of three genera and about 800 species, with many local endemics.

Many herbarium specimens are poor and incomplete, and the lack of correlation between specimens from male and female plants has led to a confused taxonomy. For a guide to collecting pandans see Stone in Ann. Miss. Bot. Gard. **70**: 137–145 (1983).

PANDANUS Parkinson

Pandanus Parkinson, Journ. Voyage South Seas: 46 (1773). —Warburg in Engler, Pflanzenr. **4**(9): 43 (1900).

Trees with candelabrum-like branching; roots adventituous, with prickles; stem and branches almost encircled by numerous leaf scars. Leaves sessile, linear, acute or attenuate at apex, armed with small prickles on margins, midrib beneath and sometimes on twin abaxial pleats; leaf-base half-embracing the stem; midrib keeled beneath; secondary veins numerous. Inflorescence terminal, capitate or spicate; bracts with prickles on margins, the ones on the peduncle green, the more distal ones coloured; flowers sessile, without a perianth. Male flowers in spicately arranged spadices; stamens many, the filaments free or connate. Female flowers in spicate or racemosely arranged bracteate spadices, or in a solitary spadix, without staminodes; ovaries free or connate, each with a single ovule; placenta sub-basal but parietal. Syncarp woody or drupe-like, the carpels closely joined for most of their length but with a ± free apex ('pileus'); pericarp thin and glabrous; mesocarp fibrous, spongy; endocarp thin and woody to thick and bony. Seed erect, the testa membranous; placenta lateral, slightly fibrous; endosperm firm.

About 600 species in the tropics and subtropics of the Old World, with ±15 in Africa. See Stone in Bot. Jahrb. Syst. **94**: 459–540 (1974) for an infrageneric classification of the genus. Much depends on authors' views of species; St. John and Huynh recognize many local species, and in their view the number of African species might be as high as 50. I agree with Stone in recognizing fewer, broader species.

P. utilis Bory from Madagascar has been cultivated in Mozambique at GI: Ravene, 80 km S of Inhambane, 3.i.1962, *St. John* 26629 (K, lvs only). This species belongs to section *Vinsonia* Warb. and differs in the stiffly erect glaucous leaves with red marginal spines. It is only known from cultivation, and its leaves are used for basketry and thatching.

Male flowers with stamens umbellately arranged at the top of a common stalk; syncarp of phalanges 5–18 mm across (subgen. *Vinsonia* sect. *Heterostigma* B.C. Stone) · **1.** *livingstonianus*
Male flowers with stamens clearly racemosely arranged; syncarp of phalanges 25–45 mm across (subgen. *Pandanus* sect. *Pandanus*) · · · · · · · · · · · · **2.** *odoratissimus*

1. **Pandanus livingstonianus** Rendle in J. Bot. **32**: 326 (1894). —Warburg in Engler, Pflanzenr. **4**, 9: 66 (1900). —Wright in F.T.A. **8**: 131 (1901). —Sim, Forest Fl. Port. E Afr.: 112, t.97 (1909). —Gomes e Sousa, Dendrol. Moçamb., Estudo Geral **1**: 177 (1966). —Huynh in Bot. Helvet. **107**: 92, figs.1–9 (1997). —M. Coates Palgrave, Trees Sthn. Africa, ed.3: 96 (2002). Type: Mozambique, Luabo, mouths of Zambesi, xii.1859, *Kirk* s.n. (K holotype). FIGURE 12.2.**39**.

   Pandanus petersii Warb. in Engl., Pflanzenr. **4**, 9: 66, fig.17/j–l (1900). Type: Mozambique, Licuare & Muansche rivers, *Peters* s.n. (B† holotype).*

   Pandanus gasicus Huynh in Bot. Helvet. **99**: 21, figs.1–12 (1989). Type: Mozambique, Gaza, Macia, Tuane, Rio Uagunumbo, 10.iii.1970, *Balsinhas* 1620 (LMU holotype).

   Pandanus globulatus Huynh in Bot. Helvet. **107**: 94, figs.10–17 (1997). Type: Mozambique, Zambezia, Pebane at km 43 on Mualama road, 15.i.1968, *Torre & Correia* 17139 (LISC holotype).

   Pandanus mosambicius Huynh in Bot. Helvet. **107**: 94, figs.18–24 (1997). Type: Mozambique, Sussundenga Dist., E end of Musapa Gap, 31.i.1962, *St. John* 26630 (BISH holotype, A, BR, K, LISC, LMA, US).

   Pandanus serrimarginalis Huynh in Bot. Helvet. **107**: 99, figs.25–29 (1997). —da Silva, Izidine & Amude, Prelim. Checklist Vasc. Pl. Mozamb.: 160 (2004). Type: Mozambique, Gaza, Macia, S. Martinho de Bilene beach, 14.xii.1961, *Lemos & Balsinhas* 309 (LMA holotype, BISH).

Tree 3–20 m high, solitary or clumped; crown narrowly pyramidal or columnar in the open to more ovoid underneath other trees, with a terminal tuft of large leaves on the stem and lateral branches ending in tufts of smaller leaves. Stem pale brown, 10–25 cm in diameter, with densely set short ascending prickles; stilt-roots sturdy, brown, 0.5–3 m high, 3–7 cm in diameter,

---

* Note: Peters gave a description of a pandan common along the banks of the Licuare (Licuari R., near Quelimane) and Munansche (not traced) rivers. Peters says the specimens were lost by the time he was writing up the botany part of the book. Nevertheless, Warburg based his description on that of Peters, and the only material he refers to is Peters'.

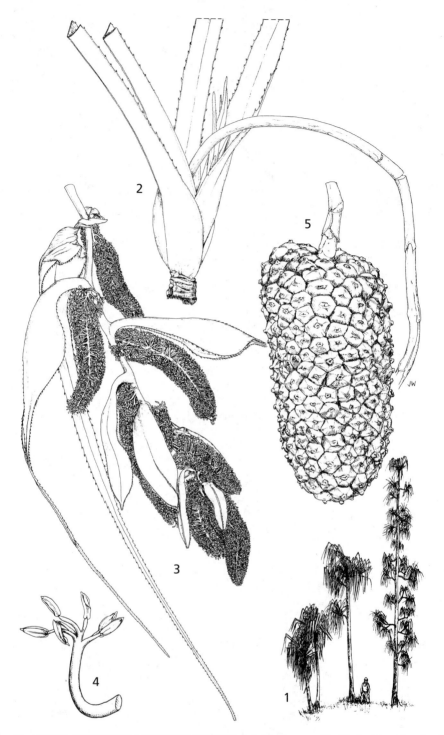

Fig. 12.2.**39**. PANDANUS LIVINGSTONIANUS. 1, habit; 2, part of leaf tuft with peduncle of syncarp (× ¹/₂); 3, male inflorescence and bracts (× ¹/₂); 4, stamen and anthers (× 3); 5, syncarp (× ¹/₂). 1–2, 5 from *Torre & Correia* 14098 & photo, 3 from *Faulkner* in Kew 301, 4 from *Torre & Correia* 16166. Drawn by Juliet Williamson.

with prickles 3–8 mm long and 2–5 mm thick; numerous short side-branches, ± in whorls, these unbranched or branching dichotomously. Leaves terminal on short lateral branches, also terminal on trunk, spiral in 3 ranks, 24–27 per tuft, dark shiny olive-green above, paler green and glaucous beneath, linear, 30–112 cm long, 1.5–3 cm wide, on young vigorous trees 50–150 cm long and 2.5–5 cm wide, hard, rigid, the midrib keeled and raised ± 1 mm, midrib beneath (abaxially) and margin with hooked prickles, apex long-attenuate; all prickles curving upwards towards the apex, rarely one or two on the midrib curved towards the base (in N Zambian population most prickles near base pointing downwards); prickles on leaf margin 0.6–3.3 mm long (longest near leaf-base, at shortest intervals near apex), on midrib 1–1.5 mm long and 0.5 mm thick at base. Male flowers in densely branched spiciform inflorescences; peduncle ± 12 cm long; axis of inflorescence 10–30 cm long with 6–10 branches, each subtended by conspicuous large off-white or cream spathe-like bracts bearing marginal prickles to 1 mm long; bracts ovate, long-attenuate at apex (lowest) to broadly ovate and acute (upper), the lowermost the largest, 29–31(89) cm long and 2.8–5 cm wide, the uppermost smallest and 3–4.3 cm long and 0.9–2.5 cm wide, midrib only visible in upper half and with minute prickles; inflorescence branches cylindrical, 3–17 cm long, 1–3 cm in diameter (only the lowermost branch sometimes has a short basal part without side branches), densely packed with many flattened side branches 2–7.5(11) mm long, each bearing at its apex (2)4–7(11) stamens; filaments (0.6)1.1–3.7 mm long, slightly flattened; anthers 1–2.8 mm long, ellipsoid, apex rounded with a small apiculum, creamy white. Female flowers many, densely set in solitary stalked heads; peduncle 16–31 cm long, 0.3–1.7 cm in diameter, curved slightly or through 180°, with 5–8 bracts or bract scars; lowermost bract the longest, 26–36(57) cm long, 2–3 cm wide, uppermost bract the smallest, 3.3–7 × 1.1–2 cm; flowering head at anthesis 3.5–6 cm long, 1.7–3 cm in diameter, with many free carpels on a central axis 1–10 mm across; carpels 4–9 mm long, diameter near apex 1–7 mm; stigma 1(3) per carpel, reniform to horseshoe-shaped, 0.9–2 mm in diameter. Syncarp ovoid with pointed apex, (6)11–15 × 6–10 cm wide; drupes/phalanges (constituent fruit parts) (4)5–6-angled, 18–34 mm long, the apical part (pileus) 2.5–5 mm long and 5–18 mm across and slightly raised with central sub-obtuse apex, 1(3)-locular and with as many stigmas 0.8–2.8 mm in diameter, minutely puberulous or covered with minute aculei.

Zambia. N: Kawambwa, ♂ fl. 23.viii.1957, *Fanshawe* 3556 (BR, K); Lumangwe Falls on Kalungwishi R., fr. 9.i.1960, *Richards* 12308 (K); same locality, ♂ fl. 30.viii.1969, *Lawton* 1563 (K). **Mozambique**. Z: Chinde Dist., Matilde, ♂ fl. 1.x.1965, *Guy* in C.K. 165326 (K, LISC, PRE, SRGH); Maganja da Costa, ♀ fl. 12.ix.1949, *Andrada* 1924 (LISC). MS: Chimanimani Mtns., Musapa Gap, ♂ fl. 15.x.1950, *Chase* 2988 (BM, LISC, PRE); same locality, ♀ fl. 19.xi.1953, *Ball* 72 (K, PRE). GI: Inchobane (S. Martinho do Bilene), ♂ fl. 8.x.1958, *Barbosa* 8333 (K). M: Maputo, Rikatla Marsh, ♂ fl. x.1918, *Junod* 415 (LISC).

Also in SE Congo. Along river-banks, in freshwater swamps, often growing gregariously; sea-level–100 m altitude, 800–900 m in Musapa Gap population, 900–1200 m in the Kawambwa/Lumangwe Falls population.

Photographs of what clearly appears to be *P. livingstonianus* from the Rio Lugenda (N Mozambique, Niassa Prov.) downstream of its confluence with the Rio Luatize have been seen, extending its distribution almost to the Tanzania border.

Conservation notes: A species that is only locally common and found in a threatened habitat; probably Vulnerable. Gomes e Sousa (1966) says that the species is fast diminishing in numbers as river-banks are being cleared. In Oldfield, Lusty & MacKinven (World List Threat. Trees: 408, 1998), *Pandanus petersii*, here reduced to synonymy, is classed as VU D2. It is said to be scattered in unprotected woodland remnants in swampy or wetland places, confined to an area stretching from Namacurra in C Mozambique to Quelimane and the Zambezi Delta.

According to Sim (1909), the timber is strong and durable, resistant to termites, and used for bearing-beams in bridges and other uses where durability and strength are required.

Livingstone (in his 'Zambesi and its tributaries', p.19) describes the populations near the Zambesi River mouth: "(the pandans) are so tall in the distance, like to remind us of the (church) steeples in our native land". A painting by Baines shows Kirk collecting this species "at Tete on the 10th of March, 1859". Kirk in his journal (Foskett: 36, 1965) recorded the event depicted by Baines on 6th June 1858 – they were in the delta at that time. In addition, no specimens have been seen from this far upstream and the habitat depicted in the illustration is wrong for the Tete area. Baines' painting clearly shows the two growth forms of *P. livingstonianus*, the young trees with only large leaves in a big terminal tuft; and older trees with a slightly smaller terminal tuft, and with the curved lateral branches bearing tufts of smaller leaves and the flowers or fruit. The same stages are visible on the photograph accompanying *Torre & Correia* 14098 (LISC), illustrated in Fig. 12.2.**39.1**.

P. petersii, which was collected close to the type locality of *P. livingstonianus* itself, has been a synonym for many years. Four other species described by Huynh are brought into synonymy here. Three of these (*P. globulatus*, *P. gasicus* and *P. serrimarginalis*; the first from the central part of Mozambique, the second and third from the southern part of Mozambique) are taxa from the coastal lowlands. The fourth (*P. mosambicius*) is the Musapa Gap population from the foothills of the Chimanimani Mountains, still within Mozambique. All four are scarcely distinguishable from the main population of *P. livingstonianus*, occurring in coastal Mozambique between Nampula and the southern border. I believe the type of *P. gasicus* has immature fruits, and Huynh's 'most remarkable' character, the almost detached upper part of the carpel seen in one specimen, I believe to be a drying artefact in immature phalanges. *P. globulatus* has relatively small fruits but is otherwise normal *P. livingstonianus*. *P. serrimarginalis* is based on fruit anatomical characters (of a rather young fruit) which, in my opinion, do not warrant specific status.

Only *P. mosambicius* gave me pause for thought. This is a well-known population at much higher altitude than normal coastal *P. livingstonianus*, but again the main reasons for separating this off are based on fruit anatomy; in my opinion the taxon falls within normal variation of *P. livingstonianus*. Huynh's use of the shape of the stigma and the presence of tiny needle-like structures thereon is not justified; the differences between reniform and hippocratiform are gradual, not absolute, and the structure of the stigma differs little from the range within *P. livingstonianus* sensu lato. The shape and degree of adherence of the pileus (the upper part of the fruit drupe) are a matter of ripeness of the fruit and the manner in which it has been dried.

Two specimens collected on the Lucite River, 'near Rhodesia in Mozambique Company territory' by Dawe in 1911–1912, are possibly from intermediate altitudes, but further locality or habitat data are lacking.

The North Zambian population, 1000 km distant from the coastal population, is very similar in all its morphological characters, although it is different in growth form and does not have the spire-like habit (J. Burrows, pers. comm.). There are more retrorse prickles on the midrib on the lower surface of the leaf, but this also occurs occasionally in other specimens, such as the cited one from *Chase*. Several specimens from across the border in Congo are almost certainly the same taxon – Luanza R., 31.viii.1966, *Malaisse* 4554 (♂ fl; BR, K); Nzilo–Kyamasumba, 19.iii.1986, *Bamps & Malaisse* 8147 (♀ fr, BR); also *Bamps & Malaisse* 8388 (♀ fr, BR), *Schaijes* 2012 (♂ fl, BR) and *Schmitz* 2948, 5934, 7772 (all ♂ fl, BR) and three more specimens – all from altitudes from 1020–1600 m.

It is possible the Angolan *P. welwitschii* Rendle in J. Bot. **32**: 324, t.347 (1894) is the same, but the description specifically says this taxon lacks stilt roots, which are always present in *P. livingstonianus*. The type (*Welwitsch* 5770 (BM holotype)), is probably from the banks of the Cuanza River, although several localities are mentioned. A drawing on one of the type sheets certainly looks like a small *P. livingstonianus*, with the terminal tuft of large leaves, and three lateral branches with smaller leaves which bear the solitary fruits, but conspicuously lacks stilt roots. All prickles on the abaxial midrib point downwards. Welwitsch mentions the trees rarely reach 6 m high, and describes the syncarps as 15 × 7.5 cm. The syncarp drupes/phalanges look like those of *P. livingstonianus* but are completely glabrous. A second specimen, km 75 from Dondo to Quibala, 220 m, *Dechamps, Murta & Silva* 1585 (BR, leaves only) is from the same locality but lacks fertile material – it has been described as *Pandanus angolensis* Huynh in Garcia de Orta, ser.Bot. **9**: 23 (1987). As these two collections are the only material of this taxon that I have seen (there are other Angolan taxa, certainly different) its status must remain slightly dubious.

2. **Pandanus odoratissimus** L.f., Suppl. Pl.: 424 (1782). —Stone in Gard. Bull. Singapore **22**: 231–257 (1967); in Ann. Miss. Bot. Gard. **60**: 266 (1973); in Rev. Fl. Ceylon **3**: 311–316, figs.6–7 (1981). Type: Sri Lanka, no locality, *Thunberg* s.n. (U).

Cultivated tree to 8 m high, apparently spreading (Mogg says 'gregarious'); only ♂ inflorescences seen. Stone (1981) states that stilt roots are generally present, and that the leaves are tristichous, 1–2(3) m long, 4–7(9) cm wide, M-shaped in section, clasping at base, with white marginal prickles. Male flowers fragrant, in dense cylindrical spikes organized into panicles; peduncle not seen; axis of inflorescence (incomplete) 14 cm long with 6 branches, each subtended by conspicuous large off-white (near base) or smaller brownish spathe-like bracts; bracts linear with long-attenuate apex (lowest) to broadly ovate and acute (upper), the lowermost the largest, 73–75 cm long and 5.3–5.6 cm wide, the uppermost smallest and 6–26 cm long and 0.4–3.6 cm wide, all bearing marginal prickles 0.2–3 mm long, midrib on lower surface with minute prickles pointing upwards; secondary inflorescence branches cylindrical, 6–8 cm long, 2–3 cm in diameter, densely packed with numerous tertiary branches to 12 mm long, each subtended by a bract 10–20 × 0.2–1.5 mm, each branch bearing along its length up to 21 stamens, ± evenly distributed or sometimes in whorls; occasionally with almost second order branching; filaments 0.5–1.2 mm long; anthers ellipsoid, 2.7–4.4 mm long, apex acuminate, creamy white. Female flowers and fruit not seen, the following description is taken from Stone (1981): ♀ inflorescence pedunculate, peduncle 10–30 cm long, bracteate with the upper bracts whitish; flowers in a cephalium, a dense globose or ellipsoid head. Syncarp globose or ellipsoid, 15–20(30) cm long, 12–18(20) cm in diameter, rounded-trigonal, of (26)50–70(143) carpellate phalanges, each 3–8 cm long and 2.5–4.5 cm across, each composed of 4–10 fused carpels which each have a V- or U-shaped stigma; ripe syncarp red-orange.

Mozambique. N: Angoche (António Enes), ♂ fl. 20.x.1965, *Mogg* 32478 (K, LISC). Originally from India, Sri Lanka and Malaysia. In freshwater coastal swamp; ± 2 m. The label on the LISC sheet says 'occasional/exotic' and 'gregarious'.

Conservation notes: Introduced species; very rare in the Flora area.

The male plants of this species are cultivated widely, e.g. on Zanzibar, for the fragrant male inflorescences; leaves are used in thatching and basketry. This is only the second record from mainland Africa.

VELLOZIACEAE

by A.E. Gonçalves

Herbaceous or ± shrubby perennials, less often ± tree-like, usually xerophytic, often unbranched at first, dichotomously (or trichotomously) branched later; stems woody-fibrous, black or dark brown, relatively slender but appearing thicker owing to the fibrous and densely overlapping persistent bases of fallen leaves and the persistent appressed adventitious fibrous roots. Leaves with parallel folds or rolled in the dry season, drought-resistant, clustered towards the ends of stems and branches; leaf sheath base broad, cylindric, always persisting; leaf blade long, narrow and grass-like, often sharply-pointed, persistent and gradually decomposing with age, or soon falling along a straight transverse line. Inflorescences terminal, becoming lateral by elongation of the axis, usually with 1 (or few clustered) flowers. Flowers growing singly in the axils of upper leaves, bisexual (seldom single-sex in *Barbaceniopsis* outside the Flora area), sometimes large and showy. Perianth blue or dark purple to white or yellow, with 6 equal to unequal segments arranged in 2 whorls, erect to reflexed at anthesis, free or often united at the base into a tube equalling or greatly exceeding the ovary, often bearing 6 joined appendages forming an outside corona usually joined to the stamens. Stamens 6, free, inserted on the perianth, separate or joined to the tube or segment bases (or in 6 bundles of 2–∞ each in *Vellozia* outside the Flora area); filaments relatively short, free or variously flattened, usually united at the base by a thin membrane forming a corona-like ring, simple and often bearing 2 lateral appendages ventrally (in *Vellozia*), joined to the corona (in *Barbacenia*) or fused with the perianth segments (in *Barbaceniopsis* and *Xerophyta*); anthers elongated, 7–8 times longer than the filaments, linear, generally 4-sporangiate, introrse to rarely extrorse, the pollen-grains often in tetrads, rarely single. Pistil with united carpels; ovary inferior, triangular in section, often depressed at the top or the walls prolonged upwards into a hypanthium, usually covered with outgrowths (trichomes), often viscous, trilocular, the anatropous ovules numerous in many rows or on 2 axile stalked lamellar placentas distally thickened or forked; style simple, mostly short, slender, filiform, usually enlarged above to a vertical columnar stigmatic region with 3 apical or subapical bands distinct to wholly confluent, linear (or with 3 stigmatoid horizontal bands in *Vellozia*). Septal nectaries present in ovary walls. Fruit a dry or hard, globose or obconical capsule, often flat or concave at the top, crowded with the remnant or scars of perianth segments, or 6-toothed, sometimes spiny, often sticky, loculicidal and irregularly dehiscent at apex, opening by 3 pores at base. Seeds numerous, small, conical, rarely rod-shaped, compressed, hard; testa wrinkled, black, rarely white-dotted; embryo small, situated near the hilum, embedded in a copious starchy, rather hard endosperm, with a terminal cotyledon and a lateral plumule.

A small and relatively well-defined family, for a long time included under Amaryllidaceae but now considered related to Haemodoraceae and Taccaceae. Essentially tropical, native to arid regions of South America (centred on Brazil and adjacent countries, extending northwards to Panama) and distributed throughout tropical and subtropical Africa and Madagascar north to the Arabian peninsula. Contains 8(9) genera of which only 2 are African, with over 200 species in the two subfamilies *Vellozioideae* and *Barbacenioideae*.

According to Menezes (Cienc. & Cult. **23**: 389–409, 1971), the subfamily *Vellozioideae* is characterized by a single leaf bundle sheath (endodermic), the flowers have anthers connected by a filament (although sometimes inconspicuous) and are without a corona; subfamily *Barbacenioideae* is characterized by a double leaf bundle sheath (endodermic and parenchymatic) and flowers with sessile anthers connected to a corona. However, in contrast to this, Smith & Ayensu state (Kew Bull. **29**: 181–205, 1974) that species from subfamily *Vellozioideae* have an adaxial epidermis not in contact with the bundle sheath, filaments always cylindrical, each one sometimes associated with a ventral coronal appendage, and anthers basifixed with pollen grains in tetrads or monads, stigmas suborbicular, horizontal to somewhat reflexed, apical and confluent. Species from subfamily *Barbacenioideae* have an adaxial epidermis which is usually in contact with the bundle sheath, filaments usually not evident or appearing broadly flattened, and anthers dorsifixed with single pollen grains, stigmas elliptic to linear in outline, apical or sometimes subapical, distinct to wholly confluent.

Xerophyta, the only genus in the Flora Zambesiaca area, belongs to the *Vellozioideae* according to Menezes (1988), and to *Barbacenioideae* according to Smith & Ayensu (1974).

No significant economic uses known.

XEROPHYTA Juss.

Xerophyta Juss., Gen. Pl.: 50 (1789). —Smith & Ayensu in Kew Bull. **29**: 184 (1974).
Vellozia sensu auctt. as regard African species, non Vand. —Greves in J. Bot. **59**: 273–284 (1921).

Small to large perennial herbs or shrubs, branched or unbranched, with a mass of densely packed adventitious roots on the stem within the leaf-bases forming a stem-like structure at the base, stem and branches covered in a thick mantle of closely packed leaf-bases. Leaves in clusters near stem apex, and on small lateral branches among the leaf bases in shrubby plants, or erect, rosette-like and sessile; leaf blade usually falling early, rarely persistent, with a distinct keel below formed by the median vein. Flowers with or without a rudimentary corona, solitary on each peduncle. Perianth segments 6, arranged in 2 whorls, persistent, blue or dark purple to white, varying between species or even on the same plant according to age; tube very short, often slightly exceeding the ovary. Stamens 6, shorter than and inserted on perianth; corona appendages ('filaments') not evident or appearing broadly flattened, ± triangular, joined to perianth tube or to segment bases; anthers erect, dorsifixed, pollen-grains single. Ovary ± globose or slightly trigonous, the surface spiny or hairy with multicellular hairs (trichomes); style short, stigma as long as or longer than style, columnar or spindle-shaped. Fruit a dry capsule. Seeds clavate or attenuate-conical, angled; testa white to yellowish-brown or black.

A genus related to *Vellozia* and distributed in central South America, tropical and subtropical Africa, Madagascar and Saudi Arabia. Generally found in and around rocky habitats. 30 species, 19 in the Flora area.

The leaves of some species are used as tying material while the stems are used to make brooms or as tooth brushes or scouring-brushes.

According to R.E. Fries (Wiss. Ergebn. Schwed. Rhod.-Kongo-Exped.: 234, 1916), *Xerophyta velutina* Baker from Angola also occurs in NW Zambia (S of Ndola, Bwana Mkubwa, rocky hill, st. 23.viii.1911, *Fries* 407, cited as *Barbacenia velutina* (Baker) Pax). Although I have not seen any specimens of *X. velutina* from the Flora area, it is included in the key and briefly described here: Shrublet with erect stem 45–75 cm high and branches 12–18 mm thick. Leaves 8–10 in terminal rosettes; sheathing-base rigid, truncate distally; blade 15–25 cm long and 6–12 mm wide, linear or lanceolate, somewhat nerved, distinctly ribbed, leathery, velvety-tomentose on both surfaces. Flowers 3, produced at stem ends; peduncles 3.8–5.5 cm long, dark-sticky upwards. Perianth segments blue, 25 mm long, lanceolate, acuminate at apex. Stamens 13 mm long; filaments very short. Ovary 6 mm long, clavate, densely dark-sticky.

African *Xerophyta* can be arranged in three groups corresponding to old Sections, based primarily on the leaf structure and which can only be seen under a low-power microscope. Xbaroid (=Section *Barbacenioides*, including species 8–19) has isolateral (undifferentiated) mesophyll, the sclerenchyma girder is Y-shaped or 3-pronged abaxially and inverted Y-shaped adaxially, and the adaxial epidermis is in contact with the bundle sheath. In Xeroph (=Section *Xerophyta*, including species 1) and Xveloid (=Section *Vellozioides*, including species 2–7) the mesophyll is dorsiventral (differentiated into palisade and spongy tissue) and the adaxial epidermis is not in contact with the bundle sheath. The latter two groups differ from each other in that in Section *Xerophyta* the leaf blade sclerenchyma girder is erect U- or W-shaped abaxially while in Section *Vellozioides* it is erect U- or Y-shaped.

Key to species

1. Branches numerous, 2–5 mm thick; leaf blades mostly 2–5 cm long, scaly on
   upper surface with small sparse scales · **1.** *pinifolia*
 – Branches very few or much stouter; leaf blades more than 5 cm long, not scaly
   above · 2
2. Trichomes (epidermal outgrowths) on ovary without glands · · · · · · · · · · · · 3

- Trichomes on ovary glandular $\cdots\cdots\cdots\cdots\cdots\cdots\cdots\cdots\cdots\cdots\cdots\cdots$ 9
3. Ovary covered with long, soft, clustered hair-like trichomes; leaf blades densely covered in trichomes, at least beneath $\cdots\cdots\cdots\cdots\cdots\cdots\cdots\cdots$ **8.** *villosa*
- Ovary covered with short hair-like, bristle-like or broad trichomes $\cdots\cdots\cdots$ 4
4. Trichomes on ovary shield-shaped, fringed with single deciduous bristles $\cdots\cdots$
$\cdots\cdots\cdots\cdots\cdots\cdots\cdots\cdots\cdots\cdots\cdots\cdots\cdots\cdots\cdots\cdots\cdots\cdots\cdots\cdots\cdots$ **9.** *eylesii*
- Trichomes on ovary or their lobes bristly or needle-shaped to forked $\cdots\cdots$ 5
5. Leaf internerves much broader than nerves; naked style base about equal to stigmatic region; trichomes on ovary simple and needle-shaped to forked $\cdots\cdots$
$\cdots\cdots\cdots\cdots\cdots\cdots\cdots\cdots\cdots\cdots\cdots\cdots\cdots\cdots\cdots\cdots\cdots\cdots\cdots$ **2.** *simulans*
- Leaf internerves much narrower than nerves; naked style base shorter than stigmatic region $\cdots\cdots\cdots\cdots\cdots\cdots\cdots\cdots\cdots\cdots\cdots\cdots\cdots\cdots\cdots$ 6
6. Trichomes on ovary simple or forked $\cdots\cdots\cdots\cdots\cdots\cdots\cdots\cdots\cdots\cdots$ 7
- Trichomes on ovary at least in part broadly stellate or orbicular with short rays
$\cdots\cdots\cdots\cdots\cdots\cdots\cdots\cdots\cdots\cdots\cdots\cdots\cdots\cdots\cdots\cdots\cdots\cdots\cdots\cdots\cdots$ 8
7. Trichomes on ovary 1–2 mm long, more or less spreading $\cdots\cdots$ **10.** *retinervis*
- Trichomes on ovary 0.2–0.6 mm long, appressed $\cdots\cdots\cdots\cdots$ **11.** *equisetoides*
8. Leaf blades glabrous or laxly covered beneath with subterete trichomes $\cdots\cdots$
$\cdots\cdots\cdots\cdots\cdots\cdots\cdots\cdots\cdots\cdots\cdots\cdots\cdots\cdots\cdots\cdots\cdots\cdots$ **12.** *suaveolens*
- Leaf blades covered beneath with flat linear-lanceolate trichomes $\cdots$ **13.** *nutans*
9. Leaf blades generally clothed at least at base $\cdots\cdots\cdots\cdots\cdots\cdots\cdots\cdots$ 10
- Leaf blades glabrous or scabrous or clothed on margins only $\cdots\cdots\cdots\cdots$ 13
10. Leaf blades pubescent, hairs mostly spreading $\cdots\cdots\cdots\cdots\cdots\cdots\cdots$ 11
- Leaf blades pilose, hairs mostly appressed $\cdots\cdots\cdots\cdots\cdots\cdots\cdots\cdots$ 12
11. Leaf sheathing base remaining whole, 3–3.5 cm long $\cdots\cdots\cdots\cdots$ **15.** *spekei*
- Leaf sheathing base rapidly disintegrating into coarse fibres, 3–7 cm long $\cdots$
$\cdots\cdots\cdots\cdots\cdots\cdots\cdots\cdots\cdots\cdots\cdots\cdots\cdots\cdots\cdots\cdots\cdots$ **16.** *zambiana*
12. Perianth segments acuminate; leaf trichomes simple $\cdots\cdots\cdots\cdots\cdots$ *velutina*
- Perianth segments rounded; leaf trichomes in bundles $\cdots\cdots\cdots$ **14.** *argentea*
13. Stem elongated, much branched, held above ground $\cdots\cdots\cdots\cdots\cdots$ 14
- Stem not over 3 cm long, sparsely branched if at all, ± subterranean $\cdots\cdots$ 18
14. Perianth segments 40–65 mm long; 'filaments' auricled $\cdots\cdots\cdots\cdots$ 15
- Perianth segments mostly under 33 mm long; 'filaments' not auricled $\cdots\cdots$ 16
15. Leaf sheathing base 4–7.5 cm long, truncate at end; branches 25 mm thick $\cdots$
$\cdots\cdots\cdots\cdots\cdots\cdots\cdots\cdots\cdots\cdots\cdots\cdots\cdots\cdots\cdots\cdots$ **18.** *splendens*
- Leaf sheathing base 1.5–4 cm long, rounded at end; branches 15 mm thick $\cdots$
$\cdots\cdots\cdots\cdots\cdots\cdots\cdots\cdots\cdots\cdots\cdots\cdots\cdots\cdots\cdots\cdots\cdots$ **17.** *kirkii*
16. Leaf blade with bristle-like point, ciliate along margins, with regularly arranged resinous glands on both surfaces $\cdots\cdots\cdots\cdots\cdots\cdots\cdots\cdots$ **3.** *capillaris*
- Leaf blade without bristle-like point, serrate or denticulate along margins, without resinous glands on both surfaces $\cdots\cdots\cdots\cdots\cdots\cdots\cdots\cdots$ 17
17. Anthers mostly 9–12 mm long; stigmatic region 10–12 mm long; leaf blades smooth $\cdots\cdots\cdots\cdots\cdots\cdots\cdots\cdots\cdots\cdots\cdots\cdots\cdots\cdots$ **4.** *squarrosa*
- Anthers mostly 12–16 mm long; stigmatic region 13–18 mm long; leaf blades rough above $\cdots\cdots\cdots\cdots\cdots\cdots\cdots\cdots\cdots\cdots\cdots\cdots\cdots$ **5.** *scabrida*
18. Plant under 10 cm high; perianth segments mostly 5–9 mm long, outermost bristle-tipped $\cdots\cdots\cdots\cdots\cdots\cdots\cdots\cdots\cdots\cdots\cdots\cdots\cdots$ **7.** *humilis*
- Plant at least 10 cm high; perianth segments 12–33 mm long, outermost not bristle-tipped $\cdots\cdots\cdots\cdots\cdots\cdots\cdots\cdots\cdots\cdots\cdots\cdots\cdots\cdots$ 19
19. Leaf blades deciduous; stigmatic region more than 4 times style base $\cdots$ **19.** *viscosa*
- Leaf blades persistent, slowly and irregularly disintegrating with age; stigmatic region shorter than style base $\cdots\cdots\cdots\cdots\cdots\cdots\cdots\cdots$ **6.** *schlechteri*

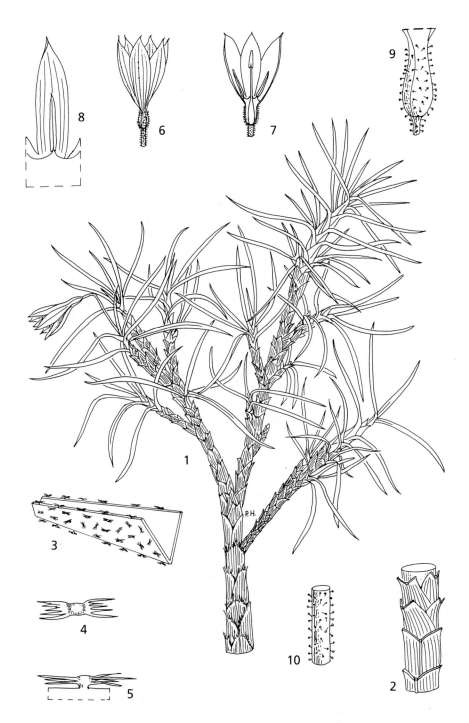

Fig. 12.2.**40**. XEROPHYTA PINIFOLIA var. PINIFOLIA. 1, habit ($\times$ $^2/_3$); 2, part of stem ($\times$ 2); 3, part of leaf (enlarged); 4 & 5, leaf hair, apical and side views (much enlarged); 6, flower ($\times$ 1); 7, longitudinal section through flower ($\times$ 1); 8, tepal with stamen ($\times$ 2); 9, ovary ($\times$ 4); 10, part of pedicel (enlarged). All from *Torre & Paiva* 10694. Drawn by Pat Halliday.

1. **Xerophyta pinifolia** Lam., Tabl. Encycl. **1**: 372, t.225 (1792). —Smith & Ayensu in Kew Bull. **29**: 187 (1974). Type: Madagascar, no locality (perhaps Fort-Dauphin), *Commerson* s.n. (P lectotype, here designated).

    Vellozia pinifolia (Lam.) Poiss., Réch. Fl. Mérid. Madagas.: 98 (1912). —Greves in J. Bot. **59**: 284 (1921).

Var. **pinifolia** – FIGURE 12.2.**40**.

Rhizomatous undershrub or shrub, 0.2–1(2.5) m high, with fibrous erect, dichotomously branched stem, sometimes with numerous rooting prostrate branches, the short erect branches 2–5 mm thick; stems and branches woody, covered with grooved scales. Leaves tufted, 3–8 towards the ends of stems, glabrous below; leaf-sheath base 0.5–2 cm long, stipule-like, without small scales on the nerves above, margins with white ligular bristle-like trichomes 4–5 mm long; blade rigid, (1)2–5(6) cm long by (1)1.5–3.5(4) mm wide, linear to lanceolate, attenuate towards apex ending in a strong point 1–4 mm long, hairy or softly tomentose above with small scales, scabrous along margins with short spinescent bristle-like trichomes; mesophyll dorsiventral. Flowers 1–2(3) at ends of stems; peduncle 0.8–2 cm long, with small glandular trichomes. Perianth white, pinkish or purplish; segments unequal, lanceolate or linear-lanceolate, outermost 8–13 × 2–3.5 mm with small glandular trichomes on outer surface, ending in a 1–2 mm long point, innermost 10–17 × 4–5.5 mm, obtuse at apex; tube (1.5)2–3(4) mm long. Filaments joined to perianth tube, with very short free part; anthers (5.5)7–9 × 0.75–0.8 mm, obtuse. Ovary 3–5 × 1.5–4 mm, ovoid-oblong, with small slender subappressed glandular trichomes; style straight, 5–6(7) mm long, stigmatic region often longer than the style. Capsule 5–7 × 2–4 mm, obovoid, somewhat attenuate towards base, hispid, opening at top. Seeds small, ovoid or subglobose, glossy, yellowish, smooth, finely reticulate.

Malawi. S: Machinga Dist., Machinga East, Mlinde Hill saddle, T.A. Nkhokwe, st. 29.iv.1982, *Patel* 878 (K). **Mozambique**. N: Marrupa Dist., 9 km from Mahua to Cuamba (Nova Freixo), c.600 m, fl.& fr. 20.ii.1964, *Torre & Paiva* 10694 (COI, EA, J, K, LISC, LMU, WAG). Z: Gurué Dist., Serra de Gurué, track to waterfall on R. Licungo, 20 km from Moçambique factory towards Régulo Mgunha, c.1300 m, fl. 8.xi.1967, *Torre & Correia* 16015 (COI, LISC, LMA, WAG).

Previously thought to be restricted to Madagascar, but now recorded from the NE part of the Flora area. Riparian forest and *Brachystegia–Combretum* woodland, sometimes on slopes, growing among rocks; 600–1700 m.

Conservation notes: Localised distribution, probably not threatened.

Var. *villosa* (H. Perret) H. Perret, restricted to Madagascar, is distinguished from var. *pinifolia* by having a villous leaf-blade.

2. **Xerophyta simulans** L.B. Smith & Ayensu in Kew Bull. **29**: 189, fig. 1 (1974); in F.T.E.A., Velloziaceae: 6, fig.1 (1975). Type: Uganda, Acholi Dist., Gulu, Kilak, 19.xi.1941, *Thomas* 4046 (K holotype).

Perennial herb, often shrubby, 0.1–2.4 m high, with a black, fibrous, cylindrical, often much-branched stem, the ultimate short erect branches up to 1 cm thick. Leaves 2–8 in tufts at end of stems; leaf-sheath base 3–5 cm long, dark red-brown, quickly disintegrating after leaf fall; blade 6–50 cm long by 1.5–7(10) mm wide, linear, attenuate towards base, apex with a strong point, pubescent towards base or laxly clothed below with straight white ± plumose trichomes, internerves usually wider than the nerves; mesophyll dorsiventral. Flowers 1–4 at stem ends, almond-scented; peduncle (1)5–12(14) cm long. Perianth blue to mauve or purple-pink shading to white; segments subequal, (12)20–40 × 4–10 mm, linear-lanceolate, acuminate towards apex; tube 1.5–4 mm long. Stamens tinged pale blue; filaments triangular, almost wholly joined to base of perianth segments with a short free part; anthers 10–22 mm long, attached 3 mm above base. Ovary 6–10 × 2.5–4 mm, narrowly ovoid to subcylindric, densely covered with ovoid simple suberect eglandular trichomes, apiculate or subulate towards apex; style 3.5–4 mm long, stigmatic region 11–17 mm long, linear. Capsule (13)15–20 × (10)15 mm, ellipsoid. Seeds suboblong, coarsely reticulate.

Zambia. N: Kasama Dist., Kasama, fl. 11.xii.1959, *Lawton* 315 (K). **Zimbabwe**. N: Makonde Dist., N of Great Dyke on Vanadium Mine road, fl. 20.x.1960, *Rutherford-Smith* 309 (SRGH). C: Harare Dist., Harare Kopje, fl. 21.xii.1908, *Rand* 1451 (BM). E: Mutare Dist., near Maranke Clinic, 48 km SW of Mutare, 975 m, fl.& fr. 22.i.1965, *Plowes* 2670 (LISC, SRGH). **Mozambique**. MS: Sussundenga Dist., Moribane, fl. x-xi.1911, *Dawe* 502a (K).

Also in Sudan, Uganda and Tanzania. On stony ground, sometimes gregarious on rocks and common on granite outcrops; 800–1400 m.

Conservation notes: Widespread, not threatened.

3. **Xerophyta capillaris** Baker in Trans. Linn. Soc. London, Bot. **1**: 264, t.36 (1878). —Smith & Ayensu in Kew Bull. **29**: 201 (1974). Type: Angola, Huíla Dist., Monino, Empalanca Plateau, fl.& fr. ii.1860, *Welwitsch* 1558 (K holotype, BM, LISU).

> *Vellozia capillaris* (Baker) Baker in F.T.A. **7**: 411 (1898). —Greves in J. Bot. **59**: 282 (1921).

Rhizomatous undershrub or shrub, 0.2–2 m high, with erect branched stem, the woody branches 1–1.2 cm thick. Leaves 4–10 in terminal rosettes; leaf-sheath base brown, 1.5–2.5 cm long, truncate; blade leathery, 5–12.5(30) cm long by 3.5–7(8) mm wide, linear, ± aristate, many-nerved; mesophyll dorsiventral. Flowers 2–3 at shoot ends; peduncle 5–6.5(7.5) cm long, with black glandular trichomes. Perianth white to violet-pink, pale lilac or blue, segments 20–25(50) × 5–6(7) mm, lanceolate, ± acuminate; tube 2–3 mm long. Filaments joined to base of perianth segments with a very short free part; anthers 10–11(13) mm long. Ovary 5–6(8) × 3 mm, obconical, densely covered with black glandular trichomes; style 6–8 mm long, stigmatic region 11–16 mm long, linear. Capsule 13–20 × 8–13(14) mm, oblong, with covering similar to ovary.

Var. **occultans** L.B. Smith & Ayensu in Kew Bull. **29**: 201 (1974). Type: Angola, Catengue, 152 m, *Pittard* 109 (BM holotype).

Shrub up to 1.5 m high. Leaf-blade pubescent with the hairs often largely hiding the resinous spots. Perianth pale mauve or violet-pink with yellow spots. Capsule not seen.

Zambia. C: Serenje Dist., above Kundalila Falls, c.13 km from Kanona, 1200 m, fl. 5.iv.1961, *Richards* 14966 (K).

Also in Angola. Rocky places, often in crevices; 1200 m (1700 m in Angola).

Conservation notes: Localised distribution in Flora area, but probably not threatened.

Var. *capillaris*, probably restricted to Angola, is separable based on having a glabrous leaf-blade and readily-visible regular resinous spots.

4. **Xerophyta squarrosa** Baker in Trans. Linn. Soc., Bot. **1**: 264 (1878). —Sölch in Merxmüller, Prodr. Fl. SW Afrika, fam.152: 2 (1969). —Coetzee in Dinteria **10**: 26 (1974). —Smith & Ayensu in Kew Bull. **29**: 202 (1974). Type: Angola, Pungo Andongo, fl.& fr. xii.1856, *Welwitsch* 1555 (BM holotype, LISU).

> *Vellozia squarrosa* (Baker) Baker in F.T.A. **7**: 410 (1898). —Greves in J. Bot. **59**: 283 (1921).

Shrub 0.9–1.5 m high, much-branched, stem 3.6–4.8 cm thick at base; stem and branches coated in mantle of closely packed rough leaf-bases and roots. Leaves 4–9, near shoot ends and on small lateral branches among leaf-bases; leaf-sheath base 1–2 cm long with fibres or bristles; blade leathery, (4)10–30(40) cm long by 1–9(12) mm wide, linear, acuminate, entire or denticulate with short spinescent trichomes on upper margins; mesophyll dorsiventral. Flowers

1–2(3) at ends of stems; peduncle 8.5–11(12.5) cm long, densely covered with spreading stalked minute glandular trichomes in upper half. Perianth pinkish white to bluish white, darker coloured towards base; segments subequal, linear to lanceolate, outermost (18)25–28(40) × 3–4 mm, attenuate-acute, with muricate glandular trichomes, innermost (20)28–30(36) × 4–7(8) mm, obtuse and slightly attenuate, not glandular; tube up to 3 mm long, thick. Filaments ligulate, joined to base of perianth segments, with short free part, sometimes with 2 lateral appendages; anthers 9–12 mm long. Ovary 6–7 × 3–4.5 mm, obconical or oblong, densely covered with spreading black glandular trichomes with spherical heads and long stalks; style 5–7 mm long, straight, stigmatic region 10–12 mm long, linear. Capsule brittle, 17 × 7 mm, ovoid or oblong, with similar covering to ovary.

Mozambique. N: Montepuez Dist., Montepuez, near centre, fl. 17.x.1942, *Mendonça* 905 (COI, LISC, LMA).

Also in Angola, Namibia and South Africa. On bare rock; c.600 m.

Conservation notes: Probably widespread and not threatened, although the only recorded population in the Flora area is very distant from others in southern Africa.

5. **Xerophyta scabrida** (Pax) T. Durand & Schinz, Consp. Fl. Afr. **5**: 271 (1894). — Smith & Ayensu in Kew Bull. **29**: 202 (1974); in F.T.E.A., Velloziaceae: 8 (1975). Type: Angola, Quango, 10°30'S, fl. ix.1876, *Pogge* 423 (B holotype).

   Barbacenia scabrida Pax in Bot. Jahrb. Syst. **15**: 144 (1892).
   Vellozia scabrida (Pax) Baker in F.T.A. **7**: 410 (1898). —Greves in J. Bot. **59**: 283 (1921).

Shrubby plant 60–90 cm high, with branched cylindrical stem 8–14 mm thick near apex. Leaves 5–8 in tufts towards stem ends; leaf-sheath base (1.5)2–3(4) cm long, remaining entire after leaf abscission, or apex partially dividing to form a fine net; blade stiff and leathery, 4–28 cm long by 1–9(10) mm wide, linear, acuminate or attenuate towards apex, grooved underneath, glabrous below except for slightly toothed midrib, ± scabrous above, finely toothed along thick yellow margins that are around twice as wide as nerves; mesophyll dorsiventral. Flowers 3–4 at ends of stems; peduncle 5.5–12.5 cm long, with densely-packed stalked glands. Perianth white, segments subequal, 18–30 × 4.5–6 mm, lanceolate, acute, outermost glandular, innermost without glands; tube 1.5–3 mm long. Filaments subtriangular, almost wholly joined to base of perianth segments, sometimes with 2 lateral appendages; anthers (8)12–16(24) mm long. Ovary 7–9(13) × (2)3–4 mm, obconical or ellipsoidal, densely covered with long-stalked, dark, glandular trichomes with spherical heads; style 6–7 mm long, stigmatic region 13–18 mm long, linear. Capsule brittle, (10)12 × (4)9 mm, oblong, with covering similar to ovary.

Mozambique. N: Ribáuè Dist., Posto Agrícola de Ribáuè, Matamane hill, fl. 5.xi.1942, *Mendonça* 1244 (COI, EA, LISC, LMU, UPS).

Also in Angola and Tanzania. Rock outcrops; c.800 m.

Conservation notes: Localised distribution in the Flora area, but possibly more widespread. Probably not threatened.

Related to *X. squarrosa* , but in *X. scabrida* the old leaf-sheath bases are fringed with fibres or bristles and the leaf blade is smooth.

6. **Xerophyta schlechteri** (Baker) N.L. Menezes in Cienc. & Cult. **23**: 422 (1971). — Smith & Ayensu in Kew Bull. **29**: 202 (1974). —Tredgold & Biegel, Rhod. Wild Fl.: 10 (1979). Type: South Africa, Free State, Donkerhoek Mt., 1600 m, *Schlechter* 4136 (Z holotype). FIGURE 12.2.**41**.

   Vellozia schlechteri Baker in Bull. Herb. Boiss., sér.2, **4**: 1003 (1904). —Greves in J. Bot. **59**: 283 (1921).
   Vellozia rosea Baker in Vierteljahrsschr. Naturf. Ges. Zürich **49**: 177 (1905). —Greves in J. Bot. **59**: 282 (1921). Type: South Africa, Gauteng, Spelonken, Shiluvane, *Junod* 969 (Z holotype).

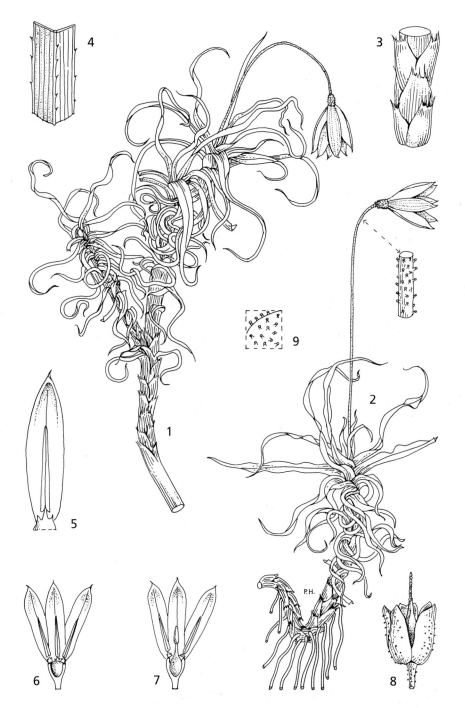

Fig. 12.2.**41**. XEROPHYTA SCHLECHTERI. 1, habit (× ²/₃); 2, habit (× ²/₃); 3, part of stem (× 2); 4, part of leaf (× 4); 5, tepal with stamen (× 2); 6, longitudinal section of flower, ovary and style removed (× 1); 7, longitudinal section of flower, 1 stamen removed (× 1); 8, fruit (× 2); 9, glands on fruit, much enlarged. 1–2 from *Philcox & Leppard* 8850, 3–9 from *Chase* 2138. Drawn by Pat Halliday.

Rhizomatous perennial herb or shrub, 10–30 cm high, erect, sticky, ± stemless. Leaves 8–16, in a rosette at top of plant at ground level; leaf-sheath base 1.5–2.5(3) cm long; blade stiff-leathery, persistent, slowly and irregularly disintegrating with age, (5)7–27(30) cm long by (1)1.5–6.5(7) mm wide, linear, acuminate, glabrous but serrate along margins; mesophyll dorsiventral. Flowers 1–2(3) at ends of stems; peduncle slender, 4.5–29 cm long. Perianth white, turning blue or purple; segments ± equal, 12–30 × 1.5–7 mm, lanceolate, acute, outermost sometimes pink or red-brown gland-dotted, white inside, innermost pink soon fading to white; tube up to 0.5 mm long. Filaments joined to perianth tube, with a very short free part; anthers 6–10 mm long, straight. Ovary 2.5–5 × 2–4 mm, subglobose, densely brown-sticky; style 4–7 mm long, stigmatic region 3–5 mm long. Capsule 8–9 × 7–8 mm.

Botswana. SE: Rocky hill, c.10 km N of Lobatse, 25°09'S; 25°41'E, 1300–1400 m, fl. 2.x.1977, *Hansen* 3204 (C, GAB, K, PRE, SRGH, UPS, WAG). **Zimbabwe**. N: Guruve Dist., Nyamunyeche Estate, fl. 21.ii.1979, *Nyariri* 711 (SRGH). W: Matobo Dist., Matopos Nat. Park, c.1300 m, fl.& fr. 26.ii.1981, *Philcox & Leppard* 8850 (K, PRE, SRGH). C: Charter Dist., Sebakwe Dam, Great Dyke, fl. 19.i.1962, *Wild* 5613 (LISC, SRGH). E: Mutare Dist., Rowa Township, Maranki Reserve, 1190 m, fl.& fr. 7.iv.1950, *Chase* 2138 (BM, COI, LISC, SRGH). S: Bikita Dist., Turwi/Dafana R. confluence, W bank of Turwi, c.305 m, fl.& fr. 5.v.1969, *Biegel* 3011 (SRGH). **Mozambique**. Z: Gurué Dist., slopes of Serra de Gurué, by Junqueiro factory, W of Picos Namuli, next to R. Malema, c.1700 m, fl. 6.xi.1967, *Torre & Correia* 15959 (COI, LISC).

Also in Namibia and South Africa (former Transvaal). Open savanna, rocky hills or slopes, dambos, shallow soil over rock, sometimes on bare granite hills or associated with the Great Dyke and serpentine soils, often forming pure colonies or close mats to 3 m wide; 300–1750 m.

Conservation notes: Widespread species; not threatened.

Closely related to *X. viscosa* which, however, has regularly deciduous leaf-blades with isolateral mesophyll, and the stigmatic region is 4–5 times longer than the style.

7. **Xerophyta humilis** (Baker) T. Durand & Schinz, Consp. Fl. Afr. **5**: 271 (1894). — Sölch in Merxmüller, Prodr. Fl. SW Afrika, fam.152: 2 (1969). —Coetzee in Dinteria **10**: 26 (1974). —Smith & Ayensu in Kew Bull. **29**: 203 (1974); in F.T.E.A., Velloziaceae: 2 (1975) as note. —Tredgold & Biegel, Rhod. Wild Fl.: 10 (1979). Type: South Africa, former Transvaal, banks of the Aapges R., *Burke* 122 (K lectotype, here designated).

Vellozia humilis Baker in J. Bot. **27**: 4 (1889); in F.C. **6**: 246 (1896); in F.T.A. **7**: 409 (1898). —Eyles in Trans. Roy. Soc. S. Africa **5**: 329 (1916). —Greves in J. Bot. **59**: 282 (1921) for most part. —Bremekamp & Obermeyer in Ann. Transv. Mus. **16**: 409 (1935). —Jacobsen in Kirkia **9**: 153 (1973).

Vellozia minuta Baker in Bull. Herb. Boiss., sér.2, **3**: 667 (1903) non *Xerophyta minuta* Baker (1875). Type: Namibia, Hereroland, Epako, fl. iii.1894, *Rautanen* s.n. (Z holotype).

Squat perennial 2–10 cm high, sometimes forming mats up to 25 cm across, brown when dry and green when moist, without or with a stem to 2.5 cm long, seldom branched, many adventitious roots on ventral side; rhizome horizontal, reaching 7 cm. Leaves grass-like, erect or slightly curved; leaf-sheath base 0.6–1.5(1.9) cm long, fraying, formed of parallel wiry strands of matted, thick, close connected fibres; blade stiff-leathery, 2–13(23) cm long by 2–6(7) mm wide, linear or linear-lanceolate, smooth, with regular longitudinal grooves below, glabrous except for minute teeth along margins and keel, especially when young; mesophyll dorsiventral. Flowers 1–2 at ends of stems, many per plant, not scented; peduncle slender, (1.5)2–10(13) cm long, purplish, rough with dense stalked black glands. Perianth pale yellow or bright blue to pink or white, often with pink on back, lilac or purple, with dark purple glands outside; segments (4)5–9(16) × 1.5–3 mm, oblong-lanceolate, outermost aristate, sometimes green with a pink keel; tube up to 0.5 mm long. Filaments joined to base of perianth segments, with a very short free part; anthers 2–4(5) mm long. Pistil slightly longer than anthers; ovary (1.2)1.5–2.5(3) × (1.2)1.5–2.5(3) mm, obconical, 3-sided, densely dotted with black short-stalked glandular trichomes; style 1.5–3 mm long, stigmatic region 1.5–2.5 mm long, linear. Capsule 2–5 × 2.5–5 mm, dotted with black glandular trichomes.

Botswana. N: Moremi Wildlife Reserve, on Maxwee–Khwai Gate road, 19°21'15"S, 23°40'30"E, fl. 3.iii.1976, *P.A.Smith* 1582 (K, LISC, SRGH). SW: 4 km S of Mamuno along track to Ncojane, 22°18'S, 20°01'E., fl. 27.xii.1977, *Skarpe* S210 (SRGH). SE: Content Farm, northern area, 24°33'S, 25°57'E, 1050 m, fl.& imm. fr. 5.x.1977, *Hansen* 3211 (BM, C, GAB, K, PRE, SRGH, UPS, WAG). **Zambia**. C: Katondwe, st. 22.ii.1968, *Fanshawe* 10295 (K, NDO). S: Livingstone Dist., Livingstone, fl. 8.xi.1973, *Chisumpa* 118 (K, NDO). **Zimbabwe**. N: Mazowe Dist., top of Iron Mask Hill, SE edge, 1615 m, fl. i.1907, *Eyles* 517 (BM, SRGH). W: Matobo Dist., Farm Besna Kobila, 1460 m, fl. xi.1957, *Miller* 4777 (COI, LISC, SRGH). C: Gweru Dist., Mlezu Govt. Agric. School Farm, 31 km SSE of Kwekwe, 19°08'S, 29°55'E, 1280 m, fl. 18.xi.1965, *Biegel* 546 (SRGH). E: Mutare Dist., Kelly's Park, Mutare commonage, fl. 26.xi.1948, *Chase* 1562 (BM, SRGH). S: Beitbridge Dist., Nulli Hills, fl. 8.iv.1967, *West* 7523 (SRGH). **Mozambique**. T: Changara Dist., 48 km from Chipembere (Chioco) to Mocubura, c.250 m, fl. 15.ii.1968, *Torre & Correia* 17669 (BR, COI, LISC, LMU, LUA, PRE, UPS). MS: Báruè Dist., Báruè road to Macossa, fl. 30.vi.1944, *Torre* 2957 (EA, J, LISC, LMA). M: Magude Dist., near Chobela, fl. 3.i.1948, *Torre* 7033 (K, LISC, SRGH, WAG).

Also in S Sudan, Kenya, Angola, Namibia and South Africa (former Transvaal and Cape Provinces). Open mopane, miombo or *Acacia* woodland and grassland, growing over rocks and on stony soils, sometimes forming dense mats, in moist to dry conditions; 150–1600 m.

Conservation notes: Widespread species, not threatened.

8. **Xerophyta villosa** (Baker) L.B. Smith & Ayensu in Kew Bull. **29**: 188 (1974). — Coetzee in Dinteria **10**: 26 (1974) non T. Durand & Schinz (1875). Type: South Africa, Limpopo Prov., Houtbosch, *Rehmann* 5792 (BM holotype).

   Vellozia villosa Baker in J. Bot. **27**: 3 (1889); in F.C. **6**: 245 (1896); in F.T.A. **7**: 408 (1898). —Greves in J. Bot. **59**: 276, 280 (1921). —Norlindh in Bot. Not. **101**: 28 (1948).

   Vellozia violacea Baker in Bull. Herb. Boiss., sér.2, **4**: 1003 (1904). —Greves in J. Bot. **59**: 276, 280 (1921). Type: South Africa, Limpopo Prov., Haenertsburg, *Junod* s.n. (Z holotype).

   Vellozia monroi Greves in J. Bot. **59**: 280 (1921). Type: Zimbabwe, Masvingo Dist., Masvingo (Fort Victoria), fl. 1909–12, *Monro* 2160 (BM lectotype, here designated).

Low perennial shrub, 0.3–0.9(1.5) m high, stemless or with an unbranched to irregularly branched stumpy stem; stem and branches stout, 2.5 cm thick at ends, woody, blackish, coated with many leaf-bases. Leaves 3–10 tufted at ends of shoots, younger ones short, 2–3 mm wide, appearing later in season, older ones long, 8.5–9.5 mm wide; leaf-sheath base 2–6 cm long, stiff, scaly, formed of parallel wiry fibres; blade stiff-leathery, 6–50 cm long, linear-lanceolate, tapering to an acute or acuminate apex, greyish-green, regular longitudinal grooves on both surfaces, the youngest with thick covering of long white or brownish trichomes, the oldest covered in short spreading whitish trichomes or glabrous below; mesophyll isolateral. Flowers 1–4 at ends of shoots, sweet-scented; peduncle slender, 4.5–19 cm long, densely covered in long white trichomes. Perianth dark blue to purple, violet, pale mauve or whitish; segments 24–50 × 4–11 mm, lanceolate, acute, outermost densely villose, with long trichomes around midrib or only at base; tube 1.5–3.5 mm long. Filaments almost or wholly joined to base of perianth segments, the free part very short or absent; anthers 6–20(25) mm long. Ovary (5)6–12(15) × 2–4 mm, narrowly obconical, densely villose with long soft spiny or clustered white or brownish eglandular hair-like trichomes; style 3–8(12) mm long, stigmatic region 11–20 mm long. Capsule 8–9 × 5–6 mm.

Zambia. N: Mbala Dist., Chitimbwa, fl.& fr. 29.xii.1955, *Nash* 245 (BM). **Zimbabwe**. N: near Tsatsi R., 22.5 km N of Concession, fl. 19.xi.1961, *Leach* 11284 (SRGH). W: Matobo Dist., Farm Besna Kobila, 1460 m, fl. xi.1954, *Miller* 2518 (LISC, SRGH). C: Goromonzi Dist., Chindamora Reserve, Ngomakurira, fl. 14.ix.1967, *Loveridge* 1750 (LISC, SRGH). E: Mutare Dist., Honzo Mt. at head of Honde Valley, fl. 5.i.1968, *Plowes* 2871 (PRE, SRGH). S: Masvingo Dist., fl. 1909, *Monro* 800 (BM, SRGH).

Also in N South Africa. Shallow soil over rocks, often on exposed granite slopes, sometimes in dense clumps; 1400–1600 m.

Conservation notes: Widespread species; not threatened.

Greves (1921) cites only two specimens for *Vellozia monroi*, *Monro* 800 and 2160 (both BM) collected in Zimbabwe at Masvingo (Fort Victoria). Smith & Ayensu (1974) refer incorrectly to *Monro* 2160 (BM) as the holotype, but I consider it to be the lectotype.

9. **Xerophyta eylesii** (Greves) N.L. Menezes in Cienc. & Cult. **23**: 422 (1971). — Smith & Ayensu in Kew Bull. **29**: 189 (1974). Type: Zimbabwe, Mazowe Dist., Iron Mask Hill, fl. x.1906, *Eyles* 440 (BM lectotype, here designated).

    Vellozia eylesii Greves in J. Bot. **59**: 281 (1921).

Dwarf shrub 15–30 cm high, with fibrous stem 2.5–5 cm thick, unbranched or forked at the top, leaves and flowers arising separately from twin apices. Leaves 4, tufted at ends; leaf-sheath base 3–5.5 cm long; blade 5–14(42) cm long by 1–10(11) mm wide, linear or linear-lanceolate, acuminate towards apex, with distinct midrib, strongly veined on both surfaces and thickly covered with short fine trichomes above, similar but less so below, softly-hairy along margins; mesophyll isolateral. Flowers 4 at shoot ends, not scented; peduncle slender, (5.5)12–16.5 cm long, hairy. Perianth pale to dark mauve or pale purple; segments lanceolate, outermost 40–50 × 4–6(7) mm, innermost similar but slightly shorter; tube 2–4 mm long. Filaments joined to base of perianth segments with a free part 1.5–3 mm long; anthers 12–19 mm long. Ovary (4)6–8(10) × (2.5)3–4 mm, oblong-cylindrical or obconical, covered with shield-like trichomes partly joined to long stiff deciduous bristles arising from shield-like bases; style 4–7 mm long, narrowly linear, stigmatic region 9–19 mm long. Capsule not seen.

Zambia. S: Kalomo Dist., Siantambo, fl. 20.xi.1962, *Mitchell* 15/55 (LISC, SRGH). **Zimbabwe**. N: Mazowe Dist., base and side of Iron Mask Hill, 1300–1530 m, fl. x.1906, *Eyles* 440 (BM, K, SRGH).

Not known elsewhere. Miombo woodland with *Loudetia* and *Hyparrhenia* grasses, mopane woodland or on copper outcrops, from apparently different habitats; 1000–1550 m.

Conservation notes: Endemic to the Flora area. Restricted distribution; probably Lower Risk near threatened.

Wild 7602 (SRGH) from W Zimbabwe (Hwange Dist., Ell Mine, near Gwayi R., st. 27.i.1967) can probably be placed here.

Greves in J. Bot. **59**: 281 (1921) cites only two specimens for *Vellozia eylesii*, *Eyles* 440 and *Clark* s.n. (both BM), both collected at Mazowe, Iron Mask Hill. Smith & Ayensu in Kew Bull. **29**: 189 (1974) refer incorrectly to *Eyles* 440 (BM) as the holotype but I consider it to be the lectotype.

10. **Xerophyta retinervis** Baker in J. Bot. **13**: 233 (1875). —Coetzee in Dinteria **10**: 26 (1974), in part. —Smith & Ayensu in Kew Bull. **29**: 190 (1974). Type: South Africa, Gauteng, Magaliesberg, *Burke* s.n. (K lectotype).

    Xerophyta clavata Baker in J. Bot. **13**: 233 (1875). —Coetzee in Dinteria **10**: 26 (1974), in part. Type: South Africa, KwaZulu-Natal, *Gerrard & Macken* 1824 (K holotype).

Vellozia retinervis (Baker) Baker in F.C. **6**: 244 (1896). —Eyles in Trans. Roy. Soc. S. Africa **5**: 329 (1916). —Greves in J. Bot. **59**: 280 (1921). —Hutchinson, Botanist Sthn. Africa: 325 (1946). —Letty, Wild Fl. Transvaal: 72, t.35 (1962). —Ross, Fl. Natal: 133 (1973).

Vellozia clavata (Baker) Baker in F.C. **6**: 244 (1896). —Greves in J. Bot. **59**: 280 (1921).

Barbacenia retinervis (Baker) Burtt Davy & Pott-Leendertz in Ann. Transv. Mus. **3**: 121, 136 (1912). —Marloth, Fl. S. Africa **4**: 125, figs.36 & 37 (1915).

Shrub 0.3–2(4) m high, with well-developed stumpy, erect stem 7.5 cm wide at base, branching just above ground, internodes short; stem and branches woody, 3.5–5 cm thick, coated with dense mass of leaf-bases immediately below top. Leaves 3–7, tufted near shoot ends creating a wispy effect; leaf-sheath base 3–6 cm long, fanlike, formed of parallel light brown to black wiry fibres connected by short threads with lattice-like interstices; blade stiff-leathery, 7.5–62 cm long by (1)2–8(18) mm wide, linear to strap-shaped, close and finely ribbed with prominent rib down centre, often with ascending spiny or bristly trichomes on keel and along thickened horny margins when young; mesophyll isolateral. Flowers 1–2(3) at ends of shoots, showy, fading quickly when picked, smelling strongly of vanilla at midday; peduncle slender, stiffly erect or flexuose, 7.5–15.5(30) cm long, glabrous, except towards tip with a few scattered bristle-like trichomes. Perianth blue to purple, violet, pale lilac or whitish; segments 20–60 × 3.5–8(12) mm, lanceolate, acuminate, outermost with covering similar to ovary; tube 1–3 mm long. Filaments joined to base of perianth segments, with relatively short free part, sometimes with 2 lateral appendages; anthers slender, 12–19 mm long. Ovary 4–8(12) × 2.5–4 mm, oblong, clavate or obconical, greenish, black when dry, often covered with white or brownish, ascending or spreading, bristle-like eglandular trichomes 1–2 mm long; style 4–5 mm long, filiform, stigmatic region (12)13–18(19) mm long, overtopping stamens. Capsule 13–19 × 9–10 mm, oblong or ellipsoid.

Botswana. SE: Mokolodi Hill, 24°45'S; 25°50'E, 1170 m, fl.& imm.fr. 23.ix.1978, *Hansen* 3465 (BM, C, GAB, K, PRE, SRGH, UPS, WAG).

Also in South Africa (former Transvaal and KwaZulu-Natal) and Swaziland. Grassland and rocky slopes, sometimes on sandy soil in open places; 1100–1200 (1700 m in South Africa).

Conservation notes: Restricted distribution within the Flora area, but more widely distributed elsewhere; not threatened.

11. **Xerophyta equisetoides** Baker in J. Bot. **13**: 233 (1875). —Smith & Ayensu in Kew Bull. **29**: 190 (1974); in F.T.E.A., Velloziaceae: 3 (1975). —Tredgold & Biegel, Rhod. Wild Fl.: 10 (1979). Type: Malawi, Zomba and E end of Lake Chilwa, fl. 20.x.1861, *Meller* s.n. (K lectotype).

Vellozia equisetoides (Baker) Baker in F.C. **6**: 245 (1896); in F.T.A. **7**: 411 (1898), in part excl. *Kirk* s.n. & *Scott-Elliot* 8524. —Greves in J. Bot. **59**: 281 (1921), in part. — Hutchinson, Botanist Sthn. Africa: 679 (1946). —Norlindh in Bot. Not. **88**: 28 (1948), in part excl. *Fries* 2211 & synonym *V. suaveolens*. —Wild in Clark, Victoria Falls Handb.: 140 (1952). —Letty, Wild Fl. Transvaal: 73 (1962). —White, F.F.N.R.: 17 (1962), in part excl. *Angus* 866 (FHO) & *White* 3680 (FHO). —Binns, Herb. Fl. Malawi: 103 (1968). — Jacobsen in Kirkia **9**: 153 (1973).

Barbacenia equisetoides (Baker) R.E. Fr., Wiss. Ergebn. Schwed. Rhod.-Kongo Exped.: 233 (1916).

Shrubby herb or small shrub, 0.5–2.6 m high with stem 2.5–3.6 cm thick, often dichotomously branched; stem and branches fibrous, coated with many pleated leaf-bases resembling *Equisetum*. Leaves 3–7(8), tufted at ends of branches; leaf-sheath base 2–6 cm long, dark red-brown, short-pubescent or bristly, formed of parallel brown or black-grey wiry strands of matted fibres connected by short threads; blade stiff-leathery, (2)4–67(70) cm long by 1–9 mm wide, subulate or 3-angled, often sharp-pointed, finely ribbed, glabrous to softly-hairy on both surfaces, often smooth along margins and keel; mesophyll isolateral. Flowers 1–5 at ends of branches, sometimes attractive, fragrant; peduncle slender, wiry or flexuous, (1)2.5–18 cm long, black on upper half, glabrous on lower half, densely covered with short or minute ascending bristle-like trichomes. Perianth blue tinged with purple or pink, pale mauve, white or yellow; segments equal or subequal, (15)20–45(50) × 3–13 mm, lanceolate to oblong, acute but appearing acuminate at apex owing to inrolling, glabrous outside, spreading; tube 1–3 mm long. Filaments triangular, joined to base of perianth segments, with a short free part; anthers slender, (6)7–20(25) mm long. Ovary 5–13 × (1)2–5.5(7) mm, ellipsoid, clavate or obconical, densely covered in short or minute brown ascending bristle-like eglandular trichomes or appressed subulate brownish bristles; style slender, 2–6 mm long, stigmatic region (5.5)7–24 mm long, overtopping stamens. Capsule brittle, 8–12 × 7–11 mm, ± globose, often flat or concave on top, with bristle-like trichomes.

1. Branches very short and stout, as long as thick; leaf-sheath bases 4–5 cm long, slightly exserted, soon splitting · 2
- Branches long and slender, much longer than thick; leaf-sheath bases to 3 cm long, much exserted, not splitting · 3
2. Leaf blades glabrous or subglabrous all over · · · · · · · · · · · · **i)** var. *equisetoides*
- Leaf blades soft-hairy on both surfaces · · · · · · · · · · · · · · · · · **ii)** var. *trichophylla*
3. Leaf blades glabrous or with only marginal trichomes · · · · **iii)** var. *pauciramosa*
- Leaf blades distinctly covered with trichomes · 4
4. Leaf blades with short stiff trichomes · · · · · · · · · · · · · · · · · · · **iv)** var. *setosa*
- Leaf blades soft-hairy · **v)** var. *pubescens*

i) Var. **equisetoides** —Smith & Ayensu in Kew Bull. **29**: 192 (1974).

> *Xerophyta melleri* Baker in J. Bot. **13**: 234 (1875). Type: Malawi, Manganja Hills, fl. ix-xi.1861, *Meller* s.n. (K holotype).
> *Xerophyta retinervis* var. *equisetoides* (Baker) Coetzee in Coetzee & van der Schijff in J. S. Afr. Bot. **39**: 269 (1973) as regard *Baines* s.n.; in Dinteria **10**: 26 (1974).

Zimbabwe. N: Makonde Dist., on Silverside road, near Mhangura, 1175 m, fl. 23.xi.1962, *Jacobsen* 1919 (PRE). W/C: 'South African Goldfields', fl. 3.x.1871, *Baines* s.n. (K). **Malawi**. S: Zomba and E end of Lake Chilwa, fl.& imm.fr. 20.x.1861, *Meller* s.n. (K).

Also in Congo, Tanzania and N South Africa. Forest and rocks, on very acid soil; up to 1200 m (1600 m in Congo).

Conservation notes: Widespread taxon; not threatened.

ii) Var. **trichophylla** (Baker) L.B. Smith & Ayensu in Kew Bull. **29**: 192 (1974); in F.T.E.A., Velloziaceae: 3 (1975). Type: Malawi, no locality, fl. 1891, *Buchanan* 854 (K lectotype).

> *Vellozia equisetoides* var. *trichophylla* Baker in F.T.A. **7**: 411 (1898). —Binns, Herb. Fl. Malawi: 103 (1968).
> *Vellozia equisetoides* sensu Watson in Gard. Chron., ser. 3, **34**: 425 (1903) based on *McClounie* collection from Malawi cultivated at Kew. —Greves in J. Bot. **59**: 281 (1921) as regard *Purves* 29. —Jacobsen in Kirkia **9**: 153 (1973) as regard *Jacobsen* 3569, non (Baker) Baker.
> *Vellozia trichophylla* (Baker) Hemsl. in Bot. Mag. **130**: t.7962 (1904) as regard *Buchanan* 162 & *Whyte* 92. —Greves in J. Bot. **59**: 281 (1921) as regard *Buchanan* 162, 854 & *Whyte* 92. —Jacobsen in Kirkia **9**: 153 (1973) as regard *Jacobsen* 1618.

Zambia. N: Mbala Dist., Plain of Death, Chilongowelo, 1440 m, fl. 21.xi.1954, *Richards* 2310 (K). W: Copperbelt, Kitwe Dist., Kitwe, fl. 16.xi.1955, *Fanshawe* 2609 (K, SRGH). C: Chisamba, fl. 5.xi.1973, *Chisumpa* 115 (K, NDO). E: Chipata Dist., Chipata Forest Reserve, fl. 6.xi.1959, *Grout* 217 (K). **Zimbabwe**. C: Harare Dist., fl. 6.xi.1951, *Greatrex* s.n. in GHS 34700 (SRGH). **Malawi**. N: From Chitipa down to Munhenja, 610 m, fl. 17.xi.1952, *Williamson* 117 (BM). C: Dedza Dist., Chongoni Forest, fl. 2.xii.1968, *Salubeni* 1235 (MAL). S: Zomba Dist., Zomba, fl. n.d., *Whyte* 92 (K).

Also in Congo, Angola and Tanzania. Forests and *Brachystegia*, *Julbernardia* or mopane woodlands, hill slopes, rocky places and shallow sandy soils; 600–1450 m.

Conservation notes: Widespread taxon; not threatened.

iii) Var. **pauciramosa** L.B. Smith & Ayensu in Kew Bull. **29**: 192, fig.2 (1974) in part excl. *Milne-Redhead* 3001 (BR, K) & *Wild* 3126 (K) from Zambia, *Mitchell* 603 (K) from Zimbabwe, and *Williamson* 117 (BM) from Malawi; in F.T.E.A., Velloziaceae: 3 (1975). Type: Zambia, Mwinilunga Dist., near Zambezi R., 6.5 km N of Kalene Hill Mission, fl. 25.ix.1952, *White* 3369 (BM holotype).

    Barbacenia wentzeliana Harms in Bot. Jahrb. Syst. **30**: 277 (1902); in Engler, Pflanzenw. Afrikas **2**: 354 (1908). —Fries, Wiss. Ergebn. Schwed. Rhod.-Kongo-Exped.: 234 (1916). Type: Tanzania, Mbeya Dist., Unyika, near Saute village on Yamba R., 1200 m, fl. xi.1899, *Goetze* 1409 (B holotype).

    Vellozia wentzeliana (Harms) Greves in J. Bot. **59**: 281 (1921) in part excl. *Rand* 155. — Brenan, Check-list For. Trees Shrubs Tang. Terr.: 629 (1949). —Wild in Clark, Victoria Falls Handb.: 140 (1952).

    Xerophyta wentzeliana (Harms) Sölch in Mitt. Bot. Staatssamml. München **4**: 74 (1961); in Merxmüller, Prodr. Fl. SW Afrika, fam.152: 2 (1969).

    Xerophyta retinervis var. *wentzeliana* (Harms) Coetzee in Coetzee & van der Schijff in J.S. Afr. Bot. **39**: 269 (1973); in Dinteria **10**: 26 (1974).

Zambia. W: Mwinilunga Dist., near Zambezi R., 6.5 km N of Kalene Hill Mission, fl. 25.ix.1952, *White* 3369 (BM, BR, COI, FHO, K).

Also in Congo, Angola, Tanzania, Namibia and N South Africa. Cracks in bare granite rocks; 1300 m.

Conservation notes: Very localised within the Flora area but with a very disjunct distribution; not threatened.

Fries (1916) records the variety from N Zambia between Mporokoso and Lufubu R., *Fries* 1189 as *B. wentzeliana*.

iv) Var. **setosa** L.B. Smith & Ayensu in Kew Bull. **29**: 194 (1974), in part excl. *Chase* 3073 (BM) & *Rand* 270 (BM). —Drummond in Kirkia **10**: 232 (1975). Type: Zimbabwe, Mutare Dist., Mutare commonage, Meneni R., fl. 22.xii.1949, *Chase* 1866 (K holotype).

    Vellozia equisetoides sensu Greves in J. Bot. **59**: 281 (1921) as regard *Cecil* 82. —sensu Wild in Kirkia **7**: t.5 (1968), non (Baker) Baker.

    Xerophyta equisetoides sensu McPherson, van der Werff & Keating in Novon **7**: 394 (1997) as regard *Plowes* 1492, non Baker.

Botswana. N: Chobe Dist., Chobe, Kasane Valley, c.1000 m, st. 20.ii.1968, *Mutakela* 186 (SRGH). **Zambia**. N: Mbala Dist., Kilungu, fl. s.d., *Glover* in *Bredo* 6167 (BR, LISU photo). S: Mazabuka Dist., Mazabuka, 15°S; 27°E., n.d., *Trapnell* s.n. (K). **Zimbabwe**. N: Mount Darwin Dist., Mavuradonha Mtns., 1460 m, fl. 12.v.1955, *Whellan* 921 (SRGH). W: Matobo Dist., Matopos Research Station, fl. 8.x.1952, *Plowes* 1492 (K, LISC, MO, SRGH). C: Chikomba Dist., near Charter, fl. xi.1899, *Cecil* 82 (K). E: Mutare Dist., Mutare commonage, Meneni R., fl. 22.xii.1949, *Chase* 1866 (BM, BR, K, PRE, SRGH).

Not known from elsewhere. Woodland on sandy soil, stream banks, granite and serpentine rock outcrops; 1000–1460 m.

Conservation notes: Endemic to the Flora area, but widespread; not threatened.

v) Var. **pubescens** L.B. Smith & Ayensu in Kew Bull. **29**: 194 (1974) in part excl. *Jeke* 43; in F.T.E.A., Velloziaceae: 4 (1975). —Drummond in Kirkia **10**: 232 (1975). Type: Zimbabwe, Mazowe Dist., near Tstatsi R., 22.5 km N of Concession, fl. 19.xi.1966, *Leach* 11283 (K holotype).

Zambia. W: Mwinilunga Dist., NE slope of Matonchi Farm, fl. 29.x.1937, *Milne-Redhead* 3001 (BR, K, PRE). **Zimbabwe**. N: Mazowe Dist., near Tsatsi R., 22.5 km N of Concession, fl. 19.xi.1966, *Leach* 11283 (K, P, SRGH).

Also in Tanzania and probably Angola. *Brachystegia* woodland, on granite rocks and rocky ground.

Conservation notes: Widespread taxon; not threatened.

Smith & Ayensu (1974) record the variety from C Zimbabwe (Harare Dist., Christon Bank, Mazowe headwaters, *Loveridge* 1193 (K)) and from N Malawi (Rumphi Dist., near Nyika Plateau conjunction, *Salubeni* 385 (K)) and C Malawi (Dedza Dist., Chongoni Forest, *Salubeni* 1285 (K)).

12. **Xerophyta suaveolens** (Greves) N.L. Menezes in Cienc. & Cult. **23**: 422 (1971). —Smith & Ayensu in Kew Bull. **29**: 194 (1974); in F.T.E.A., Velloziaceae: 6 (1975). Type: Zimbabwe, Mazowe Dist., Bernheim Hill, 1524 m, fl.& fr. x.1906, *Eyles* 439 (BM lectotype).

Perennial herb to dwarf tree, 0.15–3 m high, with blackish fibrous erect stem arising from thick base 5–20 cm across, irregularly branched. Leaves grass-like, 3–10(13) in apical tufts; leaf-sheath base 3–6 cm long, dark brown, quickly disintegrating into stiff fibres after abscission; blade rigid, 3–66 cm long by 1.5–13 mm wide, linear, filiform-attenuate towards apex, finely ribbed, glabrous below, scabrous above with fine ascending hook-like trichomes, glabrous or ciliate along margins; mesophyll isolateral. Flowers 1–6(8) tufted at ends of branches, precocious, sweet-scented, showy; peduncle slender, stiff, 2.5–16 cm long, grey-brown, glabrous or with hair-like trichomes on upper part. Perianth green, greenish-cream or white to mauve, pink, deep purple to blue, often darker above; segments subequal, (20)30–60(65) × (2.5)3–12(16) mm, narrowly lanceolate, acuminate, outermost keeled, glabrous outside, innermost with several shallow plaiting marks in centre and raised flange-like margins; tube 1–4 mm long. Filaments almost or wholly joined to base of perianth segments, with a short or no free part; anthers (9)10–20(25) mm long. Ovary (5)6–10(15) × (1.5)2–5 mm, ellipsoid or cylindrical, covered with short, flat, erect, forked or wide-stellate and shortly stiff-rayed trichomes; style 2–8 mm long, stigmatic region (7)10–20(25) mm long. Capsule (7)10–20(22) × (4)6–15(17) mm, trigonal. Seeds numerous, minute.

Leaf-blades glabrous below and scabrous above · · · · · · · · · · · · · · · **i)** var. *suaveolens*
Leaf-blades covered above or on both surfaces with long trichomes · · **ii)** var. *vestita*

i) Var. **suaveolens** —Smith & Ayensu in Kew Bull. **29**: 194 (1974); in F.T.E.A., Velloziaceae: 6 (1975). FIGURE 12.2.**42**.
  Vellozia suaveolens Greves in J. Bot. **59**: 282 (1921).

Botswana. N: Francistown, fl.& imm.fr. ii.1926, *Rand* 1 (BM). **Zambia**. N: Mpika Dist., Muchinga Escarpment, 48 km S of Ishiba Ngandu on road to Mpika, fl. 28.xi.1952, *Angus* 866 (FHO, K). W: Kitwe Dist., Kitwe, fl.& fr. 14.x.1970, *Fanshawe* 10941 (K, NDO). S: Victoria Falls, Hubert Young Drive, fl. 21.xi.1949, *Wild* 3126 (K, LISC, SRGH). **Zimbabwe**. N: Mazowe Dist., Spelonken Farm, near entrance to University College Farm, fl.& imm.fr. 1.xii.1969, *Grosvenor* 460 (LISC, SRGH). W: Hwange Dist., 32 km SE of Hwange, fl. 7.xi.1956, *Turner & Shantz* 4144 (SRGH). C: Harare Dist., Meyrick Park, fl. 12.ix.1967, *Grosvenor* 424 (LISC, SRGH). E: Mutare Dist., Mutare, NE boundary, Murahwa's Hill, 1150 m, fl. 26.xi.1968, *Chase* 8510 (LISC, PRE, SRGH). S: Beitbridge Dist., 3.2 km S of Runde R. bridge, Masvingo/ Beitbridge Road, fl. 14.vii.1961, *Leach* 11184 (K, LISC, PRE, SRGH). **Malawi**. S: Mulanje Dist., Mulanje Mt., beside Likhubula–Lichenya path, 1200 m, fl. 31.x.1986, *J.D. & E.G. Chapman* 8171 (MO, PRE). **Mozambique**. N: Cuamba Dist., Amaramba, margin of Rio Lúrio, road to Malema, fl. 14.x.1942, *Mendonça* 815 (BR, COI, EA, J, LISC, LMA, LUA,

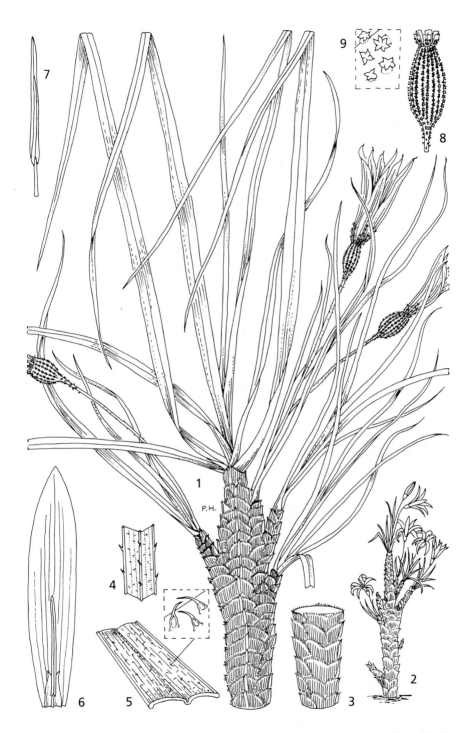

Fig. 12.2.**42**. XEROPHYTA SUAVEOLENS var. SUAVEOLENS. 1, fruiting habit (× ²/₃); 2, flowering habit (much reduced); 3, detail of stem (× ²/₃); 4, part of young leaf (enlarged); 5, part of mature leaf (enlarged); 6, tepal with stamen (× 2); 7, stigma (× 2); 8, fruit, tepals cut off (× 2); 9, hairs from fruit (much enlarged). All from *C.E.F. Allen* 41. Drawn by Pat Halliday.

SRGH, UPS, WAG); Nampula, Meconta Dist., 18 km from Meconta to Nacavala, c.250 m, fl.& fr. 26.xi.1963, *Torre & Paiva* 9332 (BR, COI, LISC, LUA, MO, PRE, WAG). MS: Báruè Dist., 13 km from Changara to Catandica (Vila Gouveia), c.400 m, fl. 24.v.1971, *Torre & Correia* 18569 (COI, EA, J, LISC, LMA, LMU, UPS).

Also in Congo, Angola and Tanzania. Mopane, acacia and miombo woodland, rocky places, sandy slopes and stream banks, often in dry situations; 250–1700 m.

Conservation notes: A widespread taxon; not threatened.

Numerous specimens cited in earlier literature are now referred to this taxon – *Angus* 866; *Baines* s.n.; *Chase* 3073; *Eyles* 25, 439, 516; *Fries* 2211; *Rand* 270; *Marloth* 2174, 3415; *Mitchell* 603; *Munro* 801; *White* 3680; *Wild* 3126; *Wild & Drummond* 6676.

This taxon is closely related to *X. eylesii* (Greves) N.L. Menezes from which it differs in the following characters: leaf-blade glabrous below and along the margins, and flowers larger with ovary covered in stellate trichomes.

ii) Var. **vestita** L.B. Smith & Ayensu in Kew Bull. **29**: 194 (1974); in F.T.E.A., Velloziaceae: 6 (1975). Type: Zambia, 48 km E of Mbala on Tunduma road, fl. 28.xi.1958, *Napper* 1161 (EA holotype).

Zambia. N: 48 km E of Mbala on Tunduma road, fl. 28.xi.1958, *Napper* 1161 (EA, K). W: Kitwe Dist., Kitwe, fl. 6.xi.1967, *Fanshawe* 10215 (K, NDO). C: Munshiwemba, fl. x.1941, *Stohr* 632 (PRE). E: Chipata Region, between Nyimba and Kachalola, 800 m, fl. 17.x.1986, *Linder* 3931 (PRE). **Zimbabwe**. N: Mazowe Dist., Spelonken Farm, fr. 5.xii.1968, *Biegel* 2695 (BR, K, LISC, SRGH). **Malawi**. N: Rumphi Dist., Rumphi-Chitipa road, fl. 9.xii.1965, *Banda* 794 (SRGH). C: Dedza Dist., Chongoni Forestry School, fl. 19.xii.1967, *Jeke* 43 (K, LISC, PRE, SRGH). S: Zomba Dist., Nkhoronje Hill, c.1000 m, fl. 4.xii.1983, *la Croix* 2424 (BM). **Mozambique**. N: Lichinga Dist., close to Metónia, 13°25'S, 35°35'E, 1200 m, fl.& fr. xi.1933, *Gomes e Sousa* 1576 (COI). Z: Ile Dist., Rio Lua, edge of Gurué with Alto-Molócuè, fl. 6.xi.1942, *Mendonça* 1299 (COI, K, LISC, SRGH, WAG). T: Tete-Songo road, between Chissua and Maroeira, c.8 km, fl. 5.xi.1973, *Macedo* 5359 (LISC, LMA).

Also in Tanzania. Grassland, savanna and *Brachystegia* woodland, rocky places, mountain slopes, stream banks and termitaria, often in dry situations; 800–1300 m.

Conservation notes: A widespread taxon; not threatened.

Various specimens cited in earlier literature are now referred to this taxon – *Jacobsen* 2166, 3569; *Jeke* 43; *West* 3453.

13. **Xerophyta nutans** L.B. Smith & Ayensu in Kew Bull. **29**: 195 (1974); in F.T.E.A., Velloziaceae: 6 (1975). Type: Tanzania, Masasi Dist., 33.5 km E of Lumesule R., 17.xii.1955, *Milne-Redhead & Taylor* 7687 (EA holotype).

Vellozia sp. 1 of White, F.F.N.R.: 17 (1962).

Shrubby plant 0.2–3 m high with a fibrous, few-branched stem, the main branches to 3 cm thick, flowering branches short. Leaves arcuate, 2–10 tufted at ends; leaf-sheath base 2–4.5 cm long, blackish-brown, soon splitting after abscission; blade 3.5–40 cm long by 1–8 mm wide, lanceolate to linear, filiform-attenuate towards apex, covered with flat linear-lanceolate grey trichomes above, glabrous below, with minute prickle-like trichomes spaced along margins; mesophyll isolateral. Flowers 1–4, in tufts at ends of small branches; peduncle 5–14 cm long, stellate-pubescent. Perianth yellowish, becoming white, mauve or light blue; segments subequal, (25)30–45 × 5–8 mm, linear-lanceolate, acute; tube 2–4 mm long. Filaments triangular, almost wholly joined to perianth segments, with a very short free part; anthers (11)12–15(17) mm long. Ovary (5)6–10(15) × 3–5 mm, ellipsoid, covered with orbicular broad-based shortly broad-rayed trichomes; style 4–8(10) mm long, stigmatic region (6)8–18(20) mm long. Capsule 8–10 × 6–8 mm, ellipsoidal. Seeds numerous, minute.

Zambia. N: Mporokoso Dist., Nsama-Mporokoso, fl.& imm.fr. 25.x.1949, *Bullock* 1375 (K). E: Chama Dist., Makutu Mtns., fl. 26.x.1972, *Fanshawe* 11542 (K, NDO). **Malawi**. N: Chitipa Dist., Mafinga Mtns., 1830–2130 m, fl. 15.iv.1963, *Chapman* 1866 (MAL, SRGH).

Also in Tanzania. Abundant in *Brachystegia* woodland forming extensive patches, common over large areas on rocky slopes and ridges; 1830–2130 m.

Conservation notes: Apparently not widespread in the Flora area, but probably not threatened.

14. **Xerophyta argentea** (Wild) L.B. Smith & Ayensu in Kew Bull. **29**: 198 (1974). — Drummond in Kirkia **10**: 232 (1975) as regards *Wild* 3612 (given in error as 4612). Type: Zimbabwe, Chimanimani Mtns, quartzite peak near Mt. Peza, 1830 m, fl. 16.x.1950, *Wild* 3612 (SRGH holotype).

Vellozia argentea Wild in Kirkia **4**: 135 (1964).
Vellozia sp. of Goodier & Phipps in Kirkia **1**: 52 (1961) as regards *Wild* 3612.

Herb to shrub, 0.3–4 m high, with erect stem to 23 mm thick, unbranched or sparsely branched, branches with mantle of closely packed leaf-bases not breaking into discrete fibres. Leaves 3–7 tufted at branch-ends; leaf-sheath base 1–2.5 cm long, scale-like, smooth; blade rigid, 4–30 cm long by 1–5 mm wide, linear, attenuate, margin entire, silvery and densely tomentose with closely compacted appressed white hairs below, striate and sparsely hairy or subglabrous above; mesophyll isolateral. Flowers 1–4 at branch-ends, sweet-scented; peduncle 2.5–16.5 cm long, brownish. Perianth blue, light purple or pinkish; segments 15–35 × 3.5–10(14) mm, oblong-elliptic, rounded, 8–10-nerved; tube 2–4 mm long. Filaments joined to base of perianth segments, with very short free part; anthers 9–13 mm long. Ovary 4–7(12) × 2.5–5(6) mm, obconical, densely purplish-glandular; style 4–8 mm long, stigmatic region 10–17 mm long, with subfleshy bands. Capsule 6–8 × 5–7 mm, ellipsoidal.

Zimbabwe. E: Chimanimani Dist., Chimanimani Mtns., Martin Forest Reserve, fl. 14.xi.1967, *Mavi* 606 (LISC, M, PRE, SRGH). **Mozambique**. MS: Sussundenga Dist., Rotanda, between Rio Mussapa and frontier, Tandara, 1750 m, fl. 31.x.1965, *Torre & Pereira* 12618 (BR, COI, J, LISC, LMA, MO).

Only known from the Chimanimani Mtns. Rocky grassland, common among quartzite crags and on cliff-ledges; 1500–2150 m.

Conservation notes: Endemic to the Chimanimani mountains; probably Lower Risk near threatened.

15. **Xerophyta spekei** Baker in J. Bot. **13**: 234 (1875). —Smith & Ayensu in Kew Bull. **29**: 198 (1974); in F.T.E.A., Velloziaceae: 7 (1975). Type: Tanzania, Tabora Dist., Boss Rock, 6°S, 1220 m, *Grant* 628 (K holotype).

Vellozia spekei (Baker) Baker in Trans. Linn. Soc. London **29**: 156 (1875); in F.T.A. **7**: 412 (1898). —Greves in J. Bot. **59**: 283 (1921). —Brenan, Check-list For. Trees Shrubs Tang. Terr.: 629 (1949).
Barbacenia tomentosa Pax in Bot. Jahrb. Syst. **15**: 144 (1892). —Harms in Engler, Pflanz. Ost-Afrikas **C**: 146 (1895). Type: Kenya, Machakos Dist., Athi R., *Fisher* 585 (B holotype, K).
Vellozia aequatorialis Rendle in J. Linn. Soc., Bot. **30**: 409 (1895). —Baker in F.T.A. **7**: 412 (1898). —Greves in J. Bot. **59**: 283 (1921). —Brenan, Check-list For. Trees Shrubs Tang. Terr.: 629 (1949). Type: Tanzania, between Zanzibar and Uyui, *W.E. Taylor* s.n. (BM holotype).
Vellozia tomentosa (Pax) Baker in F.T.A. **7**: 412 (1898). —Greves in J. Bot. **59**: 283 (1921). —Brenan, Check-list For. Trees Shrubs Tang. Terr.: 629 (1949). —Jacobsen in Kirkia **9**: 153 (1973).

Perennial shrubby herb 45–60 cm high (small tree to 5 m outside the Flora area) with stiff stems 10–13 cm thick, tapering from a stout rootstock 1.2–1.5 m across, decumbent and rooting at base, with erect woody branches 6–18 mm thick, densely covered with coarse dry leaf-bases. Leaves grass-like, 3–8(10) tufted at ends or in axils of leaf-bases on main stem; leaf-sheath base 3–3.5 cm long, scale-like, typically with middle nerve enlarged into broad smooth umbo, greyish, grey-brown or dark red-brown becoming black, tomentose to densely white-strigose towards apex; blade stiff-leathery, (5)7–30(70) cm long by (1)2–7(18) mm wide, linear, attenuate towards apex, tomentose on both surfaces or only below when young, usually bristly-serrate near apex and along margins; mesophyll isolateral. Flowers 1–5 in tufts among leaves, showy, sweet-scented; peduncle 2–17 cm long, flexuose, with hard black glands in upper half. Perianth white to blue, lilac or mauve; segments subequal, 20–30(50) × 3–4(8) mm, lanceolate or linear-lanceolate, acuminate to rounded; tube 1–2 mm long. Filaments broadly triangular, joined to base of perianth segments, with very short or no free part; anthers (10)15–18(20) mm long. Ovary 5–8(12) × 2–4 mm, globose, clavate or oblong-cylindrical, glabrous, covered with subsessile tubercular dark purple viscous glands; style rather thick, 3–5 mm long, stigmatic region 10–13 mm long. Capsule brittle, 12–13(15) × 10–11(13) mm, globose, viscous-glandular. Seeds numerous, minute.

Zambia. N: Kaputa Dist., Musesha, fl.& fr. 8.x.1958, *Fanshawe* 4875 (K).

Also in Kenya and Tanzania. Rocky slopes, rock outcrops and *Brachystegia* woodland; 450–1900 m outside Flora area.

Conservation notes: Very localised in the Flora area; probably Lower Risk near threatened.

16. **Xerophyta zambiana** L.B. Smith & Ayensu in Kew Bull. **29**: 198, fig.4 (1974). Type: Zambia, Petauke Dist., Nyimba-Luembe road, 750 m, fl. 11.xii.1958, *Robson* 898 (BM holotype).

> *Vellozia sp. 2* of White, F.F.N.R.: 17 (1962).
> *Vellozia trichophylla* sensu Jacobsen in Kirkia **9**: 153 (1973) as regard *Jacobsen* 2645, non (Baker) Hemsl.

Perennial herb or ± fibrous shrublet, 0.2–1 m high, with erect stem tapering from stout rootstock, branched or unbranched, branches short, 3–6 cm thick. Leaves 3–9 in apical tufts or in axils of upper leaf-bases, grass-like; leaf-sheath base 3–7 cm long, red-brown, glabrous, erect after abscission but quickly splitting and leaving a fibrous mass of vascular bundles; blade fleshy, pale green, 2–50 cm long by 1–13 mm wide, linear, filiform-attenuate towards apex, tomentose on both surfaces, pubescent along margins and towards base with divergent to spreading white hair-like trichomes up to 1 mm long; mesophyll isolateral. Flowers 1–4 in terminal clusters amongst leaves, sweet-scented, showy; peduncle delicate, 2–20 cm long, sparsely bristly. Perianth blue to purple, pink to pale mauve or white within and greenish outside; segments (15)25–45(50) × 4–11 mm, linear-lanceolate, attenuate towards apex, sparsely glandular towards base; tube 1–4 mm long. Filaments 3–4 mm long, almost wholly joined to perianth, bases forming a ring 1 mm high, apices narrowly triangular; anthers 10–22 mm long. Ovary 4–12 × (1.5)2–4 mm, ellipsoid, with short epigynous tube, ± densely covered with sessile or short-stalked tubercular or capitate sticky black glands; style 3–6 mm long, stigmatic region 8–16 mm long. Capsule 9–15 × 7–12 mm. Seeds appearing vitreous.

Zambia. C: Serenje Dist., Kanona, fr. 30.xi.1952, *Angus* 886 (BM, BR, FHO, K, PRE). E: Petauke Dist., Petauke, 13°55'S, 31°20'E, 1067 m, fl. 11.xii.1957, *Stewart* 93 (K). S: Mumbwa Dist., E of Kafue Hoek, road to Mumbwa, fl. 8.x.1957, *West* 3538 (LISC, SRGH). **Zimbabwe**. N: Kariba Dist., Kariba, 760 m, fl.& fr. xi.1959, *Goldsmith* 48/59 (SRGH). C: Mazowe Dist., Henderson Research Station, Iron Mask Range, fl. 12.xi.1964, *West* 6132 (SRGH). **Mozambique**. N: Montepuez Dist., Montepuez, road to Ancuabe, fl. 17.x.1942, *Mendonça* 900 (BR, COI, EA, LISC, LMA, WAG). Z: Gurué Dist., Cascata Namúli, c.1250 m, st. 3.viii.1979, *de Koning* 7563 (BM, K, LISC, LMU). T: Cahora Bassa Dist., road Chitima–Songo, c.3 km from Chitima, st. 4.ii.1972, *Macedo* 4758 (LISC, LMA).

Also in Congo. *Brachystegia* or *Uapaca* woodland, rocky places, sometimes in clumps; 500–1250 m.

Conservation notes: A widespread species; not threatened.

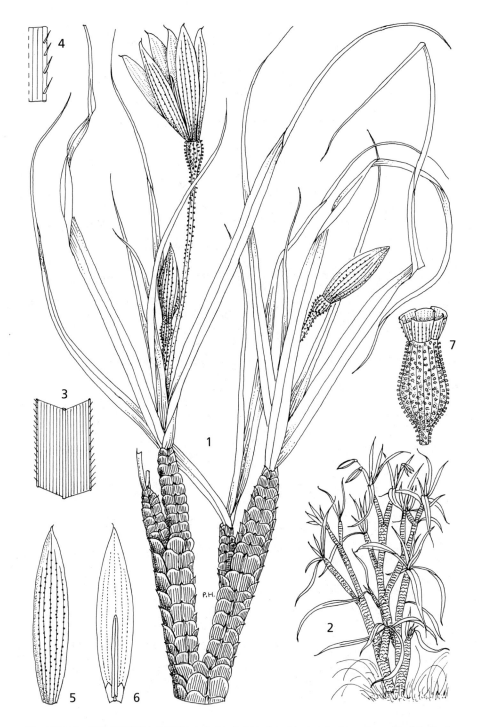

Fig. 12.2.**43**. XEROPHYTA KIRKII. 1, habit (× ²/₃); 2, habit sketch; 3, portion of leaf (× 2); 4, portion of leaf margin (much enlarged); 5, tepal, outer surface (× 1); 6, tepal with anther (× 1); 7, ovary (× 2). All from *Torre & Correia* 15903. Drawn by Pat Halliday.

17. **Xerophyta kirkii** (Hemsl.) L.B. Smith & Ayensu in Kew Bull. **29**: 201 (1974). Type: Malawi, Zomba, 1830–2150 m, fl. ix.1859, *Kirk* s.n. (K holotype). FIGURE 12.2.**43**.

Vellozia kirkii Hemsl. in Bot. Mag. **130**: t.7962 (1904). —Greves in J. Bot. **59**: 283 (1921). *Vellozia equisetoides* (Baker) Baker in F.T.A. **7**: 411 (1898) as regard *Kirk* s.n. & *Scott-Elliot* 8524.

Vellozia splendens sensu Baker in F.T.A. **7**: 412 (1898) as regard *Whyte* s.n. from Malosa Mt. —sensu Greves in J. Bot. **59**: 283 (1921) as regard *Whyte* s.n. —sensu Brenan in Mem. New York Bot. Gard. **9**: 86 (1954), non Rendle.

Subshrub to small tree (0.9)1–3(5) m high, with erect branched or unbranched fibrous stem 1.3–2.5 cm thick, with short thick primary branches and numerous short flowering shoots, thickly coated with compact leaf-bases and fibrous roots, the whole often 30–40 cm across at base. Leaves 3–10 in apical tufts, grass-like, suberect to pendulous; leaf-sheath base 1.5–4 cm long, rounded at end; blade 1.5–40(60) cm long by 1.5–12 mm wide, linear, apex acute, with distinct midrib, gland-spotted, serrate or glabrous along margins; mesophyll isolateral. Flowers (1)2–4(6) in tufts at tips, very fragrant; peduncle ± erect, 4.5–12 cm long, sticky. Perianth white or with a very pale lilac or mauve tint, reddish-lilac or dusky violet, fading to pale mauve or purple, sometimes throat green at base; segments equal, (25)40–60(65) × 5–12 mm, lanceolate, acute, outermost thickened and rough outside, strongly glandular on midline, innermost membranous; no tube. Filaments almost wholly joined to base of perianth segments; anthers subsessile, inserted 3–4 mm above base, 12–20(22) mm long, glandular. Ovary 6–11 × (2.5)3–6 mm, obconical, tubercular, strongly glandular; style 4–10(15) mm long, club-shaped, stigmatic region 15–20(22) mm long. Capsule 17 × 9 mm, sometimes covered with a viscous exudate, persistent, apparently indehiscent.

Malawi. S: Zomba Dist., Zomba Mt., 1850 m, fl. 25.i.1959, *Robson & Jackson* 1316 (BM, K, LISC, PRE). **Mozambique**. N: Ngauma Dist., Massangulo Mtns., 66 km N of Mandimba, c.1400 m, fl. 26.v.1961, *Leach & Rutherford-Smith* 11044 (SRGH). Z: Gurué Dist., slopes of Gurué Mt., via fábrica Junqueiro, W of Picos Namúli, confluence of R. Malema and R. Cocossi, c.1200 m, fl. 6.xi.1967, *Torre & Correia* 15903 (COI, EA, LISC, LMU, WAG).

Not known from elsewhere. Savanna, montane grassland and rocky outcrops; 1200–2150 m.

Conservation notes: Restricted distribution, endemic to the Flora area; probably Lower Risk near threatened.

18. **Xerophyta splendens** (Rendle) N.L. Menezes in Cienc. & Cult. **23**: 422 (1971). —Smith & Ayensu in Kew Bull. **29**: 199 (1974). Type: Malawi, Mulanje Mt., 1830 m, fl. x.1891, *Whyte* s.n. (K holotype).

Vellozia splendens Rendle in Trans. Linn. Soc. Bot. **4**: 49 (1894). —Baker in F.T.A. **7**: 412 (1898) as regard *Whyte* s.n. from Mt. Mulanje. —Greves in J. Bot. **59**: 283 (1921) in part excl. *Whyte* s.n. from Mt. Malosa. —Topham, Check-list For. Trees Shrubs Nyass. Prot.: 100 (1958). —Binns, First Check List Herb. Fl. Malawi: 103 (1968). —Chapman & White, Evergr. For. Malawi: 170 (1970).

Xerophyta sp. of Moriarty, Wild Fl. Malawi: pl.17 (1975) as regard *Moriarty* 104.

Subshrub to small tree (0.45)1.8–3(6) m high, glandular with a strong resinous smell, stems 8–12 mm thick, unbranched or forked, 1.5–3(or more) cm across at base, root system spreading; stem and branches fibrous, coated with mass of closely packed leaf-bases. Leaves 4–8 in rosettes on top of branches, glabrous, slightly sticky; leaf-sheath base 4–7.5 cm long, strongly ribbed; blade stiff-leathery, flattened, linear or linear-lanceolate, gradually tapering towards a filiform apex, outermost 30–62 cm long by 8–17 mm wide, innermost 10–20 cm long by 2–4 mm wide, with prominent midrib on both surfaces, rough along margins; mesophyll isolateral. Flowers (1)2–6(10), tufted at tips, in axils of bracts; peduncle 14–24 cm long by 1–1.5 mm wide, strong, thickly covered with raised emergences, extremely sticky upwards. Perianth lilac or white, sometimes with faint red or bluish strips; segments

(45)50–65(70) × (8)9–18 mm, oblong to ovate-lanceolate, acuminate, a small scale at base on either side of anther insertion, outermost segments with black emergences, continued up central veins nearly to apex but diminishing in size and number; tube 4–6 mm long, fleshy, covered with emergences like those on peduncle. Filaments wholly joined to perianth segments; anthers sessile, inserted c.2 mm above base, 18–28 mm long, the connective extended 1 mm beyond anther. Ovary (8)10–14 × 4–6 mm, obconical, densely sticky, covered with trichomes; style (4.5)5–8(10) mm long, stigmatic region (15)18–28(30) mm long. Capsule 25 × 20 mm, covered with viscous exudate, persistent, apparently indehiscent.

Malawi. S: Mulanje Dist., Chambe Plateau E side, 2070 m, fr. 12.ii.1979, *Blackmore, Brummitt & Banda* 369 (BM); Mulanje Forest Reserve, Chisepo where track to Sapitwa leaves Tuchila–Chambe path, fl. 11.v.1981, *Chapman* 5683 (MAL). **Mozambique**. Z: Gurué Dist., NE slopes of Mt. Namuli, 1670 m, st. 2.vi.2007, *Timberlake* 5163 (K, LMA).

Only known from Mts. Mulanje and Namuli. Abundant on rocky slopes and ledges; 1650–2450 m.

Conservation notes: Restricted distribution, but habitat not threatened; probably Lower Risk near threatened to Vulnerable.

19. **Xerophyta viscosa** Baker in J. Bot. **13**: 235 (1875). —Sölch in Merxmüller, Prodr. Fl. SW Afrika, fam.152: 2 (1969). —Coetzee in Dinteria **10**: 26 (1974). —Smith & Ayensu in Kew Bull. **29**: 202 (1974). Type: South Africa, Eastern Cape, Pondoland Region, Faku Territory, *Sutherland* s.n. (K lectotype).

> *Vellozia viscosa* (Baker) Baker in F.C. **6**: 245 (1896). —Greves in J. Bot. **59**: 283, fig.4 (1921). —Hutchinson, Botanist Sthn. Africa: 363 (1946). —Ross, Fl. Natal: 133 (1973). — Jacobsen in Kirkia **9**: 153 (1973).
>
> *Barbacenia hereroensis* Schinz in Bull. Herb. Boiss. **4**, App.3: 49 (1896). Type: Namibia, Hereroland, behind Tebris Pass, 30.viii.1896, *Fleck* 80 (Z holotype).
>
> *Vellozia hereroensis* (Schinz) Baker in F.T.A. **7**: 411 (1898). —Greves in J. Bot. **59**: 283 (1921).

Perennial shrubby herb or small shrub to 1.8(2) m high, sticky, usually with short, erect unbranched or dichotomously branched stem; stem and branches covered in remains of leaves and adventitious roots without mantle of closely packed leaf-bases. Leaves 3–7 in rosette at tops of stems; leaf-sheath base 2–3 cm long; blade stiff-leathery, 10–38 cm long by 1–5 mm wide, linear, acuminate towards apex, closely and strongly ribbed with longitudinal grooves below and prominent acute keel above, minute spine-like trichomes on keel and along margins, slightly viscous towards base with sunken scattered glands like those on margins; mesophyll isolateral. Flowers usually only 1–2; peduncle (15)17.5–22 cm long, with closely packed stalked black glands, sessile towards base. Perianth white, tinged blue to rose or pale mauve, glandular outside; segments (24)32–33 × 6–8 mm, linear or lanceolate, apiculate, outermost dotted with large black glands on back; tube 3 mm long. Filaments joined to base of perianth segments, with very short free part; anthers (12)14–16 mm long. Ovary 7–8(15) × 3.5 mm, oblong-obconical or clavate, with black, long-stalked, glandular, spherical headed trichomes; style 4 mm long, stigmatic region 18–22 mm long. Capsule oblong-obconical.

Zimbabwe. N: Makonde Dist., Lomagundi, *Jacobsen* 3681 (SRGH), cited by Jacobsen in Kirkia **9**: 153 (1973). E: Chimanimani Dist., Makurupini area, 433 m, fl. 27.viii.1969, *T.M. Wild* A18 (SRGH). **Mozambique**. Z: Gurué Dist., Mt. Namuli, slopes above Rio Licungo, 1980 m, fl. 16.xi.2007, *Harris* 326 (K, LMA).

Also in Angola, Namibia, South Africa and Lesotho. Amongst grasses on damp soil, near water; c.450 m.

Conservation notes: Localised in the Flora area, but more widespread outside; probably not threatened.

Closely related to *X. schlechteri* (Baker) N.L. Menezes which has persistent leaf-blades that slowly disintegrate, dorsiventral mesophyll and a stigmatic region shorter than the style.

COLCHICACEAE

by I. la Croix

Perennial herbs with rhizomes or a tunicated corm, occasionally stoloniferous. Stems erect or scandent, simple or branched, rarely woody, sometimes almost entirely underground. Leaves alternate, subopposite or verticillate, arising from the stem or stem base, sessile, often with a sheathing base, sometimes pseudopetiolate; lamina linear, lanceolate or ovate, occasionally subulate, with parallel veins, often with a distinct midrib, rarely with additional reticulate venation. Inflorescence a terminal raceme or cyme, sometimes pseudoumbellate, capitate or solitary, with or without bracts. Flowers hypogynous, bisexual, rarely unisexual, sessile or pedicellate, usually actinomorphic but occasionally slightly zygomorphic. Tepals usually 6, equal or only slightly unequal, free or joined for some part of their length, deciduous or persistent. Stamens 6, free or joined to petals, dehiscing by longitudinal slits. Gynoecium usually with 3 carpels (occasionally 2 or 4), syncarpous; ovary trilocular with axile placentation and few to many ovules. Styles 1 or 3, free or partly united. Fruit usually a capsule, dry or somewhat fleshy, rarely a berry. Seeds subglobose or ovoid or somewhat angled.

19 genera and about 225 species in temperate and tropical Africa, Europe, Asia, Australasia and North America. 7 genera and about 14 species occur in the Flora Zambesiaca area. There is an unconfirmed record of *Sandersonia aurantiaca* Hook. from Niassa Province in N Mozambique (da Silva, Izidine & Amude, Premin. Checklist Vasc. Plants Mozambique, 2004).

Many plants of this family are poisonous to livestock as most genera contain alkaloids, some of which are used in medicine and pharmacology.

1. Plants stemless or with a short scape ················· **1. Androcymbium**
– Plants with a stem ······································· 2
2. Flowers sessile in a spike-like cyme, without bracts ············ **2. Wurmbea**
– All flowers pedicellate in a bracteate inflorescence, or solitary ·········· 3
3. Style forming a distinct angle to the ovary ················ **3. Gloriosa**
– Style continuing line of main axis of ovary ······················ 4
4. Inflorescence single-flowered; stem erect or scandent, leaves often with tendrils; flowers large, bell-shaped ···································· **4. Littonia**
– Inflorescence several-flowered ······························ 5
5. Perianth persistent; styles 3, free ± to base ············· **5. Ornithoglossum**
– Perianth falling early; style simple or branched above base ············· 6
6. Style with 3 falcate branches; underground part a small, ovoid corm ········
   ·· **6. Iphigenia**
– Style simple; underground part a young ovoid corm attached to a knee-shaped, horizontal tuber from the previous season ··············· **7. Camptorrhiza**

1. ANDROCYMBIUM Willd.

Androcymbium Willd. in Ges. Naturf. Freunde Berlin Mag. Neuesten Entdeck. Gesammten Naturk. **2**: 21 (1808). —Krause in Notizbl. Bot. Gart. Berlin-Dahlem **7**: 512–526 (1921).

Perennial herbs with corms with firm, dark tunics; stem lacking or shortly scapose. Leaves in 2 rows or subrosulate, linear to ovate, grading into green or coloured bracts. Flowers in terminal capitulum or corymb, rarely solitary, surrounded by bracts so that the whole looks like one flower. Perianth stellate to bell-shaped; tepals 6, free, clawed. Stamens 6, arising at junction of claw and blade; anthers introrse. Ovary ovoid, trilocular; styles 3; stigmas very small. Capsule septicidal, subglobose or inversely pear-shaped; seeds minutely warty.

About 30 species in Africa and the Mediterranean (Canary Islands to the Middle East); 3 species in the Flora Zambesiaca area.

1. Leaves numerous; stamens about half the length of tepal blade · · · · · **3.** *roseum*
 – Leaves few, usually 4 or less; stamens longer than tepal blade · · · · · · · · · · · 2
2. Plant ± stemless, leaves lying on ground · · · · · · · · · · · · · · · · · **1.** *melanthioides*
 – Plant with a distinct stem with leaves set along its length · · · · · · · · · **2.** *striatum*

1. **Androcymbium melanthioides** Willd. in Ges. Naturf. Freunde Berlin Mag. Neuesten Entdeck. Gesammten Naturk. **2**: 21 (1808). —Anon. in Fl. Pl. S. Afr.: **2** t.53 (1922). —Maroyi in Kirkia **18**: 5 (2002). Type: South Africa, 'Promontori bomae spei legit prope Channakraal ad Sak river, Lichtenstein' in *Herb. Willd* 7073 (B).

 Androcymbium subulatum Baker in J. Bot. **12**: 245 (1874); in F.T.A. **7**: 559 (1898). Type: Zimbabwe, S of Umsweswe R., wooded sandbank, 15.vi.1870, *Baines* s.n. (K holotype).
 Androcymbium melanthioides var. *acaule* Baker in J. Linn. Soc., Bot. **17**: 442 (1879). Type: South Africa, Gauteng, Pretoria, *Roe* in *Bolus* 3042 (K).
 Androcymbium melanthioides var. *subulatum* (Baker) Baker in F.C. **6**: 517 (1897).

Perennial herb to 30 cm; corm conical to subglobose, 14–20 × 12–15 mm, set 10–20 cm below the ground. Stem virtually absent above ground. Leaves 2–3, 13–30 cm long or more (tips usually broken in herbarium specimens), narrowing abruptly from an ovate base c.7 cm long × 1–2.6 cm wide, to become linear, 2–4 mm wide, for most of its length. Bracts several, 6–13 × 1–5.5 cm, showy, ovate, acute or obtuse, white or purplish, striped with 15–50 green veins. Inflorescence a congested head of up to 10 flowers lying inside the bracts. Flowers purple to mauve-pink, sometimes striped violet; pedicel 3 mm long, 2.5 mm broad. Tepals 11–15 mm long with a ± tubular claw 5–9 mm long; blade concave, 6–7 mm long, 2–3 mm wide at base, ovate, acuminate, with a dark mark where the stamens are joined. Stamens attached to tepals at junction of claw and blade, 9 mm long hence longer than blade, anthers 2–3 mm long. Ovary 6 × 2.5 mm, ellipsoid; styles free, 8 mm long; stigma small, simple.

Botswana. SW: Kgalagadi Dist., Masatleng Pan, fl. 9.v.1976, *Skarpe* S-68 (K). **Zimbabwe** C: Harare (Salisbury), 1450 m, fl. vi.1919, *Eyles* 2768 (K, PRE). S: Masvingo (Fort Victoria), fl. 2.vii.1971, *Chiparawasha* 406 (K, SRGH).
 Also in Namibia and South Africa. Moist sandy soils and moist grassland; 1000–1500 m.
 Conservation notes: Moderately widespread; probably not threatened.

2. **Androcymbium striatum** A. Rich., Tent. Fl. Abyss. **2**: 336 (1851). —Baker in J. Bot. **12** : 245 (1874). —Demissew in Fl. Ethiopia **6**: 186 (1997). —Maroyi in Kirkia **18**: 6 (2002). —Hoenselaar in F.T.E.A., Colchicaceae: 2 (2005). Types: Ethiopia, Simien, Entchetkab, *Schimper* II/1338 (K syntype) & Ethiopia, Shire, *Quartin-Dillon* s.n. (P syntype).

 Androcymbium melanthioides var. *striatum* (A.Rich.) Baker in J. Linn. Soc., Bot. **17**: 442 (1879); in F.T.A. **7**: 560 (1898). —Durand & Schinz, Consp. Fl. Afr. **5**: 412 (1893).

Perennial herb to 15 cm; corm 8–15 × 7–11 mm, subglobose. Stem up to 15 cm long, above ground. Leaves 2–3, set along stem and distinct from bracts, 9–32 cm × 4–8 mm, linear, acuminate, lacking an ovate base. Bracts usually 4, 27–45 × 10–18 mm, ovate, acute, showy, white with green or green and mauve veins. Inflorescence subumbellate, lying within the bracts; flowers several, whitish; pedicel 6–10 mm long. Tepals with claw 1.5–2 mm long, blade 5–7 × 1.5–2 mm. Stamens 7–8 mm long, attached to tepals at junction of claw and lamina; anthers 1.5–2 mm long.

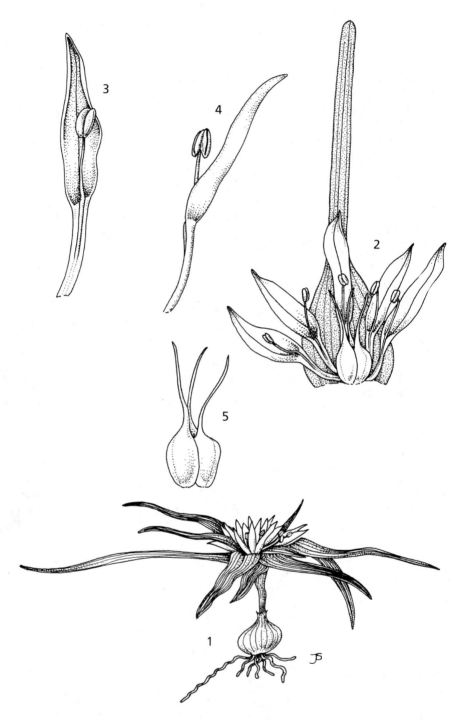

Fig. 12.2.**44**. ANDROCYMBIUM ROSEUM. 1, habit (× ²/₃); 2, part of single inflorescence with bract (× 2); 3, tepal and stamen, front view (× 3); 4, tepal and stamen, side view (× 3); 5, gynoecium (× 3). All from *Fanshawe* 6938. Drawn by Judi Stone.

Zambia. C: Lusaka SE, Kanyanja dambo, 1300 m, 29.xii.1996, *Bingham* 11299 (K, SRGH). **Zimbabwe**. N: Guruve Dist., Mupinge Pass (Impingi Poort), fl. xii.1937, *Hopkins* 6767 (K, SRGH). E: Chimanimani Dist., few miles S of Chimanimani on Orange Grove road, 1500 m, fl. xii.1968, *Goldsmith* 170/68 (K, PRE, SRGH). **Malawi**. C: Dedza Dist., Chincherere Hill, fl. 18.i.1959, *Robson* 1245 (K). S: Zomba Mt., near CCAP cottage, fl. 2.xii.1970, *Moriarty* 359 (K, MAL). **Mozambique**. N: near Lake Niassa (Nyasa), fl. 1902, *W.P. Johnson* 507 (K).

Also in Ethiopia, Kenya, Uganda, Tanzania, Angola and South Africa. In moist grassland and depressions, and on seepage rocks; 600–1700 m.

Conservation notes: Widespread species; not threatened.

3. **Androcymbium roseum** Engl. in Bot. Jahrb. Syst. **10**: 282 (1889). —Baker in F.T.A. **7**: 559 (1898). —Dyer in Fl. Pl. Afr. **31**: t.1225 (1956). Type: Namibia, Hereroland, 1100 m, v.1886, *Marloth* 1360. FIGURE 12.2.**44**.

Perennial herb; corm 12–18 × 8–16 mm, subglobose, with dark brown to black tunic, set 5–8 cm below ground. Stem ± absent above ground level. Leaves numerous, up to c.20, forming a rosette, 8–20 cm long, with an ovate base 10–22 mm wide, becoming linear, 5–7 mm wide and tapering to become long acuminate, tips often broken off in herbarium specimens. Bracts 25–40 × 7 mm, ovate-lanceolate, leaf-like, whitish-mauve, appearing pinkish when dry. Flowers c.4 in a congested head inside the bracts, lilac-pink covered with deeper pink dots near base of blade. Tepals 22–28 mm long, with claw 9–10 mm long, blade 13–16 × 2–6 mm, lanceolate, concave, sometimes notched near apex. Stamens attached at blade base and about half its length; filament 5 mm long, swollen at place of attachment, c.5 mm long, anther dorsifixed, 2 mm long, oblong. Ovary 7–9 × 5–6 mm, ovoid, with 3 rounded carpels each tapering into a slender style 8–10 mm long with a minute stigma.

Caprivi Strip. Lisikili, 21 km E of Katima Mulilo, *Codd* 7109 (PRE, WIND). **Botswana**. N: 80 km N of Kachikau, at Kabulabula road to Kasane, by Chobe R., fl. 11.vii.1937, *Ereus* 398 (K, PRE). SE: Peterhouse, Boteti delta area, NE of Mopipi, 850 m, fl, 22.iv.1973, *Thornton* 6 (K, SRGH). **Zambia**. B: 8 km N of Senanga, pan edge, 1030 m, fl. 30.vii.1952, *Codd* 7251 (K, PRE). W: Kafue Nat. Park, Mtemwa R., seasonally flooded grassland, 1080 m, fl. 16.ix.1972, *Kornaś* 2139 (K). C: Lusaka Dist., 5 km E of Lusaka, 1200 m, fl. 25.vi.1958, *Best* 139 (K, SRGH). S: Namwala Dist., between Manaungbwe fishing camp and Lubwe (Loubwe) village, edge of Kafue floodplain, fl. 6.vii.1963, *van Rensburg* 2326 (K).

Also in Angola and South Africa. Short grass on swampy or waterlogged soil; 850–1200 m.

Conservation notes: Moderately widespread species; not threatened.

2. WURMBEA Thunb.

Wurmbea Thunb., Nova Gen. Pl.: 18 (1781). —Nordenstam in Notes Roy. Bot. Gard. Edinb. **36**: 211–233 (1978).
Anguillaria R. Br., Prodr. Fl. Nov. Holl.: 273 (1810), in part.

Perennial herbs with a deep-seated corm covered with brown or blackish tunics. Stem erect, simple. Leaves inserted on stem, sessile, sheathing. Perianth actinomorphic, persistent; tepals joined at base to form a long or short tube. Anthers introrse to latrorse. Ovary syncarpous or partly so, 3-lobed at apex. Capsule septicidal or rarely loculicidal or both. Seeds globose, firm.

About 40 species in Africa (mainly the Cape) and Australia, 2 species in the Flora Zambesiaca area.

Perianth tube short (0.5–1.5 mm), $^1/_{12}$–$^1/_6$ length of perianth; flower stellate, free part of tepals narrowest at base ·····························**1.** *tenuis*
Perianth tube distinct (1.5–2.5 mm), c. $^1/_3$ length of perianth; flower cup-shaped, free parts of tepals not distinctly narrowed at base ··············**2.** *angustifolia*

1. **Wurmbea tenuis** (Hook.f.) Baker in J. Linn. Soc., Bot. **17**: 436 (1879); in F.T.A. **7**: 560 (1898). Type: Equatorial Guinea, Bioko (Fernando Po), iv.1862, *Mann* 1454 (K holotype).

> *Melanthium tenue* Hook. f. in J. Linn. Soc., Bot. **7**: 223 (1864).

Subsp. **goetzei** (Engl.) B. Nord. in Notes Roy. Bot. Gard. Edinb. **36**: 225 (1978). — Hoenselaar in F.T.E.A., Colchicaceae: 14 (2005). Type: Tanzania, Mbeya Dist., Unyika, Sunta's village, *Goetze* 1430 (B holotype, PRE). FIGURE 12.2.**45**.

> *Wurmbea goetzei* Engl. in Bot. Jahrb. Syst. **30**: 272 (1901).
> *Wurmbea homblei* De Wild. in Bull. Jard. Bot. État **5**: 8 (1915). Type: Congo, Biano plateau near Katentania, *Homblé* 814 (BR holotype).

Perennial herb to 20 cm; corm conical, ovoid or subglobose, 12–18 × 11–12 mm, with brown or blackish papery tunics; neck up to 6 cm long, disintegrating. Stem above ground 3–20 cm tall, slender. Basal leaf one, ± erect, 8–16 cm long, 0.5–2 mm wide, narrowly linear, ± semi-terete, channelled on lower surface. Stem leaves 1–2, usually not longer than spike, tapering from a sheathing base, 2–5 cm long, 3–4 mm wide at base, lower leaf always longer than upper. Inflorescence up to c.4 cm long, 1–7-flowered, flowers set 0.5–1.5 cm apart. Flowers ± stellate, 10–14 mm wide, variable in colour, white with green spots in centre of tepals, pale mauve with darker spots or pale yellow, green in centre. Tepals spreading, joined for 1–1.5 mm at base ($^1/_6$ to $^1/_{12}$ of total length), free parts ± oblanceolate, acute, 5–8 × 1–2 mm, narrowing to 0.5–1 mm wide at base. Stamens attached to tepals 1.5–2.5 mm above base, free part of filaments 2.5–4 mm long, anthers orange, 1 mm long. Ovary 2–3 mm long; styles 3, 2–4 mm long, persistent, tapering gradually to slightly capitate stigmas. Capsules erect, 8–15 × 3–5 mm, ovoid.

Zambia. N: Mbala Dist., Mbala pans, 1500 m, fl. 14.xii.1954, *Richards* 3631 (K). W: Copperbelt, Chingola Dist., Luano Forest Area, dry dambo, fl. 26.xi.1968, *Mutimushi* 2850 (K, NDO). E: Nyika Plateau, fl. 29.xii.1962, *Fanshawe* 7295 (K). **Malawi**. N: Nyika plateau, Chelinda Hill, rocky depressions, fl. 27.xii.1970, *Moriarty* 55 (K, MAL). C: Dedza Dist., foot of Chongoni Mt., 1550 m, fl. 17.i.1959, *Robson* 1235 (K, LISC). S: Mt. Mulanje, NE slopes of Namsele, on rocks, fl. 16.i.1971, *Hilliard & Burtt* 6185 (MAL). **Mozambique**. MS: Tsetserra, 2140 m, fl. 7.ii.1955, *Exell, Mendonça & Wild* 235 (LISC).

Also in Tanzania and Congo. Habitat variable, mostly montane from short wet grassland to dry sandy soil under scattered *Uapaca* trees; 1200–2300 m.

Conservation notes: Widely distributed although localised; not threatened.

Four subspecies are recognised – subsp. *tenuis* occurs in Cameroon and Nigeria, subsp. *hamiltonii* (Wendelbo) B. Nord. in Uganda, Kenya and Tanzania, and subsp. *australis* B. Nord. in South Africa and Lesotho.

2. **Wurmbea angustifolia** B. Nord. in Bot. Not. **117**: 175 (1964). —Maroyi in Kirkia **18**: 4 (2002). Type: South Africa, E Cape, Xalanga, Katberg, between Cala and Ugie, 1200–1500 m, i.1896, *Flanagan* 2660 (PRE holotype, SAM).

Perennial herb to 15 cm; corm small, 7–20 × 5–15 mm, ovoid or subglobose, covered with disintegrating brown or blackish papery tunics, neck 10–15 mm long, disintegrating. Stem 7–15 cm long, slender. Basal leaf one, usually erect, 8–18(28) cm × 1–4.5 mm, narrowly linear, margins inrolled and scabrid; stem leaves 2, with loose sheathing base 4–6 mm wide, free part

Fig. 12.2.**45**. WURMBEA TENUIS subsp. GOETZEI. 1, habit with inflorescence (× ¹/₂); 2, infructescence (× ¹/₂); 3, flower (× 5); 4, views of anthers (× 8); 5, gynoecium (× 8); 6, fruit (× 4); 7, transverse section of fruit (× 4); 8, longitudinal section of fruit (× 4); 9, seeds (× 6). 1–2 from *Richards* 17297, 3–9 from spirit material 38342. Drawn by Judi Stone.

3–18 cm long × 3.5–5 mm wide, subterete, the lower inserted about middle of stem, the upper smaller one 0.5–4 cm below the inflorescence. Inflorescence short and dense, to c.3 cm long, flowers set 3–10 mm apart. Flowers white, cream, greenish yellow to yellow, cup-shaped with suberect segments joined for 1.5–2.5 mm at base; tepals 5–7 mm long (including joined part), 1–1.5 mm wide, each with a nectary in the form of a reniform pocket at base of free part, bordering a double mauve to purple blotch. Filaments free, 1.5–2 mm long; anthers 0.5–1 mm long, oblong, bright yellow. Ovary 4–6 mm long, oblong but 3-angled; styles 3, 1–2 mm long, recurved at apex, stigma elongate 0.5 mm long.

Zimbabwe. E: Chimanimani Dist., SW slopes of Chimanimani Mtns., point 71, fl. 10.ii.1958, *Hall* 425 (BOL, SRGH).

Also in South Africa and Lesotho. Moist montane to submontane grassland, in dense turf or wet peaty soil; 1800–2300 m.

Conservation notes: Localised montane species, mostly in protected areas; probably Lower Risk near threatened.

3. GLORIOSA L.

Gloriosa L., Sp. Pl.: 305 (1753).
Clinostylis Hochst. in Flora **27**: 26 (1844).

Perennial herbs with an elongated corm, often forked. Stem scandent or erect. Leaves sessile, alternate, opposite or whorled, usually narrowing into a tendril at the apex. Flowers axillary, pedicellate or sessile, pendent or erect, red, orange or yellow, often bicoloured; tepals free, usually with undulate margins. Style filiform, 3-branched at apex, bent at an angle at base. Capsule loculicidal; seeds with a fleshy red outgrowth (strophiole).

Two species (perhaps 3) in tropical Africa and Asia.

Flowers with reflexed tepals; style forming a right angle to ovary · · · · · · · **1.** *superba*
Flowers funnel-shaped; style forming only a slight angle to ovary · · · · · **2.** *sessiliflora*

1. **Gloriosa superba** L., Sp. Pl.: 437 (1753). —Baker in F.T.A. **7**: 563 (1898). — Hepper in F.W.T.A. ed.2, **3**(1): 106 (1968). —Hoenselaar in F.T.E.A., Colchicaceae: 4 (2005). Type: India, Malabaria, *Hermann* 3: 31, no.122 (BM lectotype). FIGURE 12.2.46.

   Gloriosa simplex L., Mant. Pl. Alt.: 62 (1767). —Hepper in F.W.T.A., ed.2, **3**: 106 (1898). Type: Senegal, *Adanson* cat.nos. 4000 & 4001 (P).

   Gloriosa virescens Lindl. in Bot. Mag. **52**: t.2539 (1825). —Baker in F.T.A. **7**: 563 (1898). Type: Mozambique, 1823, *J. Forbes* s.n. (K).

   Gloriosa abyssinica A.Rich. in Tent. Fl. Abyss. **2**: 322 (1851). Types: Ethiopia, Tigray, Shire (Chiure), *Quartin-Dillon & Petit* s.n. (K syntype) & Tigray, Djeladjekanne, 31.vii.1840, *Schimper* 1437 (K syntype).

   Gloriosa superba var. *angustifolia* Baker in J. Linn. Soc., Bot. **17**: 458 (1879); in F.T.A. **7**: 563 (1898). Type: N Mozambique, Lower Rovuma R., 45 km from coast, fl. iii.1861, *C.J. Meller* s.n. (K holotype).

   Gloriosa carsonii Baker in Bull. Misc. Inf. Kew **1895**: 74 (1895); in F.T.A. **7**: 565 (1898). Type: Zambia, Lake Tanganyika, Fwambo, 1894, *Carson* 53 (K holotype).

   Gloriosa lutea Hort. in Gard. Chron. **29**: 413 (1901). Type: Zimbabwe, Nyanga, xii.1899, *Cecil* 224 (K holotype).

   Gloriosa rothschildiana O'Brien in Gard. Chron. ser.3, **33**: 322 (1903). Type: Uganda, *H.B. Rattray* s.n.

   Gloriosa grandiflora O'Brien in Gard. Chron. ser.3, **36**: 188 (1904). Type: Congo, *van Tubergen* s.n.

Gloriosa homblei De Wild. in Repert. Spec. Nov. Regni Veg. **11**: 536 (1913). Type: Congo, Katanga, Welgelegen, 1912, *Corbisier* in *Homblé* 590 (B).

Perennial herb; corm to 7 × 1.5 cm, elongated, often forked. Stem usually scandent, to c.3 m long, occasionally erect and much shorter. Leaves numerous, sessile, usually opposite or whorled, sometimes alternate, variable in size and shape, 6–15 cm long (not including tendril), 0.5–6.5 cm wide, linear to broadly ovate, usually with a tendril up to 7 cm long at apex. Flowers axillary, cream, yellow, red, red or purple-brown with a yellow margin, sometimes changing colour with age. Pedicel 9–13 cm long, bent over at apex. Tepals reflexed, oblanceolate or spathulate, apiculate, margins plain to undulate, 50–83 × 11–30 mm. Filaments free, 13–46 mm long; anthers dorsifixed, 6–10 mm long. Ovary oblong, 5–14 × 3–5 mm; style horizontal, at right angles to ovary, 25–50 mm long with 3 slender branches 3–12 mm long at tip, often not arising at the same place. Fruit 5–7 × 2–2.2 cm, splitting into 3 from base. Seeds red, fleshy.

Caprivi Strip. Sambala campsite, fl. 4.iii.2001, *Curtis* 758 (WIND). **Botswana**. N: Kwando area, mixed woodland on Kalahari sand, fl. 22.i.1976, *Williamson* 70 (K, SRGH). SE: Orapa, airport pan, fl. 31.i.1976, *Allen* 387 (K, PRE). **Zambia**. B: Mongu, fl. 22.xii.1965, *Robinson* 6743 (K). N: Mbala Dist., Saise Valley, 1500 m, fl. 22.i.1970, *Sanane* 1028 (K). W: Kitwe, in woodland, climbing or erect, fl. 6.ii.1956, *Fanshawe* 2770 (K, LISC, NDO). C: Lusaka Dist., 20 km SE of Lusaka, Jellis Farm, 1300 m, fl. 15.i.1995, *Bingham* 10283 (K). E: W of Sasare, *Brachystegia* woodland, 700 m, fl. 10.xii.1958, *Robson* 889 (K, LISC). S: Mazabuka Dist., Magoye, *Combretum/Albizia* woodland, fl. 28.xii.1962, *Angus* 3455 (K, FHO). **Zimbabwe**. N: Mt Darwin Dist., S slopes of Mavurdonna Mts., c.1000 m, fl. 17.i.1960, *Phipps* 2284 (K). W: Hwange Dist., fl. 20.xii.1934, *Eyles* 8289 (K). C: Harare, 1450 m, *Eyles* 4743 (K). E: Mutare Dist., Murahwa's Hill Commonage, 1120 m, fl. 2.i.1970, *Chase* 8578 (K, SRGH). S: Chivi Dist., Madziviri Dip, c.500 m, fl. 29.xii.1962, *Moll* 418 (K, SRGH). **Malawi**. N: Mzimba Dist, Mzuzu, Marymount, near Lunyangwa, *Brachystegia/Uapaca* woodland, 1360 m, fl. 19.xii.1968, *Pawek* 1600 (K). C: Dedza Dist., Dedza Mt., near Forestry Office, fl. 2.ii.1967, *Jeke* 60 (K, LISC, SRGH). S: Blantyre Dist., Ndirande Mt., steep rocky slopes, 1450–1570 m, fl. 3.iii.1970, *Brummitt* 8868 (K). **Mozambique**. N: Lower Rovuma R., 45 km from coast, fl. iii.1861, *Meller* s.n. (K). Z: Between Mocuba and Milange, c.18 km from Mocuba, fl. 20.v.1949, *Barbosa & Carvalho* 2745 (K, LMA). T: Tete, c.21 km from Tete–Changara crossroads, mopane woodland, c.200 m, fl. 14.ii.1968, *Torre & Correia* 17606 (LISC). MS: Inyamatshira Mts., fl. 29.i.1950, *Chase* 1952 (K, LISC, SRGH). GI: Bilene Dist., near Macia, fl. 31.iii.1959, *Barbosa & Lemos* 8424 (K, LISC, LMA). M: Maputo, Inhaca Is., c.50 m, fl. 12.xii.1984, *Groenendijk & Dunge* 1540 (K, LMU).

Widespread throughout tropical Africa and Asia. Grassland, dry bush, woodland, rainforest; 20–2230 m.

Conservation notes: A very widespread taxon; not threatened. A protected species in Zimbabwe.

Gloriosa superba is an extremely variable plant which over the years has been described as many different species, only some of which are cited here. The variation applies to all parts – habit, size and shape of leaves, size, shape and colour of flowers.

Plants described as *G. carsonii* have short, erect stems and leaves lacking tendrils. However, there are specimens labelled *G. carsonii* at Kew which fit this description but where the upper leaves have short tendrils; various collectors remark that it is twining round grasses. Plants known as *G. virescens* and *G. simplex* have tepals with plain rather than undulate margins. The name *G. rothschildiana* was applied to plants with large, red and yellow flowers. The type of G. *lutea* was a plant with large, yellow flowers with very undulate tepals (possibly a cultivated specimen), which received an Award of Merit from the Royal Horticultural Society in 1901.

All this variation is continuous and the variables do not seem to be consistently correlated. Specimens from the same area can differ considerably in flower colour

Fig. 12.2.**46**. GLORIOSA SUPERBA. 1, habit, showing flowers with plain-edged tepals and leaves without tendrils (× ¹/₂); 2, flowers with undulate tepals (× ¹/₂); 3, narrow leaves with tendrils (× ¹/₂); 4, broad leaf with tendrils (× ¹/₂); 5, anther, back view (× 2); 6, anther, front view (× 2); 7, gynoecium (× 1¹/₃); 8, fruit (× ¹/₂); 9, fruit, transverse section (× ¹/₂); 10, seed (× 2). 1 from *Pawek* 1600, 2 from spirit material 34784, 3 from *Eyles* 32, 4–7 from spirit material 34784, 8–10 from spirit material 6606. Drawn by Judi Stone.

and leaf shape, while leaf shape can often vary on a single plant. The three specimens *Lady Drewe* 4, 5 and 6, all from Kenombo in Zimbabwe, show almost the full range of variables combined in different ways.

Sometimes generalisations can be made about plants from a particular geographical area. For example, most specimens from Malawi have large flowers but these come in all colours, tepals may be wavy-edged or not, and the leaves with or without tendrils. Most Zambian specimens have rather narrow tepals with strongly undulate edges; the flowers are large but not as large as is usual in Malawi. Many (but not all) specimens from dry areas have very narrow leaves (only 5–7 mm wide), but this is not consistently linked with other characters. *Williamson* 70 from N Botswana has narrow leaves with tendrils and red flowers with rather broad, plain-edged tepals. *Meller* s.n., the type of var. *angustifolia*, also has very narrow leaves with tendrils, but the tepals are narrow and strongly undulate. Broad leaves often go with broad plain-edged tepals, but not always.

Hoenselaar (in F.T.E.A.) divides *G. superba* into two varieties. Var. *superba*, the taxon found in the Flora Zambesiaca area, is widespread, while var. *graminifolia* (Franch.) Hoenselaar is confined to Somalia and Ethiopia. In the Flora of Ethiopia, Sebsebe Demissew retains *G. baudii* (Terracc.) Chiov. (synonyms *G. graminifolia* (Franch.) Chiov., *G. abyssinica* A. Rich. var. *graminifolia* Franch. and *G. minor* Rendle) as a distinct species, while Thulin in Flora of Somalia treats it as synonymous with *G. superba*. As these taxa are only found in Ethiopia, Somalia and Kenya no decision needs to be taken here.

It seems more satisfactory to consider *Gloriosa superba* as a single, very variable species. In this it resembles the widespread orchid *Ansellia africana*, another species that shows continuous and uncorrelated variation in several characters.

The species is reputedly poisonous to livestock, and has a number of medicinal uses. It is the national flower of Zimbabwe.

2. **Gloriosa sessiliflora** Nordal & M.G. Bingham in Kew Bull. **53**: 479 (1998). Type: Zambia, Mongu, 9.xii.1995, *Bingham & Luwiika* 10752 (K holotype, MRSC, O, WAG).

Erect perennial herb 60–100 cm tall, with a stoloniferous corm. Stems unbranched, but with growth continuing from lateral buds below the inflorescence. Leaves numerous, in whorls on upper $^2/_3$ of stem, 7–10 × 1.5–2 cm, narrowly lanceolate, base clasping the stem, apex forming a tendril. Flowers reddish orange, yellowish orange towards base, borne singly or in pairs in leaf axils near stem apex, funnel- or bell-shaped, sessile or with only a short pedicel. Tepals c.4 × 1–1.5 cm, narrowly ovate, slightly undulate, with grooved nectaries towards base. Filaments yellowish, anthers orange, 6 mm long, versatile, with latrorse dehiscence. Style forming a slight angle to ovary; stigma 3-branched, each c.3 mm long.

Caprivi Strip. See note below. **Zambia**. B: Mongu, Bulozi floodplain, fl. 9.xii.1995, *Bingham & Luwiika* 10752 (K, MRSC, O, WAG).

Apparently known only from a single population at the type locality. Floodplain termite mounds in open woodland of stunted *Faidherbia albida*; c.1000 m.

Conservation notes: Very localised distribution and in an area susceptible to land use change; probably Endangered although only considered Vulnerable D2 in the Sabonet Red Data List (Golding 2002).

There is an unconfirmed record from the Caprivi Strip (Katima Mulilo, Lisikili, under trees in deep sandy soil, fl. (ex hort. Windhoek) 7.i.1985, *M. Müller* 3636 (WIND)), also cited in the Sabonet Namibia checklist (Craven 1999).

Nordal & Bingham comment in the description that this species seems to occupy an intermediate position between *Gloriosa* and *Littonia*; that it is possible that the two genera should be united, in which case the name *Gloriosa* would take priority.

4. LITTONIA Hook.

Littonia Hook. in Bot. Mag. **79**: t.4723 (1853). —Baker in J. Linn. Soc., Bot. **17**: 458–459 (1879).

Perennial herbs with tunicated corm, usually 3-lobed. Stem scandent. Leaves sessile, lower ones opposite, alternate or in whorls of 3, upper ones alternate, linear to ovate, sometimes with an apical tendril. Flowers axillary, pedicellate, nodding. Perianth bell-shaped, yellow or orange, with 6 ± equal segments, lanceolate, acute, free or shortly united, bases pouch-shaped. Stamens 6, hypogynous, filaments very slender; anthers linear-oblong, dorsifixed, extrorse, dehiscing vertically. Style erect, filiform, with 3 falcate branches. Capsule septicidal, ovoid-oblong; seeds globose with fleshy outer coat.

7 or 8 species in Africa and Arabia; 3 species in the Flora Zambesiaca area.

1. Lower leaves in whorls of 3–4 · **3.** *modesta*
 – Lower leaves opposite or alternate · 2
2. Flowers green or greenish-yellow; tepals 13–23 mm long; stem usually leafy ± to base · **1.** *littonioides*
 – Flowers orange to vermilion; tepals 27–50 mm long; leaves only in upper half of stem · **2.** *lindenii*

1. **Littonia littonioides** (Baker) K. Krause in Bot. Jahrb. Syst. **57**: 235 (1921). — Hoenselaar in F.T.E.A., Colchicaceae: 12 (2005). Type: Angola, Pungo Andongo, vi.1878, *Welwitsch* 1747 (K holotype). FIGURE 12.2.**47**.
 Sandersonia littonioides Baker in Trans. Linn. Soc., Bot. **1**: 262 (1878).
 Littonia welwitschii Benth. & Hook.f., Gen. Pl. **3**: 831 (1883). —Baker in F.T.A. **7**: 566 (1898). Type as for *L. littonioides*.

Erect perennial herb to 60 cm. Corm 10–34 mm long, 7–10 mm wide; scale-like leaf (cataphyll) extending 3–8 cm above corm. Stem 25–55 cm tall, slender, ± flexuous, leafy almost to base. Leaves 5–15, usually well spaced out but sometimes almost overlapping, alternate, 4.5–12.5 × 0.7–3.8 cm, narrowly lanceolate to ovate, apiculate or with a short tendril c.3 mm long, slightly glaucous, mid-vein raised on lower surface. Flowers solitary in leaf axils, pendent, green or greenish-yellow. Pedicel apparently fused to stem for part of its length, free part up to 6 cm long. Sepals and petals similar, joined at base for 1–2 mm, 13–20(23) × 3–5.5(6.5) mm, elliptic, acute or acuminate, sepals keeled on outside. Filaments 4–5 mm long; anthers yellow, 3–4.5 mm long. Ovary c.2 mm long, oblong, soon enlarging; style 6–10 mm long with 2–3 slender recurved branches 1.5–2 mm long.

Zambia. N: Mbala Dist., Mbala–Kambole, *Brachystegia* woodland, fl. 31.xii.1949, *Bullock* 2144 (K). W: Kitwe, plateau woodland, fl. 29.xi.1955, *Fanshawe* 2634 (K). C: Lusaka Dist., Lazy J Ranch, 20 km SE of Lusaka, 1300 m, fl. 7.i.1995, *Bingham* 10255 (K). E: Katete, St. Francis' Hospital, 1060 m, fl. 9.i.1958, *Wright* 213 (K). S: Kalomo Dist., Machili, fl. 14.xii.1960, *Fanshawe* 5976 (K). **Malawi**. N: Chitipa Dist., Kaseye Mission, 16 km E of Chitipa, 1270 m, fl. 26.xii.1977, *Pawek* 13374 (K, MAL, MO, PRE, SRGH, UC). C: Kasungu Nat. Park, *Brachystegia* woodland, 1030 m, fl. 24.xii.1970, *Hall-Martin* 1465 (PRE).

Also in Angola and Tanzania. Miombo woodland, sometimes in open areas, on sandy soils; 1000–1450 m.

Conservation notes: Widespread species; not threatened.

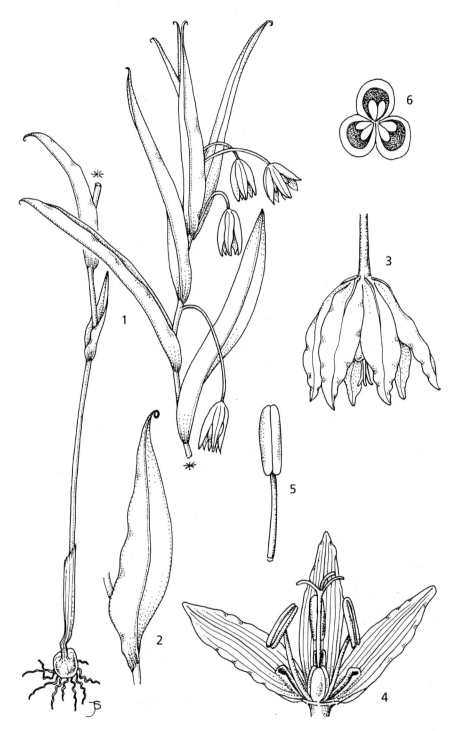

Fig. 12.2.**47**. LITTONIA LITTONIOIDES. 1, habit (× ²/₃); 2, leaf with tendril (× ²/₃); 3, flower (× 3); 4, flower with 3 tepals and 3 stamens removed (× 3); 5, stamen (× 4); 6, ovary, transverse section (× 17). 1 from *Richards* 3 (bulb from *Bullock* 2144), 2 from *Richards* 13666, 3–6 from *Richards* 3611. Drawn by Judi Stone.

2. **Littonia lindenii** Baker in F.T.A. **7**: 566 (1898). —Hoenselaar in F.T.E.A., Colchicaceae: 10 (2005). Type: Tanzania, Kigoma Dist., Ujiji, i.1884, *Linden* s.n. (K syntype) & Congo, Lake Mweru, *Descamps* s.n. (BR syntype).

Erect perennial herb to 50 cm. Corm 17–20 × 11–13 mm, ± globose to ovoid. Stem 24–48 cm tall with 2 purplish-black scale-like leaves (cataphylls) up to 6 cm above base, the lower 3–6 cm long with only 2–3 mm free, the upper with c.10 cm free. Leaves 5–7, in upper half of stem, usually starting 16–20 cm above cataphylls, 5–13.5 × 0.6–2.3 cm, lanceolate-elliptic, acute, with no tendrils, mid-vein prominent below. Pedicel 3–4 cm long, arched so flowers are pendent. Flowers orange to vermilion on outside, rich cream, yellow or greenish-yellow on inside, variable in size even on the same plant, lower flowers usually larger than upper. Sepals and petals similar, joined at base for 4–5 mm, 27–50 × 3–5.5 mm, lanceolate, acuminate, slightly keeled on outer surface, very slightly pouched at base. Filaments 8 mm long; anthers 5–6 mm long. Ovary 4 mm long; style 4–6 mm long with 3 branches c.2.5 mm long, slightly thickened at tip.

Zambia. N: Mporokoso Dist., Kalungwishi R., Lumangwe Falls, in chipya woodland, fl. 15.xi.1957, *Fanshawe* 4024 (K). W: Mwinilungwa Dist., Kabompo Gorge, 1200 m, fl. 22.xi.1962, *Richards* 17466 (K). C: Kabwe (Broken Hill) Forest Reserve, in *Brachystegia/Uapaca* woodland, fl. xi.1960, *Morze* 301 (K). **Malawi**. N: Karonga Dist., Stevenson Road, Senga stream, in *Brachystegia* woodland, 1360 m, fl. 17.xii.1969, *Pawek* 3102 (K, MAL).

Also in Congo and Tanzania. In moist miombo and similar woodland types on rocky soil, sometimes abundant; 1200–1400 m.

Conservation notes: Widespread species; not threatened.

The species seems to be particularly common in the Western Province of Zambia.

3. **Littonia modesta** Hook. in Bot. Mag. **79**: t.4723 (1853). —Baker in F.C. **6**: 527 (1897). Type: South Africa, Durban (Port Natal), x.1883, *Sanderson* s.n. (K holotype).

Perennial herb to 2 m. Corm 5–6 × c.2 cm, oblong. Stem scandent, 60–200 cm high, the basal 18 cm lacking leaves. Leaves very numerous, set close together in whorls of 3–4, growing smaller and narrower towards stem apex, variable in width even on same plant, 10–17 × 0.5–3.5 cm, lanceolate or linear-lanceolate, tip drawn out into tendril c.2 cm long (included in leaf length). Flowers axillary, bright yellow or orange. Pedicel 2–5.5 cm, bent at apex so flowers are nodding. Sepals and petals lanceolate, acute, united for 2–4 mm at base, 17–40 × 4–10 mm. Filaments c.7 mm long; anthers c.8 mm long. Ovary 8–10 × 3 mm (possibly partly developed); style 4 mm long with 3 recurved branches 2–3 mm long.

Zimbabwe. C: Harare, Mount Pleasant, 1500 m, fl. ii.1956, *Drummond* 5104 (K, LISC, PRE, SRGH). **Mozambique**. M: Namaacha, fl. 22.iii.1982, *de Koning* 9149 (LMU).

Also in South Africa and Swaziland. Woodland; 800–1500 m.

Conservation notes: Widespread and not threatened globally, but within the Flora area likely to be Vulnerable.

In the Flora area this species may occur naturally only in Mozambique; for the Zimbabwe specimen cited, the collector remarks that it was found in what seemed to be an abandoned garden and could originally have been cultivated.

5. ORNITHOGLOSSUM Salisb.

Ornithoglossum Salisb. in Parad. Lond.: t.54 (1806). —Nordenstam in Opera Bot. **64**: 1–51 (1982).

Glabrous perennial herbs, corm subterranean, subglobose to ovoid, simple or with one or two lateral lobes, papery tunics often prolonged upwards into a sheath covering lower part of stem. Stem erect, simple or branched. Leaves few to several, alternate, in 2 rows, sessile, sheathing. Flowers reflexed to pendent in bracteate terminal racemes, few to many-flowered. Flowers may not appear axillary as pedicel often fused to stem in lower half so that flowers apparently emerge between the bracts. Flowers greenish-yellow, dull brown to blackish, often bicoloured; tepals 6, free, spreading or reflexed, lower part often rolled inwards forming a claw-like structure. Nectary a channel- or pouch-like structure at junction between blade and claw or on the claw. Stamens 6, filaments inserted at base of tepals; anthers dorsifixed, versatile, extrorse, dehiscing with longitudinal slits. Styles 3, free, filiform with small capitate or oblong stigmas. Ovary syncarpous, trilocular, with few to many axile ovules. Seeds globose, firm.

8 species in southern and tropical Africa as far north as Tanzania; 1 in the Flora Zambesiaca area.

Ornithoglossum species are very poisonous to stock due to the presence of alkaloids such as colchicine and related compounds.

Ornithoglossum vulgare B. Nord. in Opera Bot. **64**: 37 (1982). —Maroyi in Kirkia **18**: 7 (2002). —Hoenselaar in F.T.E.A., Colchicaceae: 15 (2005). Type: Zimbabwe, Gweru Teachers College, 24.xi.1966, *Biegel* 1467 (SRGH holotype, K). FIGURE 12.2.**48**.

    Ornithoglossum glaucum sensu auct., non Salisb. —Baker in J. Linn. Soc., Bot. **17**: 449 (1879); in F.T.A. **7**: 561 (1898).

Perennial herb to 70 cm. Corm oblong to ovoid, sometimes lobed, 4 × 2.5 cm thick, up to 30 cm below soil surface; tunic brown, prolonged into a neck; scale-like leaf (cataphyll) 5–17 cm long, whitish to light brown, free part very short. Stem 7–70 cm tall, up to 1 cm thick at base, unbranched or with 2–3 branches. Leaves 12, usually 4–7, erect, spreading or recurved, 10–30 cm long, 0.5–2(5) cm wide, linear to lanceolate, acute, glaucous, waxy, flat or with undulate margins. Inflorescence 5–30 cm long, few to c.25-flowered. Flowers brownish-purple, dark purple to dark red, occasionally yellow; base of filaments green, upper parts and anthers brown. Pedicels erect, spreading or recurved, eventually reflexed, 1–7 cm long. Bracts about half length of pedicels, the lower ones leaf-like. Tepals spreading at first, later reflexed, (8)10–22(25) mm long, 1–3.5 mm wide when flattened, linear, with claw-like base 2–6 mm long. Filaments 5–16 mm long, slender or slightly thickened in middle part; anthers 2–6 mm long, often curved. Styles 3, free to the base (rarely joined for 0.5 mm), spreading, 6–15 mm long; ovary green, 3-lobed, c.2 × 2 mm but soon enlarging, perianth remaining as it develops. Capsule globose, 15 × 10 mm, with 3 blunt lobes at tip.

    Botswana. N: near Francistown, fl. viii.1911, *Rogers* 6035 (K). SW: Kang, 1060 m, fl. 23.vii.1976, *Mott* 992 (K, SRGH). SE: Gaborone, bare ground, 985 m, fl. 18.vii.1974, *Mott* 296 (K, UBLS). **Zambia**. N: Mbala Dist., Lufuba R., Iyendwe valley, 780 m, fl. 8.xii.1959, *Richards* 11907 (K). E: Petauke–Sasare road, *Brachystegia*/mopane woodland, 900 m, fl. 4.xii.1958, *Robson* 829 (K, LISC). S: Kalomo Dist., Machile, fl. 5.xii.1960, *Fanshawe* 5941 (K, LISC, NDO). **Zimbabwe**. N: Mutoko Dist., Ngarwe communal land, near Kotwa, fl. 6.xii.1968, *Muller & Burrows* 970 (K, SRGH). W: Matobo Dist., Matopos Hills, fl. xi.1899, *E. Cecil* 107 (K). C: Gweru Dist., grounds of Teachers' College, 1400 m, fl. 24.xi.1966, *Biegel* 1467 (K, SRGH). **Malawi**. N: Mzimba Dist., Mbawa, in miombo scrub woodland, fl. 11.xii.1952, *Jackson* 1070 (K, LISC). C: Kasungu Dist., Kasungu–Bua road, 1000 m, fl. 13.i.1959, *Robson & Jackson* 1126 (K, LISC). S: Mangochi Dist., Ciripa village, fl. 16.xii.1954, *Banda* 65 (K). **Mozambique**. N: Ribaué, 650m, fl. i.1942, *Gomes e Sousa* 2326 (PRE).

    Also in S Tanzania, Namibia and South Africa. In miombo and similar woodland types, on sandy soils; 450–1400 m.

    Conservation notes: Widespread species; not threatened.

Fig. 12.2.**48**. ORNITHOGLOSSUM VULGARE. 1, habit (× ²/₃); 2, flower (× 1); 3, immature flowers, showing variation in structure (× 2); 4, tepal and stamen (× 3); 5, tepal (× 3); 6, stamen, side view (× 3); 7, gynoecium (× 4); 8, immature fruit (× 3); 9, transverse section of fruit (× 8). 1 from *Hislop-Rhodine* 54, 2 from *Skarpe* 5-74, 3–9 from spirit material 24402. Drawn by Judi Stone.

6. IPHIGENIA Kunth

Iphigenia Kunth, Enum. Pl. **4**: 212 (1843).

Perennial herbs with tunicated corm. Stem erect, sometimes sinuous, simple. Leaves sessile, linear to filiform. Flowers in a scorpioid cyme, subumbellate or raceme-like, rarely solitary; pedicels erect, recurved or reflexed. Tepals free, linear, white, pink or brownish-red, deciduous. Anthers extrorse, basifixed. Ovary ovoid-oblong; style short with 3 falcate branches. Capsule loculicidal, ovoid to cylindrical; seeds globose-oblong, papillate.

About 15 species in Africa, Madagascar, Socotra, India, Australia and New Zealand; 2 species in the Flora Zambesiaca area.

Pedicels drooping; tepals 6–15 mm long · **1.** *oliveri*
Pedicels erect; tepals 4–5 mm long · **2.** *pauciflora*

1. **Iphigenia oliveri** Engl. in Bot. Jahrb. Syst. **15**: 467 (1892). —Baker in F.T.A. **7**: 562 (1898). —Demissew in Fl. Ethiopia **6**: 189 (1997). —Hoenselaar in F.T.E.A., Colchicaceae: 9 (2005). Type: Kenya, Taita Dist., Taveta, x.1884, *Johnston* s.n. (K). FIGURE 12.2.49.

   Iphigenia bechuanica Baker in F.T.A. **7**: 562 (1898). —Obermeyer in Kirkia **1**: 84 (1961). — Maroyi in Kirkia **18**: 8 (2002). Type: Botswana, near Kwebe, xii.1896, *Lugard* 81 (K holotype).

Slender perennial herb; corm 10–15 mm diameter, subglobose, with brown tunic. Stem 13–20 cm long, ± sinuous, with 1–2 scarious tubular scale-like leaves (cataphylls) up to c.5 cm long at base. Leaves several, spirally arranged up stem, 9–18 cm × 2–6 mm, ± erect, linear, acuminate. Flowers 3–10, nodding, arising singly from upper leaf axils, greenish or brownish. Pedicels bent over, 10–30 mm long. Tepals free to the base, spreading, narrowly linear or filiform, 6–15 mm long, falling early in fruit. Stamens with slender filaments 2 mm long; anthers extrorse. Ovary ovoid or subglobose, c.2 mm long, with numerous ovules; style very short, c.0.3 mm long, then divided into 3 branches 1–1.3 mm long, curled over. Capsules pendent, 10–14 × 7–8 mm.

Botswana. N: Ngamiland Dist., near Kwebe, fl. xii.1896, *E.J. Lugard* s.n. (K). SE: Ilahamabele–Mosu area near Sua pan, fl. 14.i.1974, *Ngoni* 325 (K, SRGH). **Zambia**. C: Lusaka, limestone outcrops, fl. 8.xii.1968, *Fanshawe* 10457 (K, NDO). S: Kalomo Dist., Kalomo, sandy riverside, 1200 m, fl. 1.i.1958, *Robinson* 2546 (K, SRGH). **Zimbabwe**. N: Makonde Dist., Darwendale, above vlei, 1360 m, fl. 18.xii.1968, *Biegel* 2711 (K, LISC, PRE, SRGH). W: Plumtree, fl. ii.1935, *McLeod* 8 (PRE). E: Mutare (Umtali), bank in grass commonage, 1100 m, fl. 12.xii.1956, *Chase* 6257 (K, SRGH). S: Chiredzi Dist., tsetse corridor, c.20 km SE of Chipinda Pools, fl. 26.i.1971, *Kelly* 445 (K, SRGH). **Mozambique**. Z: Quelimane, 30 m, fl. 1908, *Sim* 20607 (PRE).

Also in Ethiopia, Somalia, Kenya, Tanzania, Uganda, Namibia and South Africa. Short grassland, dambos, and woodland on sandy soils; 30–1400 m.

Conservation notes: Widespread species; not threatened.

In F.T.A. (1898) Baker distinguishes between *I. bechuanica* and *I. oliveri* on the basis that *I. bechuanica* has few ascending or drooping flowers, while in *I. oliveri* the flowers are many and drooping. These seem slight features on which to separate two species; on looking through herbarium specimens, I found it impossible to distinguish between them. A few plants had ± erect pedicels, but did not always have few flowers. *I. oliveri* is described by Demissew in Flora of Ethiopia (1997) as having few to several flowers, and is a widespread species on the eastern side of Africa, extending as far north as Ethiopia. Some variation is only to be expected. Obermeyer (1961), referring to *I. bechuanica*, says that she has seen no tropical African material.

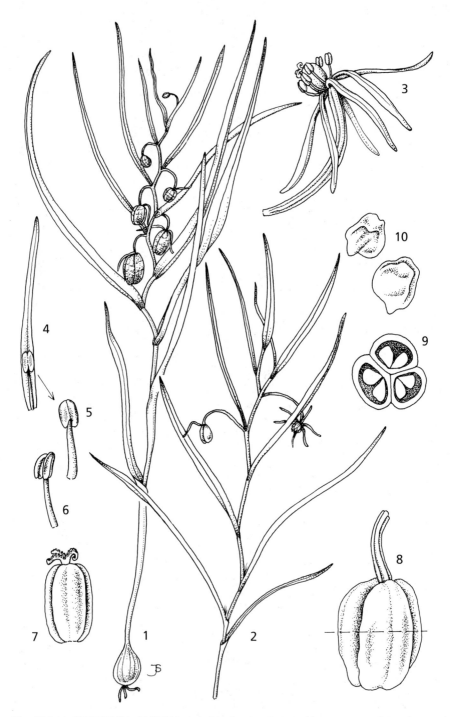

Fig. 12.2.**49**. IPHIGENIA OLIVERI. 1, habit showing fruit (× ²/₃); 2, part of habit showing flower and immature fruit (× ²/₃); 3, flower (× 3); 4, tepal and stamen (× 6); 5, stamen, back view (× 14); 6, stamen, front view (× 14); 7, gynoecium (× 14); 8, fruit (× 3); 9, transverse section of fruit (× 3); 10, seeds (× 6). 1 from *Ngoni* 325, 2 from *Whitehouse* 2, 3–10 from *Wild* 1610. Drawn by Judi Stone.

2. **Iphigenia pauciflora** Martelli, Fl. Bogos.: 86 (1886). —Demissew in Fl. Ethiopia **6**: 187 (1997). —Hoenselaar in F.T.E.A., Colchicaceae: 7 (2005). Type: Eritrea, Keren, *Beccari* 248 (FT holotype).

Iphigenia abyssinica Chiov. in Ann. Bot. (Roma) **9**: 147 (1911). Type: Ethiopia, Tigray/Gondar region, Mai-Aini in Tzelemti, *Chiovenda* 641 (FT holotype).

Slender perennial herb; corm c.8 × 6 mm, subglobose. Stem 10–15 cm tall. Leaves usually 4, 8–15 cm × 1–2 mm wide, narrowly linear, acute, suberect. Pedicel 13–21 mm long, erect, straight, subtended by paired bracts (occasionally 3), 4–8 mm long. Flowers yellowish. Tepals spreading, 4–5 mm long, 0.5–0.7 mm wide. Stamens c.2 mm long, the filament slightly flattened; anthers 0.7 mm long. Ovary 3 × 1.5 mm long, broadest near apex; style very short with 3 branches, c.1 mm long. Capsule erect, 9–10 × 6–7 mm.

Malawi. N: Rumphi Dist., c.22 km N of Rumphi, in dambo, 1100 m, fr. 26.ii.1978, *Pawek* 13921 (K, MO); Mzimba Dist., c.11 km S of Eutini, fl.& fr. 31.i.1976, *Pawek* 19783 (K, UC).

Widespread in tropical Africa from Ivory Coast and Cameroon to Ethiopia, Sudan, Kenya, Tanzania and N Malawi. Short grass near rivers, dambos and in *Brachystegia* woodland; 1100–1800 m.

Conservation notes: Widespread species, not threatened globally, but with a restricted distribution in the Flora area; probably Vulnerable.

7. CAMPTORRHIZA Phill.

Camptorrhiza E. Phillips in Fl. Pl. S. Afr. **15**: t.575 (1935). —Obermeyer in Kirkia **1**: 86 (1961).

Iphigeniopsis Buxb. in Bot. Arch. **38**: 228 (1936).

Perennial herbs with tunicate corm. Leaves sessile, linear, sheathing. Flowers greenish or pink, in a raceme-like scorpioid cyme; rachis sinuous; pedicels recurved; tepals free, soon reflexed. Anthers latrorse, filaments terete or swollen in the middle. Ovary subglobose to ovoid; style simple, erect. Capsule loculicidal, globose to elliptic; seeds globose, firm.

Camptorrhiza differs from *Iphigenia* in the simple style, persistent perianth segments and laterally dehiscing anthers.

Two species, one in India the other in southern Africa.

Camptorrhiza strumosa (Baker) Oberm. in Kirkia **1**: 86 (1961). —Maroyi in Kirkia **18**: 9 (2002). Type: Botswana, Kwebe, xii.1896, *Lugard* 59 (K holotype, SRGH). FIGURE 12.2.**50**.

Iphigenia strumosa Baker in F.T.A. **7**: 562 (1898).

Iphigenia schlechteri Engl. in Bot. Jahrb. Syst. **32**: 89 (1902). Type: Mozambique, Maputo (Lourenço Marques), 29.xi.1897, *Schlechter* 11525 (BM holotype, K, PRE).

Iphigenia guineensis Baker in J. Linn. Soc., Bot. **17**: 451 (1879). Types: Angola, *Welwitsch* 1625, 1626 (BM, K).

Iphigenia junodii Schinz in Mém. Herb. Boiss. **10**: 28 (1900). Type: Mozambique, Delagoa Bay, hills at Rikatla, *Junod* 128 (Z holotype, BR).

Iphigenia flexuosa Baker in Bull. Herb. Boiss. ser.2, **4**: 996 (1904). Type: Namibia, between Etiro and Karibib, *Rautanen* 435 (Z holotype).

Iphigenia dinteri Dammer in Bot. Jahrb. Syst. **48**: 361 (1912). Types: Namibia, Windhoek, Brakwater, *Dinter* 1556 (B syntype?) & Okahandja, *Dinter* 395 (B syntype?, NBG).

Camptorrhiza schlechteri (Engl.) E. Phillips in Fl. Pl. S. Afr. **15**: t.575 (1935).

Iphigeniopsis flexuosa (Baker) Buxb. in Bot. Archiv. **38**: 228, 264 (1936).

Iphigeniopsis junodii (Schinz) Buxb. in Bot. Archiv. **38**: 229, 264 (1936).

Iphigeniopsis schlechteri (Engl.) Buxb. in Bot. Archiv. **38**: 264 (1936).

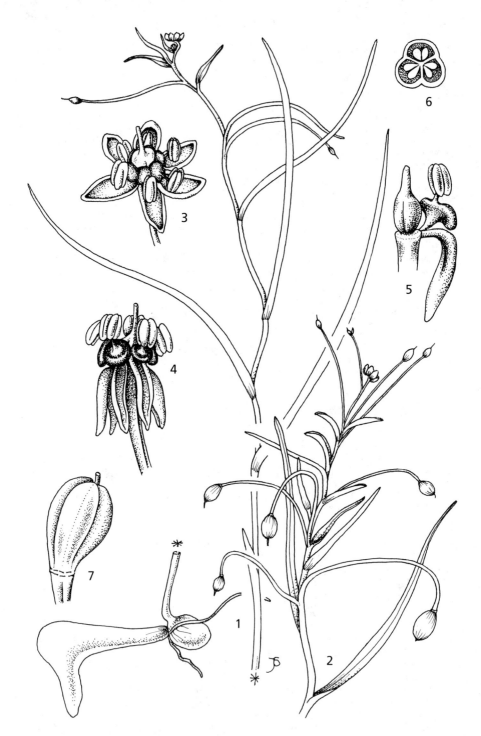

Fig. 12.2.**50**. CAMPTORRHIZA STRUMOSA. 1, habit (× 1); 2, habit with maturing fruits (× 1); 3, immature flower (× 5); 4, mature flower (× 5); 5, gynoecium, stamen and petal (× 6); 6, ovary, transverse section (× 10); 7, fruit (× 4). 1 from *C.A. Smith* 6306, 2 from *Barbosa & Lemos* 7865, 3–7 from spirit material 2381. Drawn by Judi Stone.

Iphigeniopsis strumosa (Baker) Buxb. in Bot. Archiv. **38**: 228, 264 (1936) .
Camptorrhiza flexuosa (Baker) Sterling in Bot. J. Linn. Soc. **68**: 284 (1974).
Camptorrhiza junodii (Schinz) Sterling in Bot. J. Linn. Soc. **68**: 284 (1974).

Perennial herb to 40 cm high. New corm small, ovoid, attached to knee-shaped tuber from the previous season, c.25 × 18 mm. Stem 22–35 cm long, slender, sinuous with basal, membranous scale-like leaf, free part 0.5–2 cm long. Leaves several to numerous, starting ± halfway up stem, grading into bracts, 7–14 cm long, 1–8 mm wide, narrowly linear, folded, acute, suberect. Inflorescence a raceme-like scorpioid cyme. Pedicels arising opposite leaves, 2–3 cm long, very slender, becoming recurved and usually sinuous in fruit, elongating to c.4 cm. Flowers described variously as green and russet, green with a purple-black swelling at filament base, yellow-green, or brown and green. Tepals 2.5–4 × 1–1.5 mm, oblong, concave with tip turned in, ± spreading but reflexing and persistent as ovary swells. Filaments 1 mm long including a fleshy base 0.5 mm long and wide; anthers 1.5 mm long, lying in hollow of concave tepal. Ovary c.2 mm long; style simple, 1 mm long. Capsule pendent, globose or pear-shaped, 10–20 × 9–12 mm.

Caprivi Strip. West Caprivi, Manketti village, ?1993, *Maggs* 913 (WIND). **Botswana**. N: Kwebe, Ngamiland, fl. xii.1896, *Lugard* 59 (K). SW: Ghanzi, Hide Stone area, 940 m, fl. 18.i.1970, *Brown* 7950 (K). **Zambia**. S: Kalomo, dry pan, fr. 11.ii.1965, *Fanshawe* 9158 (K, NDO). **Zimbabwe**. N: Gokwe Dist., Sengwa Research Station, *Brachystegia* woodland, fl. 6.xii.1968, *Jacobsen* 319 (K, SRGH). W: Hwange Dist., near Makalolo 1 pan, c.80 km SE of Main Camp, 1015 m, fl, 11.xii.1968, *Rushworth* 1351 (K, LISC, SRGH). E: Mutare Dist., Sabi Drift, 818 m, fl. 2.xii.1954, *Wild* 4659 (K, LISC, PRE, SRGH). S: Masvingo, 1909–1912, *Monroe* 2188 (BM). **Mozambique**. M: Maputo, waste ground, fr. xii. 1989, *Groenendijk* 2242 (K, LMU).

Also in South Africa and Namibia. In various types of dry woodland on sand soils; 50–1100 m.

Conservation notes: Widespread species; not threatened.

LILIACEAE

by P. Wilkin

Perennial herbs with leaves cauline to rosette-forming, underground organs bulbous or rhizomatous. Stems erect, bearing 1 to many, alternate to verticillate or apparently opposite leaves. Inflorescence usually racemose, sometimes umbelliform, often bracteate. Flowers usually actinomorphic, perfect; tepals free, in 2 whorls of 3, usually weakly differentiated, often spotted or otherwise contrastingly marked, bearing nectaries and forming a tube; stamens 6, free, anthers usually dorsifixed; ovary superior, usually 3-locular. Fruit a capsule or berry. Seeds often flattened and arranged on top of each other in fruit, less commonly rounded to angular.

A family of 17 genera and c. 600 species with a predominantly northern temperate distribution. The highest diversity is in North America and Eastern Asia.

Previously Liliaceae included many genera that have now been included in numerous other families in Liliales and Asparagales. Here the family is regarded in its strictest sense, following numerous recent publications (e.g. Chase, Monocot relationships: an overview, Amer. J. Bot. **91**: 1645–1655, 2004).

LILIUM L.

Lilium L., Sp. Pl.: 302 (1753). —Baker in J. Linn. Soc., Bot. **14**: 225 (1875). — Woodcock & Stearn, Lilies World: 113 (1950).

Erect herbs from multi-scaled bulbs which lack a tunic. Leaves cauline, appearing whorled to

Fig. 12.2.**51**. LILIUM FORMOSANUM. 1, habit (× ²/₃); 2, flower, outer peraianth segment and stamen (× ²/₃); 3, flower, inner perianth segment and stamen (× ²/₃); 4, style and ovary (× ²/₃). 1 from *Coveny* 6107, 2–4 from *Coveny* 6060. Drawn by Juliet Williamson. From F.T.E.A.

alternate, sometimes with axillary bulbils. Inflorescence usually racemose, terminal, sometimes reduced to a solitary flower. Flowers variable in shape and colour, pedicellate; tepals erect, forming a tube to recurved, with a nectary in the form of a narrow channel near base of inner surface, sometimes surrounded by hairs or papillae. Filaments usually erect or spreading, sometimes pubescent, anthers dorsifixed. Style borne at apex of 3-locular ovary, elongate, stigma capitate to 3-lobed. Fruit a capsule, containing seeds with narrow wings.

A genus of c.100 species predominantly native to the temperate northern hemisphere. Species diversity is highest in Eastern Asia, especially in montane areas. One widely cultivated species of *Lilium* appears to have become naturalised in Zimbabwe.

Lilium formosanum (Baker) Wallace in Garden (London) **40**: 442 (1891). —Stapf in Bot. Mag. **154**: t.9205 (1930). —Woodcock & Stearn, Lilies World: 219 (1950). —Grimshaw in F.T.E.A, Liliaceae: 3 (2005). —Maroyi in Kirkia **18**: 187 (2006). Types: Taiwan (Formosa), Tamsuy, 1864, *Oldham* 565 (K syntypes) & Taiwan, no locality, received 1862, *Swinhoe* s.n. (K syntype). FIGURE 12.2.**51**.

> *Lilium longiflorum* Thunb. var. *formosanum* Baker in Gard. Chron., n.s. **14**: 524 (1880).
> *Lilium philippinense* Baker var. *formosanum* Grove in Gard. Chron., ser.3 **70**: 63 (1921). Types: Taiwan (but stated to be from Japan when distributed), obtained as seed in 1918 and cultivated in UK and Ireland in 1919–1920, *Wilson* 10961 (K 3 syntypes) and *Wilson* 10963 (not seen).
> *Lilium zairei* Mynett & Mackiewicz in Acta Hort. **177**: 667 (1986), invalid name.

Single-stemmed erect herb from 80–170 cm tall. Bulb 2–4 cm wide, subglobose to broadly ellipsoid. Stem 4–12 mm in diameter, usually maroon or purple to brown-purple. Leaves 40–182 × 2–7(13) mm, linear to narrowly lanceolate, smaller and less dense towards stem apex. Flowers usually 1–3, orientated horizontally, white, pleasantly scented, usually on ± erect pedicels bearing leaf-like bracts. Tepals erect in basal two-thirds forming a narrow tube, reflexed to recurved in apical third, outer whorl 10.7–19.6 × 1.8–3.2 cm, oblanceolate, with maroon, purple or brown-purple hue on outer surface, inner whorl 11.1–22.5 × 3.3–4.2 cm, spathulate. Reproductive organs exserted from tube, filaments 6.1–12.3 cm long, anthers 8–24 × 2–4.7 mm, ovary 3.4–8.1 cm long; style 8–9 cm long, stigma 8–13 mm wide, 3-lobed, lobes recurved. Capsule 7–9 × 2 cm, cylindric, erect (not seen in FZ specimens).

Zimbabwe. C: Marondera Dist., Marondera (18°10'N, 31°32'W), fl. 10.i.1998, *Poilecot* 7540 (G, K). E: Chimanimani Dist., 'Westward Ho' on road between Nyahodi Valley and main Chimanimani–Chipinge road, 20.i.1973, *Grosvenor* 791 (K, LISC, SRGH); Nyanga Dist., 20 km S of Juliasdale on road to Mutare, fl. 11.ii.1997, *Brummitt & Pope* 19581 (K, SRGH).

The species is endemic to Taiwan, but has become naturalised in Zimbabwe (and possibly other FZ countries) in and around plantations. It is also naturalised elsewhere in eastern, southern and central Africa (e.g. *L. zairei*, based on bulbs collected in Kisangani), Australasia (where it is weedy) and the Neotropics. It is usually encountered in anthropogenic habitats at higher altitudes; 1000–1600 m.

Conservation notes: Introduced species.

Lilium formosanum differs from the closely related *L. longiflorum* Thunb. (the Easter Lily) which has 8–18 mm wide, lanceolate or oblong-lanceolate leaves. *L. longiflorum* also has tepals in which both whorls are oblanceolate and white, the outer being slightly narrower with a green hue towards the base on the outer surface, and lacks red or purple pigmentation; in *L. formosanum* the tepal whorls are not differentiated. The vegetative and floral dimensions of *L. formosanum* Grove also mentions a specimen collected by W.R. Price in Formosa (Taiwan) in 1912, but it is not clear which of the six specimens at Kew alone are being referred to.

SMILACACEAE

by M.A. Diniz

Climbing, creeping or scandent subshrubs or shrubs, somewhat robust, usually glabrous, rarely pubescent, often spiny, with a woody rhizome, usually dioecious. Leaves usually alternate, less often opposite, simple, entire, petiolate; blade thin to leathery, palmately 3–7-curvinerved. Stipules modified to form 2 tendrils arising from a persistent leaf-sheath at base of petiole. Inflorescence a peduncle of 1–many-flowered axillary or terminal umbels, simple or arranged in a raceme or spike. Flowers unisexual, 3-merous, actinomorphic, pedicellate; perianth segments (tepals) 6, free at base, rarely united, reflexed or not. Stamens 3–6(15) in male flowers; filaments sometimes united into a column; anthers 2-thecous, entire, basifixed. Staminodes 1–6, filiform in female flowers. Ovary superior, sessile, 3-locular with axile placentation, 1–2 ovules per locule; stigmas 3, recurved. Fruit a berry, globose to obovoid, 1–3-seeded. Seeds globose or flattened, smooth, with hard endosperm.

A family with 4 genera, of which only *Smilax* occurs in tropical Africa.

SMILAX L.

Smilax L., Sp. Pl.: 1028 (1753); Gen. Pl. ed.5: 455 (1754).

Subshrubs or shrubs with branches climbing or pendent, sometimes spiny. Leaves usually alternate, shortly petiolate, often with a pair of spiral tendrils near base of petiole as extensions of persistent leaf-sheath; blade coriaceous or subcoriaceous, with reticulate venation. Inflorescence axillary or terminal, solitary or simple umbels or arranged in racemes or spikes. Flowers small, unisexual, yellowish to greenish, sometimes aromatic; bracts small; perianth with 6 segments free at the base, caducous. Male flowers usually with 6 stamens with free filaments, or rarely inserted at base of perianth segments, erect or erect-patent; anthers basifixed, introrse. Female flowers with 3–6 staminodes; ovary ovoid with 1 ovule per locule, pendulous. Fruit a berry, subglobose, black, reddish, violet or purplish when mature. Seeds 1–3, globose, pendulous; embryo small, enveloped by hard endosperm.

A genus of c.375 species distributed in tropical and subtropical regions and in temperate areas. Only 2 species found in the Flora Zambesiaca area.

Leaf lamina broadly oblong to subcircular; leaf base rounded or subcordate; flowers in pedunculate axillary umbels · **1.** *anceps*
Leaf lamina triangular, ovate to lanceolate; leaf base cordate, hastate, truncate or saggitate; flowers in sessile umbels inserted along a rachis · · · · · · · · · · **2.** *aspera*

1. **Smilax anceps** Willd., Sp. Pl. **4**: 782 (1806). —A. de Candolle in A. & C. de Candolle, Monogr. 1, Smilaceae: 169 (1878). —Baker in F.C. **6**: 256 (1896). — Cowley in F.T.E.A., Smilacaceae: 2 (1989). —Obermeyer in F.S.A. **5**(3): 85, fig.17 (1992). Type: Mauritius, *W. Heyne* (B-W 18393 holotype). FIGURE 12.2.**52.**

   Smilax kraussiana Meisn. in Flora **28**: 312 (1845).—A. de Candolle in A. & C. de Candolle, Monogr. 1, Smilaceae: 171 (1878). —Baker in F.T.A. **7**: 423 (1898). —Wood, Natal Plants **4**: fig.339 (1906). Type: South Africa, KwaZulu-Natal, *Drège* 4503 (K isotype).

   Smilax mossambicensis Garcke in Peters, Naturw. Reise Mossamb., Bot.: 518 (1862). Type: Mozambique, Inhambane, *Peters* s.n. (B† holotype).

Climbing or scandent glabrous subshrub or shrub with stems up to 5 m long. Stems slender, terete, ribbed or angled, flexuose, somewhat zigzag with scattered sharp prickles 1–3 mm long, more densely arranged near base. Leaves alternate with petiole 0.5–2.5 cm long, canaliculate, thickened, usually with a pair of long spiral tendrils near base; lamina 4–14 × 1.5–13 cm, circular, ovate to elliptic, acute to obtuse, usually apiculate at apex, slightly subcordate to rounded, rarely cuneate at base, with 2–3 nerves from base at each side of midrib, prominent below, net-veined, margins entire, coriaceous. Inflorescence an axillary globose many-flowered umbel; peduncle 0.8–3 cm long with two ovate, amplexicaul, obtuse bracts at or near middle, up to 5 mm long, persistent. Pedicels 0.4–1.2 cm long, elongate up to 2.2 cm in fruit with minute basal bracteoles. Perianth segments 3–5 mm long, linear to oblong, obtuse at apex, recurved, greenish white, brownish or yellowish, the outer ones broader than inner. Male flower with 6 stamens inserted at base of perianth segments; filaments 3–5 mm long; anthers up to 1 mm long, oblong, basifixed, erect. Female flower with 3 staminodes shorter than perianth segments. Ovary 2–3 mm

Fig. 12.2.**52**. SMILAX ANCEPS. 1, flowering ♂ branch (× ¹/₂); 2, flowering ♀ branch (× ¹/₂); 3, ♂ flower (× 2¹/₂); 4 stamen (× 3¹/₂); 5, ♀ flower (× 2¹/₂); 6, fruiting branch (× ¹/₂); 7, fruit in cross-section (× ¹/₂). 1, 3–4 from *Barbosa* 8436, 2, 5 from *Milne-Redhead* 3314, 6–7 from *Moura* 36. Drawn by G.C. Matos.

long, ovoid, sessile; ovule solitary in each locule; stigma subsessile, 3-branched, arms c. $^1/_3$ as long as ovary, linear, papilate recurved. Berry 8–10 mm in diameter, globose, purplish or black, 1–3 seeded; seeds c.4 × 2 mm, kidney-shaped, flattened, purplish grey.

Zambia. B: Kalabo, near resthouse, fl. 16.xi.1959, *Drummond & Cookson* 6524 (LISC, SRGH). N: Lake Chila (Abercorn Lake), fr. 19.vii.1930, *Hutchinson & Gillett* 3891 (BM, K). W: Mwinilunga Dist., SW of Dobeka bridge near Dobeka Plain, ♀ fl. 19.xi.1937, *Milne-Redhead* 3314 (BM, K, LISC). C: Mkushi Dist., fl. 3.i.1958, *Robinson* 2592 (K). S: Victoria Falls, ♂ fl. 1909, *Monro* 1993a (BM). **Zimbabwe**. E: Vumba, E slopes of forest, ♂ fl. 11.xii.1945, *Wild* 496 (K, LISC, SRGH). S: Mberengwa Dist., Buhwa Mt., ♂ fl. 10.xii.1953, *Wild* 4326 (K, LISC, SRGH). **Malawi**. N: Chitipa Dist., Misuku Hills, descent to Sokora, ♀ fl. 7.vii.1973, *Pawek* 7078 (K, MAL, MO). C: Dedza Dist., near Mphunzi, Kachere, imm.fr. 22.i.1959, *Robson* 1305 (BM, LISC). S: Zomba, Bwaira stream, fr. i.vii.1954, *Banda* 37 (BM). **Mozambique**. N: Cabo Delgado, 2 km from Mueda to Negomano, close to Missão de Santo António, c.850 m, imm.fr. 3.i.1964, *Correia* 91 (LISC). Z: Gurué, ♂ fl. 10.ii.1932, *Vincent* 171 (BM). MS: Cheringoma, 110 km from Beira, serração de Durundi, fr. 26.v.1942, *Torre* 4221 (LISC). GI: Gaza, Chirindzene, 6 km on road to Chiconela, fr. 20.ix.1980, *Nuvunga, Boane & Conjo* 366 (BM, K, LMU). M: Marracuene, 18 km from Marracuene (Vila Luiza) to Bobole (Bobone), ♂ fl. 15.xii.1944, *Mendonça* 3423 (LISC).

Widespread from Senegal eastwards to Sudan, Ethiopia, Uganda, Kenya and Tanzania, south to South Africa and Swaziland. Also in Madagascar, Comoro Islands, Mauritius, Seychelles and Reunion. Locally common on forest edges, clearings and secondary bushland and thicket, also along rivers, lakesides and marshes; sea-level to 2400 m.

Conservation notes: Widespread species; not threatened.

The species has been used medicinally. In E and S Africa a decoction of the root has been used as a tonic, diuretic, febrifuge, antirheumatic, antisyphilitic, and as a remedy for gonorrhoea and ophthalmic conditions. The stems are often used to make granaries, stretchers and baskets.

2. **Smilax aspera** L., Sp. Pl.: 1028 (1753). —A. de Candolle in A. & C. de Candolle, Monogr. 1, Smilaceae: 163 (1878). —Cowley in F.T.E.A., Smilacaceae: 2 (1989). Type: Unknown origin; from a plant grown in Clifford's garden in Netherlands (BM-CLIFF).

Smilax goetzeana Engl. in Bot. Jahrb. Syst. **30**: 275 (1901). Type: Tanzania, Njombe Dist., Kinjilaa Dinde (Kingika), Mt. Pikurugwe ridge, *Goetze* 1251 (B† holotype).

Climbing or creeping glabrous shrub with stems up to 15 m long. Stems slender, 4–6-angled or terete, flexuous, unarmed or with a few short prickles. Leaves alternate, petiole 1–4.7 cm long, not thickened towards apex, with a pair of long tendrils from near the base; lamina 3–15 × 2–8 cm, triangular, oblong, ovate, lanceolate, or sometimes reniform, acute to acuminate at apex, cordate, hastate, truncate or sagittate at base, coriaceous or subcoriaceous, sometimes with white blotches, 2–4 nerves from base on each side of midrib, net-veined, with occasional scattered prickles 1–3 mm long on midrib and nerves. Inflorescence axillary, 1.5–15 cm long, with sessile umbels spirally arranged along rachis; bracts minute. Flowers on pedicels 1.5–9 mm long. Perianth segments 2–4 × 1 mm, linear-oblong to elliptic, obtuse at apex, slightly recurved, white, greenish, pink, yellowish or brownish, aromatic, the inner equal or shorter and narrower than the outer ones. Male flower with 6 stamens shorter than perianth segments; anthers up to 1 mm long, shorter than filaments, yellow-greenish. Female flower with 6 staminodes shorter than perianth segments. Ovary c.2 mm long, ovoid, greenish; ovule solitary in each locule; style subsessile, stigmas 3, oblong, recurved. Berry 5–8 mm in diameter, globose, red or black when mature, 3-seeded; seeds c.3 × 2 mm, oblong.

Zambia. E: Nyika Plateau, 9.6 km SW of Resthouse, 2150 m, st. 27.x.1958, *Robson & Angus* 417 (BM, K).

Known from Ethiopia, Uganda, Kenya, Tanzania, Zaire, and Zambia. Also found in India (along the Himalayas, Assam, Madras) and Sri Lanka. Common in most parts of Western Europe, E Mediterranean, and North Africa. On evergreen forest edges; 2150 m.

Conservation notes: Widespread species, although very local in the Flora area.

Cowley (1989) mentions that the species also occurs in Malawi, but I have not seen any material from that country.

INDEX TO BOTANICAL NAMES

AFROTHISMIA Schltr. **141**
  insignis 143
  winkleri 143
  zambesiaca Cheek sp. nov. **141,142,**148
ALISMA L. **1,2**
  humile Kunth 12
  plantago-aquatica L. **2,3**
  reniforme D. Don 6
ALISMATACEAE **1**
Althenia 96,97
Amorphophallus abyssinicus (A. Rich.)
    N.E. Br. 140
Amphibolis 99
ANDROCYMBIUM Willd. **176**
  melanthioides Willd. **177**
    var. *acaule* Baker 177
    var. *subulatum* (Baker) Baker 177
  roseum Engl. 177,**178,179**
  striatum 177
  subulatum Baker 177
Anguillaria R. Br. 179
APONOGETON L. f. **65**
  abyssinicus A. Rich. 66,**71**
    var. abyssinicus 71
    var. albiflorus 71
    var. cordatus 71
    var. glanduliferus 71
    var. graminifolius Lye 71
  afroviolaceus Lye 66,**69**
    var. afroviolaceus **69**
    var. angustifolius Lye 69
  boehmii Engl. 71
  braunii K. Krause 71
  desertorum A. Spreng. 66,**70**
  dinteri Engl. & K. Krause 70
  eylesii Rendle 70
  gracilis Schinz 68
  gramineus Lye 68,69
  hereroensis Schinz 72
  holubii Oliv. 70
  junceus Schltdl. 66,**71**
    subsp. *junceus* 72
    subsp. *rehmannii* (Oliv.) Oberm. 71,72
  junceus sensu Obermeyer 72
  kraussianum C. Krauss 70
  natalensis 66
  oblongus Peter 71
  ranunculiflorus 65

rehmannii Oliv. 66,**72**
spathaceus E. Mey. 72
stuhlmannii Engl. 66,**68**
vallisnerioides Baker **66,67,**69
violaceus Lye 69
APONOGETONACEAE **65**
Barbacenia
  equisetoides (Baker) R.E. Fr. 165
  hereroensis Schinz 175
  retinervis (Baker) Burtt Davy & Pott-
      Leendertz 165
  scabrida Pax 160
  tomentosa Pax 171
  velutina (Baker) Pax 155
  wentzeliana Harms 167
BLYXA Thouars 19,**42**
  aubertii Rich. **43**
    var. aubertii **43**
    var. echinosperma (C.B. Clarke) Cook
        & Lüönd 43
  hexandra Cook & Lüönd 43,**44,45,**46
  radicans Ridl. 43,**44**
Boottia Wall. 29
  aschersoniana Gürke 39
  cylindrica T.C.E. Fr. 38
  exserta Ridl. 32
  fischeri Gürke 35
  kunenensis Gürke 40
  macrantha C.H. Wright 33
  mossambicensis Peter 33
  muricata C.H. Wright 39
  stratiotes T.C.E. Fr. 38
BURMANNIA L. 141,**143**
  bicolor Mart.
    var. *africana* Ridl. 145
    var. *micrantha* Engl. & Gilg 145
  blanda Gilg 145
  inhambanensis Schltr. 145
  madagascariensis Mart. **144,145**
  madagascariensis Baker 145
  welwitschii Schltr. 145
BURMANNIACEAE **141**
BURNATIA Micheli 1,**2**
  enneandra Micheli **4,5**
BUTOMOPSIS Kunth **16**
  lanceolata (Roxb.) Kunth 16
  latifolia (D. Don) Kunth, **16,17**
  Butomus latifolius D. Don 16

CALDESIA Parl. 1,**6**
  parnassifolia (Bassi) Parl. 8
  reniformis (D. Don) Makino **6,7**
CAMPTORRHIZA Phill. **193**
  flexuosa (Baker) Sterling 176,195
  junodii (Schinz) Sterling 195
  strumosa (Baker) Oberm. **193,194**
  schlechteri (Engl.) E. Phillips 193
Caulinia Willd. 60
  ovalis R. Br., 51
  serrulata R. Br. 105
Clinostylis Hochst. 182
COLCHICACEAE **176**
Coleogeton 77
  pectinatus (L.) Les & Haynes 79
CYMODOCEA K.D. Koenig 99,**104**
  acaulis Peter 105
  angustata 104
  ciliata (Forssk.) Asch. 107
  isoetifolia Asch. 100
  rotundata (Ehrenb. & Hempr.) Aschers.
    104
  serrulata (R. Br.) Asch. & Magnus 104,
    105,106
CYMODOCEACEAE 96, **99**
Damasonium Schreb. 29
  ulvifolium Planch. 31
DIOSCOREA L. **110**
  abyssinica Kunth 128
  alata L. 109,110,112,116
  andongensis Rendle 130
  angustiflora Rendle 127
  anthropophagorum A. Chev. 119
  apiculata De Wild. 131
  asteriscus Burkill 111,**113**,115,**116**
  baya De Wild. 112,**113**,115,116,**117**
    var. baya **117**
    var. kimpundi De Wild. 118
    var. *subcordata* De Wild. 117
  beccariana Martelli 131
  brevipes Burtt-Davy 136
  buchananii Benth. 111,112,**114**,116,**118**
    var. *ukamensis* R. Knuth 118
  buchholziana Engl. 123
  bulbifera L. 111,115,117,**119**
    var. *anthropophagorum* (A. Chev.)
      Summerh. 119
  cayenensis Lam. 109,110,112,115,116,128
    var. *praehensilis* (Benth.) A. Chev. 127
  cochleari-apiculata De Wild. 111, 112,
    114, 115, **120, 121**
  cotinifolia Kunth 112,**114**,115,116,**122**
  crinita Hook.f. 131,133
  dinteri Schinz 131
  dregeana (Kunth) T. Durand & Schinz
    111,**114**,115,**123**
    var. *hutchinsonii* Burkill 123

dumetorum (Kunth) Pax 111,**114**,115,
  122,**123**
elephantipes 137
esculenta (Lour.) Burkill 109
excisa R. Knuth 131
forbesii Baker 131
hirtiflora Benth. 111,**113**,115,116,**125**,131,
  136
  subsp. hirtiflora 125
  subsp. orientalis Milne-Redh. **125**
  subsp. pedicellata Milne-Redh. **126**
  var. *nyassica* Burkill 126
  var. *trapnellii* Burkill 125
hockii De Wild. 135
hylophila Harms 130
liebrechtsiana De Wild. 127,128
lindiensis R. Knuth 125
macroura Harms 134
malifolia Baker 122
marlothii R. Knuth 136
montana (Burch.) Spreng.
  var. *glauca* R. Knuth 136
  var. *lobata* Weim. 136
  var. *sagittata* Suess. 136
odoratissima Pax 127, 128
polyantha Rendle 125
praehensilis Benth. 112,**113**,115,116,
  127,129
preussii Pax 111,112,115,**130**
  subsp. hylophila (Harms) Wilkin **113**,
    130
  subsp. preussii **130**, 131
    var. *glabra* Burkill 130
quartiniana A. Rich 111,112,**114**,115,**131**
  var. *apiculata* (De Wild.) Burkill 132
  var. *dinteri* (Schinz) Burkill 132
  var. *excisa* (R. Knuth) Burkill 132
  var. *quartiniana* 132
  var. *schliebenii* (R. Knuth) Burkill 132
  var. *stuhlmannii* (Harms) Burkill 132
rehmanni Baker 136
retusa Mast. 133
retusa sensu Mogg 132
rotundata Poir. 109
rubiginosa Benth. 125
rupicola Kunth. 119
sagittifolia Pax. 128
sansibarensis Pax 111,112,**113**,115,**133**
sativa L. sensu Rendle 116
sativa sensu Baker 119
schimperiana Kunth 111,**113**,115,116,**134**
  var. *vestita* Pax 135
schliebenii R. Knuth 132
stellato-pilosa De Wild. 135
stolzii R. Knuth 120
stuhlmannii Harms 131
sylvatica Eckl. 112,**114**,115,116,**136**

triphylla L.
  var. *dumetorum* (Kunth) R. Knuth 124
  var. *tomentosa* Rendle 124
  welwitschii Rendle 134
DIOSCOREACEAE **109**
Diplanthera Thouars 102
  uninervis (Forssk.) F.N. Williams 102
Echinodorus sensu Wright 11
  humilis (Kunth) Buchenau 12
  schinzii Buchenau 4
Egeria densa Planch. 18
ENHALUS Rich. 19,**49**
  acoroides (L.f.) Royle **49,50**
Freycinetia 148
GLORIOSA L. 176,**182**
  abyssinica A.Rich. 182
    var. *graminifolia* Franch. 185
  baudii (Terracc.) Chiov. 185
  carsonii Baker 182,183
  graminifolia (Franch.) Chiov. 185
  grandiflora O'Brien 182
  homblei De Wild. 183
  lutea Hort. 182,183
  minor Rendle 185
  rothschildiana O'Brien 182,183
  sessiliflora Nordal & M.G. Bingham 182, **185**
  simplex L. 182,183
  superba L. **182,184**
    var. *angustifolia* Baker 182,185
    var. graminifolia (Franch.) Hoenselaar 185
    var. superba 185
  virescens Lindl. 182,183
Groenlandia 76
GYMNOSIPHON Blume 141,143,**146**
  afro-orientalis Cheek sp. nov. **146,147**
  usambaricus 148
HALODULE Endl. 99,**102**
  uninervis **103**
  wrightii Asch. 95,104
  wrightii sensu Hartog 102
HALOPHILA Thouars 18,19,**51**
  linearis Hartog 54
  minor (Zoll.) Hartog 51,**54**
  ovalis (R. Br.) Hook. f. **51**
    subsp. linearis **52**
    subsp. ovalis **52,53**
  ovalis 'lanceolate form' of Macnae 54
  ovalis 'normal form' of Macnae 52
  ovata sensu C.H. Wright 52
  ovata sensu Hartog 54
  stipulacea (Forssk.) Asch. 51,**54**,95
Helmia
  dregeana Kunth 123
  dumetorum Kunth 123
Heterozostera 93

HYDRILLA Rich. 19,**27**
  dregeana C. Presl 24
  muscoides (Harv.) Planch. 24
  verticillata (L. f.) Royle **27,28**
  verticillata sensu Gibbs 25
    var. *brevifolia* Casp. 27
Hydrocharis 18
HYDROCHARITACEAE **18**
Hydrogeton heterophyllus Lour. 86
IPHIGENIA Kunth 176,**191**,193
  abyssinica Chiov. 193
  bechuanica Baker 191
  dinteri Dammer 193
  flexuosa Baker 193
  guineensis Baker 193
  junodii Schinz 193
  oliveri Engl. **191,192**
  pauciflora Martelli 191,**193**
  schlechteri Engl. 193
  strumosa Baker 193
Iphigeniopsis Buxb. 193
  flexuosa (Baker) Buxb. 193
  junodii (Schinz) Buxb. 193
  schlechteri (Engl.) Buxb. 193
  trumosa (Baker) Buxb. 195
JUNCAGINACEAE **73**
LAGAROSIPHON Harv. 19,**20**
  cordofanus Casp. 20,**21**
  crispus Rendle 21
  fischeri Gürke 21
  ilicifolius Oberm. 20,22,**25,26**,29
  major (Ridl.) Moss 20,**24**
  muscoides Harv. 20,**22**
    forma *brevifolia* 24
    forma *longifolia* 24
    forma *typica* 24
  muscoides Harv.
    var. *major* Ridl. 24
    var. *major* sensu Schinz & Junod 22
    var. *major* sensu Schinz 25
  nyassae Ridl. 21
  schweinfurthii Casp. 24
  sp. sensu Hall-Martin & Drummond 21
  tenuis Rendle 21
  tsotsorogensis Bremek. & Oberm. 21
  verticillifolius Oberm. 20,**22,23**
Lemnopsis Zoll. 51
  minor Zoll. 54
Leontice leontopetaloides L. 138
Lepilaena 96,97
LILIACEAE **195**
LILIUM L. **195**
  formosanum (Baker) Wallace **196,197**
  longiflorum Thunb. 197
    var. *formosanum* Baker 197
  philippinense Baker
    var. *formosanum* Grove 197

zairei Mynett & Mackiewicz 197
Limnobium 18
LIMNOCHARITACEAE **15**
LIMNOPHYTON Miq. 1,**8**
  angolense Buchenau 8,**10**
  fluitans Graebn **8**
  obtusifolium (L.) Miq. **8**
  obtusifolium sensu auct. mult. 11
LITTONIA Hook. 176,185,**186**
  lindenii Baker 186,**188**
  littonioides (Baker) K. Krause **186,187**
  modesta Hook. 186,**188**
  welwitschii Benth. & Hook.f. 186
Maidenia 18
Maundia 73
Melanthium tenue Hook. 180
NAJADACEAE **58**
NAJAS L. 18,**58**
  subgen. Caulinia (Willd.) Rendle **60**
  subgen. Najas 59
  baldwinii Horn 64
  graminea Delile 59,**64**
    var. graminea. **64**
  graminea sensu Triest & Symoens 63
  hagerupii Horn 64
  horrida Magnus 59,**60,62**,63
  interrupta K. Schum. 60,63
  marina L. **59**
    subsp. armata **60**
    subsp. *delilei* (Rouy) Oberm. 60
    var. *angustifolia* sensu R.E. Fries 61
    var. *muricata* (Delile) K. Schum. 60
  meiklei Horn 63
  minor All. 64
  minor sensu Bennett 61
  pectinata (Parl.) Magnus 63
  pectinata sensu Durand & Schinz 61
  pectinata sensu Richards & Morony 60
  schweinfurthii Magnus 64
  testui Rendle 59,**63**,64
  welwitschii sensu Richards & Morony 61,63
Nanozostera Toml. & Posl. 93
  capensis (Setch.) Toml. & Posl. 95
Nechamandra 18
ORNITHOGLOSSUM Salisb. 176,**188**
  glaucum sensu auct. 189
  vulgare B. Nord. **189,190**
OTTELIA Pers. 19,**29**
  alismoides (L.) Pers. 30
  australis Bremek. 31
  cylindrica (T.C.E. Fr.) Dandy 31,**38**
    var. *stratiotes* (T.C.E. Fr.) Cook, Symoens
      & Urmi-König 38
  exserta (Ridl.) Dandy 30,**32,34**
  fischeri (Gürke) Dandy 30,33,**35**
  gigas T.C.E. Fr. 42
  kunenensis (Gürke) Dandy 31,**40**

lancifolia A. Rich. 31
  var. *fluitans* Ridl. 31
  latifolia De Wild. 31
  lisowskii Symoens sp. nov. 30,**38**
  luapulana Symoens sp. nov. 30,**36,37**
  macrantha (C.H. Wright) Dandy 33
  muricata (C.H. Wright) Dandy 30,**39,41**
  obtusifolia T.C.E. Fr. 31,32
  plantagine Ridl. 31
  stratiotes (T.C.E. Fr.) Dandy 38,39
  ulvifolia (Planch.) Walp. 30,**31**
  verdickii Gürke 30,**42**
    var. *elanga* De Wildeman 42
  vernayi Bremek. & Oberm. 31,32
  vesiculata Ridl. 31
PANDANACEAE **148**
PANDANUS Parkinson **148**
  subgen. Pandanus sect. Pandanus 149
  subgen. Vinsonia sect. Heterostigma B.C.
    Stone 149
  angolensis Huynh 153
  gasicus Huynh 149,152
  globulatus Huynh 149,152
  livingstonianus Rendle **149,150**
  mosambicus Huynh 149,152
  odoratissimus L.f. 149,**153**
  petersii Warb. 149,151,152
  serrimarginalis Huynh 149,152
  utilis Bory 149
  welwitschii 152
Phucagrostis rotundata Kuntze 104
Phylospadix 93
Physkium Lour. 46
POTAMOGETON L. 77,**79,81**
  sect. *Coleophylli* Koch 79
  subgen. *Coleogeton* (Rcbh.) Raunk. 77,79
  subgen. *Coleophylli* Koch 77
  subsect. Javanici 84
  americanus Cham. & Schltdl.
    var. *thunbergii* (Cham. & Schltdl.) A.
      Benn. 87,89
  berchtoldii Fieber 84
  capensis Hagstr. 87
  crispus L. 82,**85**
    var. *najadoides* Graebn. 85
  filiformis sensu Bennett 79
  fluitans sensu Bennett 87
  friesii sensu Bennett 83
  javanicus Hassk. 86
  livingstonei A. Benn. 79,81
  longifolius sensu Burkill 87
  lucens 89
  lucens sensu K. Schum. 87
  natans L.
    var. *capensis* T. Durand & Schinz 89
  natans sensu Bennett 91
  natans sensu Thunberg 89

nodosus Poir. 82,83,**89**,91
  var. *billotii* (F.W. Schultz) Hagstr.
  forma *angustissimus* Hagstr. 89
nodosus sensu Dandy 91
octandrus Poir. 82,**86**
panormitanus Guss. 83
pectinatus L. 79
preussii A. Benn. 86
pusillus L. 82,**83**,85
  var. *africanus* A. Benn. 83
  ?repens Hagstr. 87
richardii Solms 82,89,**91**,92
richardii sensu Wild 90
schweinfurthii A. Benn. 82,83,**87**,90
stagnorum Hagstr. 89
subjavanicus Hagstr. 83, 84
thunbergii sensu Obermeyer 89,91
thunbergii Cham. & Schltdl. 89
trichoides Cham. & Schltdl. 82,**84**
  × apertus Miki 84
  × bunyonyiensis Denny & Lye 89
POTAMOGETONACEAE 76,96
Pseudalthenia 96,97
RANALISMA Stapf 1,**11**
  humile (Kunth) Hutch. **12**
  rostratum Stapf 11
Rautanenia schinzii (Buchenau) Buchenau 4
Sagittaria humilis (Kunth) Kuntze 12
  obtusifolia L. 10
Sandersonia aurantiaca Hook. 176
  littonioides Baker 186
Sararanga 148
Schizotheca Solms 55
  hemprichii Ehrenb. 56
Serpicula verticillata L.f. 27
SMILACACEAE **198**
SMILAX L. **198**
  anceps Willd. **198**,199
  aspera L. 198,**200**
  goetzeana Engl. 200
  kraussiana Meisn. 198
  mossambicensis Garcke 198
Statiotes acoroides L.f., 49
Stenomeris Planch. 109
STUCKENIA Börner 77, **79**
  filiformis (Pers.) Börner 81
  pectinata (L.) Börner **79**,80
SYRINGODIUM Kütz. 99,**100**
  filiforme Kütz. 100
  isoetifolium (Asch.) Dandy **100**,101
TACCA J.R. & G. Forst. 109,**137**
  abyssinica Baker 138
  involucrata Schumach. & Thonn. 138
  leontopetaloides (L.) Kuntze **138**,139
  pinnatifida J.R. & G. Forst. 138
    var. *acutifolia* H. Limpr. 138
  quanzensis Welw. 138
Taccaceae 109

Tenagocharis Hochst. 16
  latifolia (D. Don) Buchenau 16
Testudinaria Burch. 110
  montana 137
  multiflora Marloth 136
  paniculata Dummer 136
  sylvatica Kunth 136
Tetroncium 73
THALASSIA K.D. König 18,19,**55**
  hemprichii (Ehrenb.) Asch. 55,57,105
  hemprichii (Solms) Asch. **56**
  testudinum K.D. König 55
Thalassia sp. sensu Cohen 56
THALASSODENDRON Hartog 99,**107**
  ciliatum (Forssk.) Hartog 56,**107**,108
  pachyrhizum 107
Trichopus Gaertn. 109
TRIGLOCHIN L. 73,**74**
  bulbosa L. **74**,75
    subsp. bulbosa **74**
  milnei Horn 74
  striata Ruiz & Pav. 74,**76**
Udora cordofana Hochst. 21
VALLISNERIA L. 18,19,**46**
  aethiopica Fenzl 48
  spiralis L. **46**
    forma aethiopica (Fenzl) T. Durand &
      Schinz **47**,48
    var. denseserrulata 48
Vellozia aequatorialis Rendle 171
  argentea Wild 171
  capillaris (Baker) Baker 159
  clavata (Baker) Baker 165
  equisetoides (Baker) Baker 165, 174
  equisetoides sensu Greves 167
  equisetoides sensu Watson 166
  equisetoides var. *trichophylla* Baker 166
  eylesii Greves 164
  hereroensis (Schinz) Baker 175
  humilis Baker 162
  kirkii Hemsl. 174
  minuta Baker 162
  monroi Greves 163,164
  pinifolia (Lam.) Poiss. 158
  retinervis (Baker) Baker 165
  rosea Baker 160
  scabrida (Pax) Baker 160
  schlechteri Baker 160
  sp. 1 of White 170
  sp. 2 of White 172
  sp. of Goodier & Phipps 171
  spekei (Baker) Baker 171
  splendens Rendle 174
  splendens sensu Baker 174
  squarrosa (Baker) Baker 159
  suaveolens Greves 168
  tomentosa (Pax) Baker 171
  trichophylla (Baker) Hemsl. 166

trichophylla sensu Jacobsen 172
villosa Baker 163
violacea Baker 163
viscosa (Baker) Baker 175
wentzeliana (Harms) Greves 167
VELLOZIACEAE **153**
  subfamily Vellozioideae 154
  subfamily Barbacenioideae 154
Vleisia 97
WIESNERIA Micheli 1,**13**
  filifolia Hook. f. 13,**14,15**
  schweinfurthii Hook. f. **13**
  sparganiifolia Graebn. 13
Wisneria 13
WURMBEA Thunb. 176,**179**
  angustifolia B. Nord. **180**
  goetzei Engl. 180
  homblei De Wild. 180
  tenuis (Hook.f.) Baker **180**
    subsp. australis B. Nord. 180
    subsp. goetzei (Engl.) B. Nord. **180,181**
    subsp. hamiltonii (Wendelbo) B. Nord. 180
    subsp. tenuis 180
XEROPHYTA Juss. 154,**155**
  section Barbacenioides 155
  section Vellozioides 155
  section Xerophyta 155
  argentea (Wild) L.B. Smith & Ayensu 156,**171**
  capillaris Baker 156,**159**
    var. capillaris 159
    var. occultans L.B. Smith & Ayensu 159
  clavata Baker 164
  equisetoides Baker 156,**165**
    var. equisetoides **166**
    var. pauciramosa 166
    var. pauciramosa L.B. Smith & Ayensu 166,**167**
    var. pubescens L.B. Smith & Ayensu 166,**167**
    var. setosa L.B. Smith & Ayensu 166,**167**
    var. suaveolens **168**
    var. trichophylla (Baker) L.B. Smith & Ayensu 166

    var. vestita L.B. Smith & Ayensu 168,**170**
  equisetoides sensu McPherson, van der Werff & Keating 167
  eylesii (Greves) N.L. Menezes 156,**164**,170
  humilis (Baker) T. Durand & Schinz 156,**162**
  kirkii (Hemsl.) L.B. Smith & Ayensu 156,**173,174**
  melleri Baker 166
  nutans L.B. Smith & Ayensu 156,**170**
  pinifolia Lam. 155,**158**
    var. pinifolia **157,158**
    var. villosa (H. Perret) H. Perret 158
  retinervis Baker 156,**164**
    var. *wentzeliana* (Harms) Coetzee 167
    var. *equisetoides* (Baker) Coetzee 166
  scabrida (Pax) T. Durand & Schinz 156,**160**
  schlechteri (Baker) N.L. Menezes 156,**160,161,**175
  simulans L.B. Smith & Ayensu 156,**158**
  sp. of Moriarty 174
  spekei Baker 156,**171**
  splendens (Rendle) N.L. Menezes 156,**174**
  squarrosa Baker 156,**159**
  suaveolens (Greves) N.L. Menezes 156,**168,169**
  velutina Baker 155,156
  villosa (Baker) L.B. Smith & Ayensu 156,**163**
  viscosa Baker 156,**175**
  wentzeliana (Harms) Sölch 167
  zambiana L.B. Smith & Ayensu 156,**172**
ZANNICHELLIA L. 96,**97**
  palustris L. **97,98**
ZANNICHELLIACEAE **96**,99
Zoostera ciliata Forssk. 107
ZOSTERA L. **93**
  capensis Setch. **94,95**
  marina var. *angustifolia* sensu A. Bennett 95
  nana sensu A. Bennett 95
  stipulacea Forssk. 54,95
  uninervis Forssk. 102
ZOSTERACEAE **93**

FAMILIES OF VASCULAR PLANTS REPRESENTED IN
THE FLORA ZAMBESIACA AREA

PTERIDOPHYTA
(Flora Zambesiaca families and family number. Published 1970)

Actiniopteridaceae		Gleicheniaceae	9	Parkeriaceae	
see Adiantaceae	18	Grammitidaceae	20	see Adiantaceae	18
Adiantaceae	18	Hymenophyllaceae	15	Polypodiaceae	21
Aspidiaceae	27	Isoetaceae	4	Psilotaceae	1
Aspleniaceae	23	Lindsaeaceae	19	Pteridaceae	
Athyriaceae	25	Lomariopsidaceae	26	see Adiantaceae	18
Azollaceae	13	Lycopodiaceae	2	Salviniaceae	12
Blechnaceae	28	Marattiaceae	7	Schizaeaceae	10
Cyatheaceae	14	Marsileaceae	11	Selaginellaceae	3
Davalliaceae	22	Oleandraceae		Thelypteridaceae	24
Dennstaedtiaceae	16	see Davalliaceae	22	Vittariaceae	17
Dryopteridaceae		Ophioglossaceae	6	Woodsiaceae	
see Aspidiaceae	27	Osmundaceae	8	see Athyriaceae	25
Equisetaceae	5				

GYMNOSPERMAE
(Flora Zambesiaca families and family number. Volume 1(1) 1960)

Cupressaceae	3	Cycadaceae	1	Podocarpaceae	2

ANGIOSPERMAE
(Flora Zambesiaca families, volume and part number and year of publication)

Acanthaceae			Asphodelaceae	12(3)	2001
tribes 1–5	8(5)	2013	Avicenniaceae	8(7)	2005
tribes 6–7	8(6)	2015	Balanitaceae	2(1)	1963
Agapanthaceae	13(1)	2008	Balanophoraceae	9(3)	2006
Agavaceae	13(1)	2008	Balsaminaceae	2(1)	1963
Aizoaceae	4	1978	Barringtoniaceae	4	1978
Alangiaceae	4	1978	Basellaceae	9(1)	1988
Alismataceae	12(2)	2009	Begoniaceae	4	1978
Alliaceae	13(1)	2008	Behniaceae	13(1)	2008
Aloaceae	12(3)	2001	Berberidaceae	1(1)	1960
Amaranthaceae	9(1)	1988	Bignoniaceae	8(3)	1988
Amaryllidaceae	13(1)	2008	Bixaceae	1(1)	1960
Anacardiaceae	2(2)	1966	Bombacaceae	1(2)	1961
Anisophylleaceae			Boraginaceae	7(4)	1990
see Rhizophoraceae	4	1978	Brexiaceae	4	1978
Annonaceae	1(1)	1960	Bromeliaceae	13(2)	2010
Anthericaceae	13(1)	2008	Buddlejaceae		
Apocynaceae	7(2)	1985	see Loganiaceae	7(1)	1983
Aponogetonaceae	12(2)	2009	Burmanniaceae	12(2)	2009
Aquifoliaceae	2(2)	1966	Burseraceae	2(1)	1963
Araceae	12(1)	2012	Buxaceae	9(3)	2006
Araliaceae	4	1978	Cabombaceae	1(1)	1960
Arecaceae			Cactaceae	4	1978
see Palmae	13(2)	2010	Caesalpinioideae		
Aristolochiaceae	9(2)	1997	see Leguminosae	3(2)	2006
Asclepiadaceae	-	-	Campanulaceae	7(1)	1983
Asparagaceae	13(1)	2008	Canellaceae	7(4)	1990

Cannabaceae	9(6)	1991
Cannaceae	13(4)	2010
Capparaceae	1(1)	1960
Caricaceae	4	1978
Caryophyllaceae	1(2)	1961
Casuarinaceae	9(6)	1991
Cecropiaceae	9(6)	1991
Celastraceae	2(2)	1966
Ceratophyllaceae	9(6)	1991
Chenopodiaceae	9(1)	1988
Chrysobalanaceae	4	1978
Colchicaceae	12(2)	2009
Combretaceae	4	1978
Commelinaceae	-	-
Compositae		
tribes 1–5	6(1)	1992
tribes 6–12	-	-
Connaraceae	2(2)	1966
Convolvulaceae	8(1)	1987
Cornaceae	4	1978
Costaceae	13(4)	2010
Crassulaceae	7(1)	1983
Cruciferae	1(1)	1960
Cucurbitaceae	4	1978
Cuscutaceae	8(1)	1987
Cymodoceaceae	12(2)	2009
Cyperaceae	-	-
Dichapetalaceae	2(1)	1963
Dilleniaceae	1(1)	1960
Dioscoreaceae	12(2)	2009
Dipsacaceae	7(1)	1983
Dipterocarpaceae	1(2)	1961
Dracaenaceae	13(2)	2010
Droseraceae	4	1978
Ebenaceae	7(1)	1983
Elatinaceae	1(2)	1961
Ericaceae	7(1)	1983
Eriocaulaceae	13(4)	2010
Eriospermaceae	13(2)	2010
Erythroxylaceae	2(1)	1963
Escalloniaceae	7(1)	1983
Euphorbiaceae	9(4)	1996
Euphorbiaceae	9(5)	2001
Flacourtiaceae	1(1)	1960
Flagellariaceae	13(4)	2010
Fumariaceae	1(1)	1960
Gentianaceae	7(4)	1990
Geraniaceae	2(1)	1963
Gesneriaceae	8(3)	1988
Gisekiaceae		
see Molluginaceae	4	1978
Goodeniaceae	7(1)	1983
Gramineae		
tribes 1–18	10(1)	1971
tribes 19–22	10(2)	1999
tribes 24–26	10(3)	1989
tribe 27	10(4)	2002
Guttiferae	1(2)	1961
Haloragaceae	4	1978
Hamamelidaceae	4	1978
Hemerocallidaceae	12(3)	2001
Hernandiaceae	9(2)	1997
Heteropyxidaceae	4	1978
Hyacinthaceae	-	-
Hydnoraceae	9(2)	1997
Hydrocharitaceae	12(2)	2009
Hydrophyllaceae	7(4)	1990
Hydrostachyaceae	9(2)	1997
Hypericaceae		
see Guttiferae	1(2)	1961
Hypoxidaceae	12(3)	2001
Icacinaceae	2(1)	1963
Illecebraceae	1(2)	1961
Iridaceae	12(4)	1993
Irvingiaceae	2(1)	1963
Ixonanthaceae	2(1)	1963
Juncaceae	13(4)	2010
Juncaginaceae	12(2)	2009
Labiatae		
see Lamiaceae & Verbenaceae		
Lamiaceae		
Viticoideae, Pingoideae	8(7)	2005
Lamiaceae		
Scutellaroideae-		
Nepetoideae	8(8)	2013
Lauraceae	9(2)	1997
Lecythidaceae		
see Barringtoniaceae	4	1978
Leeaceae	2(2)	1966
Leguminosae,		
Caesalpinioideae	3(2)	2007
Mimosoideae	3(1)	1970
Papilionoideae	3(3)	2007
Papilionoideae	3(4)	2012
Papilionoideae	3(5)	2001
Papilionoideae	3(6)	2000
Papilionoideae	3(7)	2002
Lemnaceae		
see Araceae	12(1)	2012
Lentibulariaceae	8(3)	1988
Liliaceae sensu stricto	12(2)	2009
Limnocharitaceae	12(2)	2009
Linaceae	2(1)	1963
Lobeliaceae	7(1)	1983
Loganiaceae	7(1)	1983
Loranthaceae	9(3)	2006
Lythraceae	4	1978
Malpighiaceae	2(1)	1963
Malvaceae	1(2)	1961
Marantaceae	13(4)	2010
Mayacaceae	13(2)	2010
Melastomataceae	4	1978
Meliaceae	2(1)	1963
Melianthaceae	2(2)	1966

Menispermaceae	1(1)	1960	Ptaeroxylaceae	2(2)	1966
Menyanthaceae	7(4)	1990	Rafflesiaceae	9(2)	1997
Mesembryanthemaceae	4	1978	Ranunculaceae	1(1)	1960
Mimosoideae			Resedaceae	1(1)	1960
see Leguminosae	3(1)	1970	Restionaceae	13(4)	2010
Molluginaceae	4	1978	Rhamnaceae	2(2)	1966
Monimiaceae	9(2)	1997	Rhizophoraceae	4	1978
Montiniaceae	4	1978	Rosaceae	4	1978
Moraceae	9(6)	1991	Rubiaceae		
Musaceae	13(4)	2010	subfam. Rubioideae	5(1)	1989
Myristicaceae	9(2)	1997	tribe Vanguerieae	5(2)	1998
Myricaceae	9(3)	2006	subfam. Cinchonoideae	5(3)	2003
Myrothamnaceae	4	1978	Rutaceae	2(1)	1963
Myrsinaceae	7(1)	1983	Salicaceae	9(6)	1991
Myrtaceae	4	1978	Salvadoraceae	7(1)	1983
Najadaceae	12(2)	2009	Santalaceae	9(3)	2006
Nesogenaceae	8(7)	2005	Sapindaceae	2(2)	1966
Nyctaginaceae	9(1)	1988	Sapotaceae	7(1)	1983
Nymphaeaceae	1(1)	1960	Scrophulariaceae	8(2)	1990
Ochnaceae	2(1)	1963	Selaginaceae		
Olacaceae	2(1)	1963	see Scrophulariaceae	8(2)	1990
Oleaceae	7(1)	1983	Simaroubaceae	2(1)	1963
Oliniaceae	4	1978	Smilacaceae	12(2)	2009
Onagraceae	4	1978	Solanaceae	8(4)	2005
Opiliaceae	2(1)	1963	Sonneratiaceae	4	1978
Orchidaceae	11(1)	1995	Sphenocleaceae	7(1)	1983
Orchidaceae	11(2)	1998	Sterculiaceae	1(2)	1961
Orobanchaceae			Strelitziaceae	13(4)	2010
see Scrophulariaceae	8(2)	1990	Taccaceae		
Oxalidaceae	2(1)	1963	see Dioscoreaceae	12(2)	2009
Palmae	13(2)	2010	Tecophilaeaceae	12(3)	2001
Pandanaceae	12(2)	2009	Tetragoniaceae	4	1978
Papaveraceae	1(1)	1960	Theaceae	1(2)	1961
Papilionoideae			Thymelaeaceae	9(3)	2006
see Leguminosae, Papilionoideae			Tiliaceae	2(1)	1963
Passifloraceae	4	1978	Trapaceae	4	1978
Pedaliaceae	8(3)	1988	Turneraceae	4	1978
Periplocaceae			Typhaceae	13(4)	2010
see Asclepiadaceae	-	-	Ulmaceae	9(6)	1991
Philesiaceae			Umbelliferae	4	1978
see Behniaceae	13(1)	2008	Urticaceae	9(6)	1991
Phormiaceae			Vacciniaceae		
see Hemerocallidaceae	12(3)	2001	see Ericaceae	7(1)	1983
Phytolaccaceae	9(1)	1988	Vahliaceae	4	1978
Piperaceae	9(2)	1997	Valerianaceae	7(1)	1983
Pittosporaceae	1(1)	1960	Velloziaceae	12(2)	2009
Plantaginaceae	9(1)	1988	Verbenaceae	8(7)	2005
Plumbaginaceae	7(1)	1983	Violaceae	1(1)	1960
Podostemaceae	9(2)	1997	Viscaceae	9(3)	2006
Polygalaceae	1(1)	1960	Vitaceae	2(2)	1966
Polygonaceae	9(3)	2006	Xyridaceae	13(4)	2010
Pontederiaceae	13(2)	2010	Zannichelliaceae	12(2)	2009
Portulacaceae	1(2)	1961	Zingiberaceae	13(4)	2010
Potamogetonaceae	12(2)	2009	Zosteraceae	12(2)	2009
Primulaceae	7(1)	1983	Zygophyllaceae	2(1)	1963
Proteaceae	9(3)	2006